Student's Solutions Manual

Elementary Algebra
Second Edition

Student's Solutions Manual

Elementary Algebra
Second Edition

Mark Dugopolski

Southeastern Louisiana University

ADDISON-WESLEY PUBLISHING COMPANY
Reading, Massachusetts • Menlo Park, California • New York
Don Mills, Ontario • Wokingham, England • Amsterdam • Bonn
Sydney • Singapore • Tokyo • Madrid • San Juan • Milan • Paris

Reproduced by Addison-Wesley from camera-ready copy supplied by the authors.

Copyright © 1996 Addison-Wesley Publishing Company, Inc.

ISBN   0-201-89514-5

3 4 5 6 7 8 9 10 ML 999897

To the student

This manual contains a complete solution to every odd-numbered exercise in the accompanying text. Solutions are given for all exercises in the Warm-ups, Chapter Tests, and Tying It All Together sections. This manual should be used as a reference only after you have attempted to solve a problem on your own.

TABLE OF CONTENTS

Student's Solutions Manual

Elementary Algebra
Second Edition

1.1 WARM-UPS

1. True, because we can multiply the numerator and denominator of a fraction by any nonzero number and get an equivalent fraction.
2. True, because they both reduce to 2/3.
3. False, because 8/12 in lowest terms is 2/3.
4. True, because if we multiply we get 2/6, which reduces to 1/3.   **5.** True, because of the definition of multiplication of fractions.
6. True, because of the definition of multiplication of fractions.

7. True, because $\frac{1}{2} \div 3 = \frac{1}{2} \cdot \frac{1}{3} = \frac{1}{6}$.

8. True, because $5 \div \frac{1}{2} = 5 \cdot 2 = 10$.

9. False, because $\frac{1}{2} + \frac{1}{4} = \frac{2}{4} + \frac{1}{4} = \frac{3}{4}$.

10. True, because $2 - \frac{1}{2} = \frac{4}{2} - \frac{1}{2} = \frac{3}{2}$.

1.1 EXERCISES

1. $\frac{3}{4} = \frac{3 \cdot 2}{4 \cdot 2} = \frac{6}{8}$

3. $\frac{8}{3} = \frac{8 \cdot 4}{3 \cdot 4} = \frac{32}{12}$

5. $5 = 5 \cdot \frac{2}{2} = \frac{10}{2}$

7. $\frac{3}{4} = \frac{3 \cdot 25}{4 \cdot 25} = \frac{75}{100}$

9. $\frac{3}{10} = \frac{3 \cdot 10}{10 \cdot 10} = \frac{30}{100}$

11. $\frac{5}{3} = \frac{5 \cdot 14}{3 \cdot 14} = \frac{70}{42}$

13. $\frac{3}{6} = \frac{3 \cdot 1}{3 \cdot 2} = \frac{1}{2}$

15. $\frac{12}{18} = \frac{6 \cdot 2}{6 \cdot 3} = \frac{2}{3}$

17. $\frac{15}{5} = \frac{5 \cdot 3}{5} = 3$

19. $\frac{50}{100} = \frac{50 \cdot 1}{50 \cdot 2} = \frac{1}{2}$

21. $\frac{200}{100} = \frac{100 \cdot 2}{100} = 2$

23. $\frac{18}{48} = \frac{6 \cdot 3}{6 \cdot 8} = \frac{3}{8}$

25. $\frac{26}{42} = \frac{2 \cdot 13}{2 \cdot 21} = \frac{13}{21}$

27. $\frac{84}{91} = \frac{7 \cdot 12}{7 \cdot 13} = \frac{12}{13}$

29. $\frac{2}{3} \cdot \frac{5}{9} = \frac{10}{27}$

31. $\frac{1}{3} \cdot 15 = \frac{15}{3} = 5$

33. $\frac{3}{4} \cdot \frac{14}{15} = \frac{3}{2 \cdot 2} \cdot \frac{2 \cdot 7}{3 \cdot 5} = \frac{7}{10}$

35. $\frac{2}{5} \cdot \frac{35}{26} = \frac{2}{5} \cdot \frac{5 \cdot 7}{2 \cdot 13} = \frac{7}{13}$

37. $3\frac{1}{2} \cdot \frac{6}{5} = \frac{7}{2} \cdot \frac{2 \cdot 3}{5} = \frac{21}{5}$

39. $4\frac{1}{2} \cdot 3\frac{1}{3} = \frac{9}{2} \cdot \frac{10}{3} = \frac{90}{6} = 15$

41. $\frac{3}{4} \div \frac{1}{4} = \frac{3}{4} \cdot \frac{4}{1} = 3$

43. $\frac{1}{3} \div 5 = \frac{1}{3} \cdot \frac{1}{5} = \frac{1}{15}$

45. $5 \div \frac{5}{4} = 5 \cdot \frac{4}{5} = 4$

47. $\frac{6}{10} \div \frac{3}{4} = \frac{2 \cdot 3}{2 \cdot 5} \cdot \frac{2 \cdot 2}{3} = \frac{4}{5}$

49. $2\frac{3}{16} \div \frac{5}{2} = \frac{35}{16} \cdot \frac{2}{5} = \frac{7}{8}$

51. $\frac{1}{4} + \frac{1}{4} = \frac{2}{4} = \frac{2 \cdot 1}{2 \cdot 2} = \frac{1}{2}$

53. $\frac{5}{12} - \frac{1}{12} = \frac{4}{12} = \frac{1}{3}$

55. $\frac{1}{2} - \frac{1}{4} = \frac{1 \cdot 2}{2 \cdot 2} - \frac{1}{4} = \frac{2}{4} - \frac{1}{4} = \frac{1}{4}$

57. $\frac{1}{3} + \frac{1}{4} = \frac{1 \cdot 4}{3 \cdot 4} + \frac{1 \cdot 3}{4 \cdot 3} = \frac{4}{12} + \frac{3}{12} = \frac{7}{12}$

59. $\frac{3}{4} - \frac{2}{3} = \frac{3 \cdot 3}{4 \cdot 3} - \frac{2 \cdot 4}{3 \cdot 4} = \frac{9}{12} - \frac{8}{12} = \frac{1}{12}$

61. $\frac{1}{6} + \frac{5}{8} = \frac{1 \cdot 4}{6 \cdot 4} + \frac{5 \cdot 3}{8 \cdot 3} = \frac{4}{24} + \frac{15}{24} = \frac{19}{24}$

63. $\frac{5}{24} - \frac{1}{18} = \frac{5 \cdot 3}{24 \cdot 3} - \frac{1 \cdot 4}{18 \cdot 4} = \frac{15}{72} - \frac{4}{72} = \frac{11}{72}$

65. $\frac{23}{6} + \frac{5}{16} = \frac{23 \cdot 8}{6 \cdot 8} + \frac{5 \cdot 3}{16 \cdot 3} = \frac{184}{48} + \frac{15}{48} = \frac{199}{48}$

67. $\frac{3}{5} = \frac{3 \cdot 20}{5 \cdot 20} = \frac{60}{100} = 60\%, \ \frac{3}{5} = \frac{3 \cdot 2}{5 \cdot 2} = \frac{6}{10} = 0.6$

69. $9\% = \frac{9}{100} = 0.09$

71. $0.08 = \frac{8}{100} = 8\%, \ 0.08 = \frac{8}{100} = \frac{2 \cdot 4}{25 \cdot 4} = \frac{2}{25}$

73. $\frac{3}{4} = \frac{3 \cdot 25}{4 \cdot 25} = \frac{75}{100} = 0.75, \ \frac{3}{4} = \frac{75}{100} = 75\%$

75. $2\% = \frac{2}{100} = \frac{2 \cdot 1}{2 \cdot 50} = \frac{1}{50}, \ 2\% = \frac{2}{100} = 0.02$

77. $0.01 = \frac{1}{100} = 1\%$

79. $\frac{3}{8} \div \frac{1}{8} = \frac{3}{8} \cdot \frac{8}{1} = 3$

81. $\frac{3}{4} \cdot \frac{28}{21} = \frac{3}{4} \cdot \frac{4 \cdot 7}{3 \cdot 7} = 1$

83. $\frac{7}{12} + \frac{5}{32} = \frac{7 \cdot 8}{12 \cdot 8} + \frac{5 \cdot 3}{32 \cdot 3} = \frac{56}{96} + \frac{15}{96} = \frac{71}{96}$

85. $\frac{5}{24} - \frac{1}{15} = \frac{5 \cdot 5}{24 \cdot 5} - \frac{1 \cdot 8}{15 \cdot 8} = \frac{25}{120} - \frac{8}{120} = \frac{17}{120}$

87. $3\frac{1}{8} + \frac{15}{16} = \frac{25}{8} + \frac{15}{16} = \frac{50}{16} + \frac{15}{16} = \frac{65}{16}$

89. $7\frac{2}{3} \cdot 2\frac{1}{4} = \frac{23}{3} \cdot \frac{9}{4} = \frac{69}{4}$

91. $\frac{1}{2} + \frac{1}{3} + \frac{1}{4} = \frac{1 \cdot 6}{2 \cdot 6} + \frac{1 \cdot 4}{3 \cdot 4} + \frac{1 \cdot 3}{4 \cdot 3}$

$\qquad = \frac{6}{12} + \frac{4}{12} + \frac{3}{12} = \frac{13}{12}$

93. $\frac{1}{2} \cdot \frac{1}{2} \cdot \frac{1}{2} = \frac{1}{8}$

95. $A = 20 \cdot 3\frac{1}{2} \cdot 25 \cdot 1\frac{1}{2} = 20 \cdot \frac{7}{2} \cdot 25 \cdot \frac{3}{2}$

$= 2625 \text{ in}^2$

$V = 2625 \cdot 92\frac{5}{8} = 2625 \cdot \frac{741}{8} = \frac{1,945,125}{8} \text{ in}^3$

$= \frac{1,945,125}{8} \text{ in}^3 \cdot \frac{1 \text{ ft}^3}{1728 \text{ in}^3} \approx 140.7 \text{ ft}^3$

97. $54\frac{3}{4} + \frac{3}{16} - \frac{1}{8} + \frac{5}{16} - \frac{1}{4}$

$= 54\frac{12}{16} + \frac{3}{16} - \frac{2}{16} + \frac{5}{16} - \frac{4}{16} = 54\frac{14}{16} = 54\frac{7}{8}$

The stock gained $\frac{1}{8}$ for the week.

$\frac{1}{8} \div 54\frac{3}{4} = \frac{1}{8} \cdot \frac{4}{219} = \frac{1}{438} \approx 0.002 = 0.2\%$

Gain was 0.2% for the week.

1.2 WARM-UPS

1. True, because we refer to the numbers 1, 2, 3, 4, . . . as either the natural numbers or the counting numbers.   **2.** True, because of the definition of counting numbers.   **3.** False, because the smallest counting number is 1. **4.** False, because zero can be expressed as a ratio of two integers, for example $0 = 0/4$. **5.** True, because 3 and -3 are opposites of each other.   **6.** False, because the absolute value of a number is the number's distance from 0, and 4 is 4 units away from 0. So the absolute value of 4 is 4.   **7.** True, because the opposite of -9 is 9.   **8.** True, because the opposite of the opposite of a number is the number. **9.** False, because -6 is to the left of -3 on the number line and so -6 is less than -3. **10.** False, because -5 is to the left of 0 on the number line and 4 and 6 are to the right of 0 on the number line.

1.2 EXERCISES

1. 6, because 6 is located to the right of -3 on the number line.

3. 0, because 0 is located to the right of -6 on the number line.

5. -2, because -2 is located to the right of -3 on the number line.

7. -12, because -12 is located to the right of -15 on the number line.

9. The counting numbers smaller than 6 are 1, 2, 3, 4, and 5. We show them on a number line as follows.

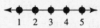

11. The whole numbers smaller than 5 are 0, 1, 2, 3, and 4. We show them on a number line as follows.

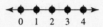

13. The whole numbers are 0, 1, 2, 3, and so on. The whole numbers between -5 and 5 are 0, 1, 2, 3, and 4.   Note that there are no negative whole numbers.

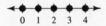

15. The counting numbers larger than -4 are 1, 2, 3, 4, 5, and so on. All of the counting numbers are larger than -4.

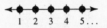

17. The integers larger than 1/2 are 1, 2, 3, 4, 5, and so on. We show them on a number line as follows.

19. True, because every integer is a ratio of two integers. For example, $-3 = -3/1$.

21. False, because the smallest counting number is 1.

23. True, because the ratio the circumference and diameter of a circle is π and π is irrational.

25. True, because the ratio of a whole number and 1 is a ratio of integers. For example, 2 is a whole number and $2 = 2/1$ expresses 2 as a ratio of integers.

27. True, because the positive integers are natural numbers.

29. False, because 0 is rational ($0 = 0/1$).

31. $|-6| = -(-6) = 6$

33. $|0| = 0$, because 0 is 0 units from 0 on the number line.

35. $|7| = 7$, because 7 is positive.

37. $|-9| = -(-9) = 9$

39. $|-45| = -(-45) = 45$

41. $\left|\frac{3}{4}\right| = \frac{3}{4}$

43. $|-5.09| = -(-5.09) = 5.09$

45. -16, because -16 is to the left of 9 on the number line.

47. First change the first fraction to fourths.

$$-\frac{5}{2} = -\frac{5 \cdot 2}{2 \cdot 2} = = -\frac{10}{4}$$

On the number line $-10/4$ is to the left of $-9/4$. So the smaller number is $-10/4$ or $-5/2$.

49. Because $|-3| = 3$, the smaller number is 2.

51. Because $|-4| = 4$, the smaller number is 3.

53. Because $|-5| = 5$ and $|-9| = 9$, -9 has the larger absolute value.

55. Because $|16| = 16$ and $|-9| = 9$, 16 has the larger absolute value.

57. True, because $|-3| = 3$ and $|-5| = 5$, and if we add the absolute values we do get 8.

59. True, because the absolute value of any negative number is a positive number and any positive number is greater than 0.

61. True, because $|-9| = 9$ and $|6| = 6$.

1.3 WARM-UPS

1. True, because the difference between 9 and 8 is 1, and we use the negative sign because -9 has a larger absolute value than 8.

2. True, because to add two negative numbers we add their absolute values $(2 + 4 = 6)$ and then attach a negative sign to the result.

3. True, because $0 - 7 = 0 + (-7) = -7$.

4. False, because $5 - (-2) = 5 + 2 = 7$.

5. False, because $-5 - (-2) = -5 + 2 = -3$.

6. False, because the additive inverse of -3 is 3.

7. True, because the variable b can represent a negative or a positive number, and when b is a negative number $-b$ is positive.

8. False, because the sum of a positive number and a negative number may be either positive or negative depending on which number has the larger absolute value.

9. True, because all subtraction can be expressed as addition of the opposite $(b - a = b + (-a))$.

10. False, because $-5 - (-7) = -5 + 7 = 2$.

1.3 EXERCISES

1. $3 + 10 = 13$

3. $(-3) + (-10) = -(3 + 10) = -13$

5. $-0.25 + (-0.9) = -(0.25 + 0.90) = -1.15$

7. $-\frac{1}{3} + \left(-\frac{1}{6}\right) = -\frac{1 \cdot 2}{3 \cdot 2} + \left(-\frac{1}{6}\right) = -\frac{2}{6} + \left(-\frac{1}{6}\right)$

$\qquad = -\frac{3}{6} = -\frac{1 \cdot 3}{2 \cdot 3} = -\frac{1}{2}$

9. Since -8 and 8 are additive inverses, their sum is 0, $-8 + 8 = 0$.

11. $-\frac{17}{50} + \frac{17}{50} = 0$

13. Since 9 has the larger absolute value, the result is positive, $-7 + 9 = 9 - 7 = 2$.

15. Since -13 has the larger absolute value, the result is negative, $-(13 - 7) = -6$.

17. Since 8.6 has the larger absolute value, the result is positive, $8.6 - 3.0 = 5.6$.

19. $3.9 + (-6.8) = -(6.8 - 3.9) = -2.9$

21. Since $-1/2$ has the larger absolute value, the result is negative.

$\frac{1}{4} + \left(-\frac{1}{2}\right) = \frac{1}{4} + \left(-\frac{2}{4}\right) = -\left(\frac{2}{4} - \frac{1}{4}\right) = -\frac{1}{4}$

23. Since subtraction is equivalent to addition of the opposite, $8 - 2 = 8 + (-2)$.

25. Since subtraction is equivalent to addition of the opposite, $4 - 12 = 4 + (-12)$.

27. Since subtraction is equivalent to addition of the opposite, $-3 - (-8) = -3 + 8$.

29. Since subtraction is equivalent to addition of the opposite, $8.3 - (-1.5) = 8.3 + 1.5$.

31. $6 - 10 = 6 + (-10) = -4$

33. $-3 - 7 = -3 + (-7) = -10$

35. $5 - (-6) = 5 + 6 = 11$

37. $-6 - 5 = -6 + (-5) = -11$

39. $\frac{1}{4} - \frac{1}{2} = \frac{1}{4} + \left(-\frac{2}{4}\right) = -\frac{1}{4}$

41. $\frac{1}{2} - \left(-\frac{1}{4}\right) = \frac{2}{4} + \frac{1}{4} = \frac{3}{4}$

43. $10 - 3 = 10 + (-3) = 7$

45. You must line up the decimal points when you subtract decimal numbers.
$1 - 0.07 = 1.00 - 0.07 = 0.93$

47. Remember to line up the decimal points.
$7.3 - (-2) = 7.3 + 2.0 = 9.3$

49. $-0.03 - 5 = -0.03 + (-5.00) = -5.03$

51. $-5 + 8 = 8 - 5 = 3$

53. $-6 + (-3) = -(6 + 3) = -9$

55. $-80 + (-40) = -(80 + 40) = -120$

57. $61 - (-17) = 61 + 17 = 78$

59. $(-12) + (-15) = -(12 + 15) = -27$

61. The result of this addition is negative because -20 has a larger absolute value than 13.
$13 + (-20) = -(20 - 13) = -7$

63. $-102 + (-99) = -(102 + 99) = -201$

65. $-161 + (-161) = -(161 + 161) = -322$

67. The result of this addition is negative because -16 has the larger absolute value.
$-16 + 0.03 = -(16.00 - 0.03) = -15.97$

69. $0.08 - 3 = 0.08 + (-3.00) = -(3.00 - 0.08)$
$$= -2.92$$

71. $-3.7 + (-0.03) = -(3.70 + 0.03) = -3.73$

73. $-2.3 - (-6) = -2.3 + 6 = 6 - 2.3 = 3.7$

75. $\frac{3}{4} + \left(-\frac{3}{5}\right) = \frac{3 \cdot 5}{4 \cdot 5} + \left(-\frac{3 \cdot 4}{5 \cdot 4}\right)$
$$= \frac{15}{20} + \left(-\frac{12}{20}\right) = \frac{15}{20} - \frac{12}{20} = \frac{3}{20}$$

77. $-\frac{1}{12} - \left(-\frac{3}{8}\right) = -\frac{2}{24} + \frac{9}{24} = \frac{7}{24}$

79. $45.87 + (-49.36) = -3.49$

Consult your calculator manual if you have trouble getting the answer with your calculator.

81. $0.6578 + (-1) = -0.3422$

83. $-3.45 - 45.39 = -48.84$

85. $-5.79 - 3.06 = -8.85$

87. $97.86 - 27.89 - 42.32 - 25.00 - 3.50 - 8.00$
$$= -\$8.85$$

89. $14° - 20° = -6°F$

1.4 WARM-UPS

1. True, because in algebra the times symbol can be omitted between a number and a variable.   **2.** False, the product of -2 and 5 is -10.   **3.** True.

4. False, because $0 \div 6 = 0$.

5. True, because the quotient of two negative numbers is a positive number.   **6.** True, because the quotient of a positive number and a negative number is a negative number.

7. True.

8. False, because $(-0.2)(0.2) = -0.04$.

9. True, because

$$\left(-\frac{1}{2}\right) \div \left(-\frac{1}{2}\right) = \left(-\frac{1}{2}\right) \cdot \left(-\frac{2}{1}\right) = 1.$$

10. False, because division by zero is undefined.

1.4 EXERCISES

1. $-3 \cdot 9 = -27$

3. $(-12)(-11) = 132$

5. $-\frac{3}{4} \cdot \frac{4}{9} = \left(-\frac{3}{4}\right)\left(\frac{4}{3 \cdot 3}\right) = -\frac{1}{3}$

7. $0.5(-0.6) = -0.30 = -0.3$

9. $(-12)(-12) = 144$

11. $-3 \cdot 0 = 0$

13. $8 \div (-8) = -1$

15. $(-90) \div (-30) = 90 \div 30 = 3$

17. $\frac{44}{-66} = -\frac{2}{3}$

19. $\left(-\frac{2}{3}\right) \div \left(-\frac{4}{5}\right) = \left(-\frac{2}{3}\right) \cdot \left(-\frac{5}{4}\right)$

$\qquad = \left(-\frac{2}{3}\right) \cdot \left(-\frac{5}{2 \cdot 2}\right) = \frac{5}{6}$

21. $\frac{-125}{0}$ is undefined.

23. $0 \div \left(-\frac{1}{3}\right) = 0$

25. $(40) \div (-0.5) = (400) \div (-5) = -80$

27. $-0.5 \div (-2) = 0.25$

29. $(25)(-4) = -100$

31. $(-3)(-9) = 27$

33. $-9 \div 3 = -3$

35. $20 \div (-5) = -4$

37. $(-6)(5) = -30$

39. $(-57) \div (-3) = 19$

41. $(0.6)(-0.3) = -0.18$

43. $(-0.03)(-10) = 0.30 = 0.3$

45. $(-0.6) \div (0.1) = (-6) \div (1) = -6$

47. $(-0.6) \div (-0.4) = (-6) \div (-4) = 1.5$

49. $-\frac{12}{5}\left(-\frac{55}{6}\right) = \frac{2 \cdot 6}{5} \cdot \frac{5 \cdot 11}{6} = 22$

51. $-\frac{11}{4} \div \frac{33}{4} = -\frac{11}{4} \cdot \frac{4}{33} = -\frac{1}{3}$

53. $(0.45)(-365) = -164.25$

55. $(-52) \div (-0.034) = 1529.41$

57. $(-4)(-4) = 16$

59. $-4 + (-4) = -(4 + 4) = -8$

61. $-4 + 4 = 0$

63. $-4 - (-4) = -4 + 4 = 0$

65. $0.1 - 4 = -(4.0 - 0.1) = -3.9$

67. $(-4) \div (0.1) = (-40) \div (1) = -40$

69. $(-0.1)(-4) = 0.4$

71. $|-0.4| = 0.4$

73. $\frac{-0.06}{0.3} = \frac{-6}{30} = -0.2$

75. $\frac{3}{-0.4} = -\frac{30}{4} = -7.5$

77. $-\frac{1}{5} + \frac{1}{6} = -\frac{1 \cdot 6}{5 \cdot 6} + \frac{1 \cdot 5}{6 \cdot 5} = -\frac{6}{30} + \frac{5}{30} = -\frac{1}{30}$

79. $\left(-\frac{3}{4}\right)\left(\frac{2}{15}\right) = \left(-\frac{3}{2 \cdot 2}\right)\left(\frac{2}{3 \cdot 5}\right) = -\frac{1}{10}$

81. $\frac{45.37}{6} = 7.562$

83. $(-4.3)(-4.5) = 19.35$

85. $\frac{0}{6.345} = 0$

87. $199.4 \div 0$ is undefined.

1.5 WARM-UPS

1. False, because $(-3)^2 = (-3)(-3) = 9$.

2. False, because $5 - 3 \cdot 2 = 5 - 6 = -1$.

3. True, because we do operations within parentheses first. **4.** False, because $|5 - 6| = |-1| = 1$ and $|5| - |6| = -1$.

5. False, because $5 + 6 \cdot 2 = 5 + 12 = 17$ and $(5 + 6) \cdot 2 = 11 \cdot 2 = 22$. **6.** False, because $(2 + 3)^2 = 5^2 = 25$ and $2^2 + 3^2 = 4 + 9 = 13$.

7. False, because $5 - 3^3 = 5 - 27 = -22$.

8. True, because $(5 - 3)^3 = 2^3 = 2 \cdot 2 \cdot 2 = 8$.

9. False, because $6 - \frac{6}{2} = 6 - 3 = 3$ and $\frac{0}{2} = 0$.

10. True, because $\frac{6 - 6}{2} = \frac{0}{2} = 0$.

1.5 EXERCISES

1. $(4 - 3)(5 - 9) = (1)(-4) = -4$

3. $|3 + 4| - |-2 - 4| = 7 - 6 = 1$

5. $\frac{7 - (-9)}{3 - 5} = \frac{16}{-2} = -8$

7. $(-6 + 5)(7) = (-1)(7) = -7$

9. $(-3 - 7) - 6 = (-10) - 6 = -16$

11. $-16 \div (8 \div 2) = -16 \div (4) = -4$

13. The exponent 4 indicates that 4 is used as a factor 4 times: $4 \cdot 4 \cdot 4 \cdot 4 = 4^4$.

15. To indicate that -5 is used as a factor 4 times we use the exponent 4.

$(-5)(-5)(-5)(-5) = (-5)^4$

17. The factor $-y$ appears 3 times so we use an exponent of 3: $(-y)(-y)(-y) = (-y)^3$.

19. $\frac{3}{7} \cdot \frac{3}{7} \cdot \frac{3}{7} \cdot \frac{3}{7} \cdot \frac{3}{7} = \left(\frac{3}{7}\right)^5$

21. $5^3 = 5 \cdot 5 \cdot 5$

23. $b^2 = b \cdot b$

25. $\left(-\frac{1}{2}\right)^5 = \left(-\frac{1}{2}\right)\left(-\frac{1}{2}\right)\left(-\frac{1}{2}\right)\left(-\frac{1}{2}\right)\left(-\frac{1}{2}\right)$

27. $(0.22)^4 = (0.22)(0.22)(0.22)(0.22)$

29. $3^4 = 3 \cdot 3 \cdot 3 \cdot 3 = 9 \cdot 9 = 81$

31. $0^9 = 0$

33. $(-5)^4 = (-5)(-5)(-5)(-5) = 625$

35. $(-6)^3 = (-6)(-6)(-6) = -216$

37. $(10)^5 = 100{,}000$

39. $(-0.1)^3 = -0.001$

41. $\left(\frac{1}{2}\right)^3 = \frac{1}{2} \cdot \frac{1}{2} \cdot \frac{1}{2} = \frac{1}{8}$

43. $\left(-\frac{1}{2}\right)^2 = \left(-\frac{1}{2}\right)\left(-\frac{1}{2}\right) = \frac{1}{4}$

45. $-8^2 = -64$

47. $-(-8)^4 = -4096$

49. $-(7-10)^3 = -(-3)^3 = 27$

51. $(-2^2) - (3^2) = -4 - 9 = -13$

53. $3^2 \cdot 2^2 = 9 \cdot 4 = 36$

55. $-3 \cdot 2 + 4 \cdot 6 = -6 + 24 = 18$

57. $(-3)^3 + 2^3 = -27 + 8 = -19$

59. $-21 + 36 \div 3^2 = -21 + 36 \div 9$
$= -21 + 4 = -17$

61. $-3 \cdot 2^3 - 5 \cdot 2^2 = -3 \cdot 8 - 5 \cdot 4 = -24 - 20$
$= -44$

63. $\frac{-8}{2} + 2 \cdot 3 \cdot 5 - 2^3 = -4 + 30 - 8 = 18$

65. $(-3 + 4^2)(-6) = (13)(-6) = -78$

67. $(-3 \cdot 2 + 6)^3 = (-6 + 6)^3 = 0^3 = 0$

69. $2 - 5(3 - 4 \cdot 2) = 2 - 5(3 - 8)$
$\qquad = 2 - 5(-5) = 2 + 25 = 27$

71. $3 - 2 \cdot |5 - 6| = 3 - 2 \cdot |-1| = 3 - 2 \cdot 1 = 1$

73. $(3^2 - 5)|3 \cdot 2 - 8| = (9 - 5)|6 - 8|$
$= (4)|-2| = 4 \cdot 2 = 8$

75. $\frac{3 - 4 \cdot 6}{7 - 10} = \frac{3 - 24}{-3} = \frac{-21}{-3} = 7$

77. $\frac{7 - 9 - 3^2}{9 - 7 - 3} = \frac{7 - 9 - 9}{2 - 3} = \frac{-11}{-1} = 11$

79. $3 + 4[9 - 6(2 - 5)] = 3 + 4[9 - 6(-3)]$
$= 3 + 4[9 - (-18)] = 3 + 4[27] = 3 + 108 = 111$

81. $6^2 - [(2 + 3)^2 - 10] = 36 - [5^2 - 10]$
$\qquad\qquad\qquad = 36 - [15] = 21$

83. $4 - 5 \cdot |3 - (3^2 - 7)| = 4 - 5 \cdot |3 - (2)|$
$= 4 - 5 \cdot |1| = 4 - 5 = -1$

85. $-2|3 - (7 - 3)| - |-9| = -2|3 - (4)| - 9$
$= -2|-1| - 9 = -2 \cdot 1 - 9 = -2 - 9 = -11$

87. $1 + 2^3 = 1 + 8 = 9$

89. $(-2)^2 - 4(-1)(3) = 4 - (-12) = 16$

91. $4^2 - 4(1)(-3) = 16 - (-12) = 16 + 12 = 28$

93. $(-11)^2 - 4(5)(0) = 121 - 0 = 121$

95. $-5^2 - 3 \cdot 4^2 = -25 - 3 \cdot 16 = -25 - 48 = -73$

97. $[3 + 2(-4)]^2 = [3 + (-8)]^2 = [-5]^2 = 25$

99. $|-1| - |-1| = 1 - 1 = 0$

101. $\frac{4 - (-4)}{-2 - 2} = \frac{8}{-4} = -2$

103. $3(-1)^2 - 5(-1) + 4 = 3 \cdot 1 - (-5) + 4$
$\qquad\qquad\qquad = 3 + 5 + 4 = 12$

105. $5 - 2^2 + 3^4 = 5 - 4 + 81 = 82$

107. $-2 \cdot |9 - 6^2| = -2 \cdot |9 - 36| = -2 \cdot 27$
$= -54$

109. $-3^2 - 5[4 - 2(4 - 9)] = -9 - 5[4 - 2(-5)]$
$= -9 - 5[4 + 10] = -9 - 5[14] = -79$

111. $1 - 5|5 - (9 + 1)| = 1 - 5|5 - (10)|$
$= 1 - 5|-5| = 1 - 5 \cdot 5 = 1 - 25 = -24$

113. $3.2^2 - 4(3.6)(-2.2) = 41.92$

115. $(5.63)^3 - [4.7 - (-3.3)^2] = 184.643547$

117. $\frac{3.44 - (-8.32)}{6.89 - 5.43} \approx 8.0548$

119. $246.3(1.0086)^{11} \approx 270.6$

Population will be 270.6 million in 2000.

1.6 WARM-UPS

1. True, because in $2x + 3y$ the last operation to be performed is addition. **2.** False, because in $5(y - 9)$ the last operation to be performed is multiplication. **3.** True, because the last operation to be performed is multiplication. **4.** False, because the last operation to be performed is addition. **5.** True, because the last operation to be performed is multiplication.
6. False, because $2(-2) + 4 = 0$.
7. False, $(-3)^3 - 5 = -27 - 5 = -32$.
8. False, because $2(5) - 3 = 13$ is incorrect.
9. True.
10. False, because $2(x + 7)$ is read as the product of 2 and $x + 7$.

1.6 EXERCISES

1. In the expression $a^3 - 1$, the subtraction is performed last. So the expression is referred to as a difference.
3. In the expression $(w - 1)^3$, the subtraction is done first and the exponent is evaluated last. So the expression is called a cube.
5. In the expression $3x + 5y$, multiplication is done first and addition last. So the expression is called a sum.
7. Since the subtraction is performed after the division, the expression is a difference.
9. In the expression $3(x + 5y)$ the last operation is multiplication. So the expression is a product.
11. In the expression squaring is done last. So the expression is a square.
13. The difference of x^2 and a^2
15. The square of $x - a$
17. The quotient of $x - 4$ and 2
19. The difference of $x/2$ and 4
21. The cube of ab
23. Since the word sum indicates addition, the expression is $2x + 3y$.
25. The difference of 8 and $7x$ is $8 - 7x$.

27. The square of $a + b$ is $(a + b)^2$.

29. The product of $x + 9$ and $x + 12$ is $(x + 9)(x + 12)$.

31. The quotient of $x - 7$ and $7 - x$ is $\frac{x - 7}{7 - x}$.

33. Use $a = -1$ and $b = 2$ in the expression.

$$-(a - b) = -(-1 - 2) = -(-3) = 3$$

35. Use $b = 2$ in the expression.

$$-2^2 + 7 = -4 + 7 = 3$$

37. Use $c = -3$ in the expression.

$$c^2 - 2c + 1 = (-3)^2 - 2(-3) + 1 = 9 - (-6) + 1$$
$$= 9 + 6 + 1 = 16$$

39. Use $a = -1$ and $b = 2$ in the expression.

$$a^3 - b^3 = (-1)^3 - (2)^3 = -1 - 8 = -9$$

41. Use $a = -1$ and $b = 2$ in the expression.

$$(a - b)(a + b) = (-1 - 2)(-1 + 2) = (-3)(1)$$
$$= -3$$

43. Use $a = -1$, $b = 2$, and $c = -3$:

$$b^2 - 4ac = 2^2 - 4(-1)(-3) = 4 - 12 = -8$$

45. Use $a = -1$, $b = 2$, and $c = -3$:

$$\frac{a - c}{a - b} = \frac{-1 - (-3)}{-1 - 2} = \frac{2}{-3} = -\frac{2}{3}$$

47. Use $a = -1$, $b = 2$, and $c = -3$:

$$\frac{2}{a} + \frac{6}{b} - \frac{9}{c} = \frac{2}{-1} + \frac{6}{2} - \frac{9}{-3} = -2 + 3 - (-3) = 4$$

49. Use $a = -1$ in the expression.

$$a \div |a| = -1 \div |-1| = -1 \div 1 = -1$$

51. Use $a = -1$ and $b = 2$ in the expression.

$$|b| - |a| = |2| - |-1| = 2 - 1 = 1$$

53. Use $a = -1$ and $c = -3$ in the expression.

$$-|-a - c| = -|-(-1) - (-3)|$$
$$= -|1 + 3| = -|4| = -4$$

55. Use $a = -1$ and $b = 2$ in the expression.

$$(3 - |a - b|)^2 = (3 - |-1 - 2|)^2$$
$$= (3 - |-3|)^2 = (3 - 3)^2 = (0)^2 = 0$$

57. Replace x by 2 in the equation to see if the equation is satisfied.

$$3x + 7 = 13$$
$$3(2) + 7 = 13$$
$$6 + 7 = 13 \qquad \text{Correct}$$

So 2 is a solution to the equation.

59. Replace x by -2 in the equation.

$$\frac{3x-4}{2} = 5$$

$$\frac{3(-2)-4}{2} = 5$$

$$\frac{-10}{2} = 5 \qquad \text{Incorrect.}$$

So -2 is not a solution to the equation.

61. Replace x by -2 in the equation.

$$-x+4 = 6$$
$$-(-2)+4 = 6$$
$$2+4 = 6 \qquad \text{Correct}$$

The number -2 satisfies the equation.

63. Replace x by 4 in the equation.

$$3x-7 = x+1$$
$$3(4)-7 = 4+1$$
$$5 = 5 \qquad \text{Correct}$$

The number 4 satisfies the equation.

65. Replace x by 3 in the equation.

$$-2(x-1) = 2-2x$$
$$-2(3-1) = 2-2(3)$$
$$-4 = -4 \qquad \text{Correct}$$

The number 3 satisfies the equation.

67. Replace x by 1 in the equation.

$$x^2+3x-4 = 0$$
$$1^2+3(1)-4 = 0$$
$$1+3-4 = 0 \qquad \text{Correct}$$

The number 1 satisfies the equation.

69. Replace x by 8 in the equation.

$$\frac{x}{x-8} = 0$$

$$\frac{8}{8-8} = 0$$

$$\frac{8}{0} = 0 \qquad \text{Incorrect}$$

Since 0 appears in the denominator, 8 does not satisfy this equation.

71. Replace x by -6 in the equation.

$$\frac{x+6}{x+6} = 1$$

$$\frac{-6+6}{-6+6} = 1$$

$$\frac{0}{0} = 1 \qquad \text{Incorrect}$$

Since $0/0$ is undefined, -6 is not a solution to the equation.

73. $5x+3x = 8x$

75. $3(x+2) = 12$

77. $\frac{x}{3} = 5x$

79. $(a+b)^2 = 9$

81. $b^2 - 4ac = (6.7)^2 - 4(4.2)(1.8) = 14.65$

83. $b^2 - 4ac = (3.2)^2 - 4(-1.2)(5.6) = 37.12$

85. Replace T by 36.5:
$$81.7 + 2.4(36.5) = 169.3$$
The height of the person is about 169.3 cm.
A male with height 180 cm has a tibia with a length of about 41 cm

87. For Seattle, $GB = \dfrac{(31-27)+(36-32)}{2} = 4$

For California, $GB = \dfrac{(31-27)+(39-32)}{2} = 5.5$

For Oakland, $GB = \dfrac{(31-21)+(43-32)}{2} = 10.5$

89. Since 100 yds = 300 ft,

$$2L + 2W = 2(300) + 2(160) = 920 \text{ feet}$$

1.7 WARM-UPS

1. False, because $24 \div (4 \div 2) = 24 \div 2 = 12$ and $(24 \div 4) \div 2 = 6 \div 2 = 3$.
2. False, because $1 \div 2 = 0.5$ and $2 \div 1 = 2$.
3. True, because $6 - 5 = 1$ and $-5 + 6 = 1$.
4. False, because $9 - (4-3) = 9 - 1 = 8$ and $(9-4) - 3 = 5 - 3 = 2$.
5. True, because of the commutative property of multiplication.
6. True, because of the distributive property.
7. True, because $(0.02)(50) = 1$.
8. True, because of the distributive property.
9. True, because of the distributive property.
10. True, because $0 + 0 = 0$.

1.7 EXERCISES

1. $9 + r = r + 9$
3. $3(2+x) = 3(x+2)$
5. Since $4 - 5x = 4 + (-5x)$, we can write $4 - 5x = -5x + 4$.
7. $x \cdot 6 = 6x$.
9. $(x-4)(-2) = -2(x-4)$
Note that the commutative property of multiplication only allows us to rewrite the multiplication part of the expression.

11. By the commutative property of multiplication we can write $4 - y \cdot 8 = 4 - 8y$.

13. $(4w)(w) = 4(w \cdot w) = 4w^2$

15. $3a(ba) = 3a(ab) = (3a \cdot a)b = (3a^2)b = 3a^2b$

17. $(x)(9x)(xz) = 9 \cdot x \cdot x \cdot x \cdot z = 9x^3z$

19. $8 - 4 + 3 - 10 = 11 - 14 = -3$

21. $8 - 10 + 7 - 8 - 7 = 15 - 25 = -10$

23. $-4 - 11 + 7 - 8 + 15 - 20 = -43 + 22 = -21$

25. $-3.2 + 2.4 - 2.8 + 5.8 - 1.6 = -7.6 + 8.2$
$$= 0.6$$

27. $3.26 - 13.41 + 5.1 - 12.35 - 5$
$$= 8.36 - 30.76 = -22.4$$

29. $3(x - 5) = 3x - 3 \cdot 5 = 3x - 15$

31. $2m + 12 = 2m + 2 \cdot 6 = 2(m + 6)$

33. $a(2 + t) = a2 + at = 2a + at$

35. $-3(w - 6) = -3w - (-3)(6) = -3w - (-18)$
$$= -3w + 18$$

37. $-4(5 - y) = -4(5) - (-4y) = -20 + 4y$

39. $4x - 4 = 4x - 4 \cdot 1 = 4(x - 1)$

41. $-1(a - 7) = -1a - (-1)(7) = -a - (-7)$
$$= -a + 7$$

43. $-1(t + 4) = -1(t) + (-1)(4) = -t + (-4)$
$$= -t - 4$$

45. $4y - 16 = 4y - 4 \cdot 4 = 4(y - 4)$

47. $4a + 8 = 4a + 4 \cdot 2 = 4(a + 2)$

49. The multiplicative inverse of $\frac{1}{2}$ is 2 because $\frac{1}{2} \cdot 2 = 1$.

51. The multiplicative inverse of -5 is $-\frac{1}{5}$ because $-5 \cdot \left(-\frac{1}{5}\right) = 1$.

53. The multiplicative inverse of 7 is $\frac{1}{7}$ because $7 \cdot \frac{1}{7} = 1$.

55. The multiplicative inverse of 1 is 1 because $1 \cdot 1 = 1$.

57. The multiplicative inverse of -0.25 is -4 because $-0.25(-4) = 1$.

59. The multiplicative inverse of $2.5 = \frac{5}{2}$ is $\frac{2}{5}$ because $\frac{5}{2} \cdot \frac{2}{5} = 1$.

61. The statement $3 \cdot x = x \cdot 3$ is true because of the commutative property of multiplication.

63. The statement $2(x - 3) = 2x - 6$ is correct because of the distributive property.

65. The statement $-3(xy) = (-3x)y$ is correct because of the associative property.

67. The statement $4 + (-4) = 0$ is true because of the inverse property.

69. The statement $x^2 \cdot 5 = 5x^2$ is correct because of the commutative property of multiplication.

71. The statement $1 \cdot 3y = 3y$ is correct because of the identity property. One is the multiplicative identity.

73. The statement $2a + 5a = (2 + 5)a$ is correct because of the distributive property.

75. The statement $-7 + 7 = 0$ is correct because of the inverse property. The numbers 7 and -7 are additive inverses of each other.

77. The statement $(2346)0 = 0$ is correct because of the multiplication property of 0.

79. The statement $ay + y = y(a + 1)$ is correct because of the distributive property.

81. We can use the commutative property to reverse the addition and write $a + y = y + a$.

83. We can use the associative property to move the parentheses and write $5(aw) = (5a)w$.

85. Use the distributive property to write $\frac{1}{2}x + \frac{1}{2} = \frac{1}{2}(x + 1)$.

87. Use the distributive property to write $6x + 15 = 3 \cdot 2x + 3 \cdot 5 = 3(2x + 5)$.

89. Since 4 and 0.25 are multiplicative inverses of each other, we can write $4(0.25) = 1$.

91. By the multiplication property of 0, we can write $0 = 96(0)$.

93. The multiplicative inverse of $0.33 = \frac{33}{100}$ is $\frac{100}{33}$. So we can write $0.33\left(\frac{100}{33}\right) = 1$.

95. $\frac{1}{0.04} + \frac{1}{0.05} = 45$ bricks/hour

97. $\frac{1 \text{ person}}{0.34483 \text{ sec}} \approx 2.9$ people/sec

$\frac{2.9 \text{ people}}{1 \text{ sec}} \times \frac{604,800 \text{ sec}}{\text{week}}$

$= 1,753,920$ people/week

1. True, because of the distributive property.
2. False, because if $x = 1$ then the equation $-3x + 9 = -3(x + 9)$ becomes $6 = -30$, which is incorrect. **3.** True, because of the distributive property. **4.** True, by combining like terms. **5.** False, because if $a = 2$ then $(3a)(4a) = 12a$ becomes $48 = 24$. **6.** False, because $3(5 \cdot 2) = 30$ and $15 \cdot 6 = 90$.
7. False, because if $x = 1$ then $x + x = x^2$ becomes $2 = 1$. **8.** False, because if $x = 1$ then $x \cdot x = 2x$ becomes $1 = 2$. **9.** False, because if $x = 2$ then $3 + 2x = 5x$ becomes $7 = 10$.
10. True, because of the distributive property.

1.8 EXERCISES

1. $35(200) = (35 \cdot 2)100 = 70 \cdot 100 = 7000$

3. $\frac{4}{3}(0.75) = \frac{4}{3} \cdot \frac{3}{4} = 1$

5. $256 + (78 + 22) = 256 + 100 = 356$

7. $35 \cdot 3 + 35 \cdot 7 = 35(3 + 7) = 35(10) = 350$

9. $18 \cdot 4 \cdot 2 \cdot \frac{1}{4} = 18 \cdot 2 \cdot 4 \cdot \frac{1}{4} = 18 \cdot 2 \cdot 1 = 36$

11. $(120)(300) = (120 \cdot 3)100 = 360 \cdot 100$
$$= 36{,}000$$

13. $12 \cdot 375(-6 + 6) = 12 \cdot 375(0) = 0$

15. $78 + 6 + 8 + 4 + 2 = 78 + 10 + 10 = 98$

17. $5w + 6w = (5 + 6)w = 11w$

19. $4x - x = 4x - 1x = 3x$

21. $2x - (-3x) = 2x + 3x = 5x$

23. $-3a - (-2a) = -3a + 2a = -1a = -a$

25. $-a - a = -1a - 1a = (-1 - 1)a = -2a$

27. The terms in $10 - 6t$ are not like terms. So they cannot be combined.

29. $3x^2 + 5x^2 = (3 + 5)x^2 = 8x^2$

31. The terms in $-4x + 2x^2$ are not like terms and so they cannot be combined.

33. $5mw^2 - 12mw^2 = (5 - 12)mw^2 = -7mw^2$

35. $3(4h) = (3 \cdot 4)h = 12h$

37. $6b(-3) = -18b$

39. $(-3m)(3m) = -3 \cdot 3 \cdot m \cdot m = -9m^2$

41. $(-3d)(-4d) = (-3)(-4)d \cdot d = 12d^2$

43. $(-y)(-y) = (-1y)(-1y) = (-1)(-1)yy = y^2$

45. $-3a(5b) = (-3 \cdot 5)ab = -15ab$

47. $-3a(2 + b) = -3a(2) + (-3a)(b)$
$$= -6a - 3ab$$

49. $-k(1 - k) = -k(1) - (-k)(k) = -k + k^2$

51. $\frac{3y}{3} = \frac{1}{3}(3y) = \left(\frac{1}{3} \cdot 3\right)y = 1 \cdot y = y$

53. $\frac{-15y}{5} = \frac{1}{5}(-15)y = -3y$

55. $2\left(\frac{y}{2}\right) = 2 \cdot \frac{1}{2} \cdot y = 1 \cdot y = y$

57. $8y\left(\frac{y}{4}\right) = \frac{8}{4}yy = 2y^2$

59. $\frac{6a - 3}{3} = \frac{1}{3}(6a - 3) = 2a - 1$

61. $\frac{-9x + 6}{-3} = -\frac{1}{3}(-9x + 6) = 3x - 2$

63. $x - (3x - 1) = x - 3x + 1 = -2x + 1$

65. $5 - (y - 3) = 5 - y + 3 = 8 - y$

67. $2m + 3 - (m + 9) = 2m + 3 - m - 9$
$$= 2m - m + 3 - 9 = m - 6$$

69. $-3 - (-w + 2) = -3 + w - 2 = w - 5$

71. $3x + 5x + 6 + 9 = 8x + 15$

73. $-2x + 3 + 7x - 4 = 7x - 2x + 3 - 4 = 5x - 1$

75. $3a - 7 - (5a - 6) = 3a - 7 - 5a + 6$
$$= 3a - 5a - 7 + 6 = -2a - 1$$

77. $2(a - 4) - 3(-2 - a) = 2a - 8 + 6 + 3a$
$$= 2a + 3a - 8 + 6 = 5a - 2$$

79. $-5m + 6(m - 3) + 2m$
$$= -5m + 6m - 18 + 2m = 3m - 18$$

81. $5 - 3(x + 2) - 6 = 5 - 3x - 6 - 6 = -3x - 7$

83. $x - 0.05(x + 10) = x - 0.05x - 0.5$
$$= 1.00x - 0.05x - 0.5 = 0.95x - 0.5$$

85. $4.5 - 3.2(x - 5.3) - 8.75$
$$= 4.5 - 3.2x + 16.96 - 8.75 = -3.2x + 12.71$$

87. $3x - (4 - x) = 3x - 4 + x = 4x - 4$

89. $y - 5 - (-y - 9) = y - 5 + y + 9 = 2y + 4$

91. $7 - (8 - 2y - m) = 7 - 8 + 2y + m$
$$= 2y + m - 1$$

93. $\frac{1}{2}(10 - 2x) + \frac{1}{3}(3x - 6) = 5 - x + x - 2 = 3$

95. $0.2(x + 3) - 0.05(x + 20)$

$= 0.2x + 0.6 - 0.05x - 1 = 0.15x - 0.4$

97. $2k + 1 - 3(5k - 6) - k + 4$
$$= 2k + 1 - 15k + 18 - k + 4 = -14k + 23$$

99. $-3m - 3[2m - 3(m + 5)]$

$= -3m - 3[-m - 15] = -3m + 3m + 45 = 45$

101. $0.15(38,000) + 0.28(x - 38,000)$

$= 0.28x - 4940$

$0.28(52,000) - 4940 = \$9620$

Income of \$150,000 pays approximately \$40,000

in tax.

103. $2(x) + 2(x + 40) = 4x + 80$

$4(30) + 80 = 200$ feet

CHAPTER 1 REVIEW

1. $\frac{1}{3} + \frac{3}{8} = \frac{1 \cdot 8}{3 \cdot 8} + \frac{3 \cdot 3}{8 \cdot 3} = \frac{8}{24} + \frac{9}{24} = \frac{17}{24}$

3. $\frac{3}{5} \cdot 10 = \frac{3}{5} \cdot 5 \cdot 2 = 6$

5. $\frac{2}{5} \cdot \frac{15}{14} = \frac{2}{5} \cdot \frac{3 \cdot 5}{2 \cdot 7} = \frac{3}{7}$

7. $4 + \frac{2}{3} = 4 \cdot \frac{3}{3} + \frac{2}{3} = \frac{12}{3} + \frac{2}{3} = \frac{14}{3}$

9. $\frac{1}{2} + \frac{1}{3} + \frac{1}{4} = \frac{1 \cdot 6}{2 \cdot 6} + \frac{1 \cdot 4}{3 \cdot 4} + \frac{1 \cdot 3}{4 \cdot 3}$
$$= \frac{6}{12} + \frac{4}{12} + \frac{3}{12} = \frac{13}{12}$$

11. Of the numbers listed, the numbers 0, 1, 2, and 10 are whole numbers.

13. Of the numbers listed, the numbers −2, 0, 1, 2, and 10 are integers.

15. The only irrational numbers listed are $-\sqrt{5}$ and π. Note that any terminating decimal is rational.

17. True, because every whole number can be expressed as a ratio of integers. For example, $9 = 9/1$.

19. False, because the counting numbers between −4 and 4 are 1, 2, and 3.

21. False, because the national debt is finite.

23. True, because the integers greater than −1 are 0, 1, 2, 3, and so on.

25. $-5 + 7 = 7 - 5 = 2$

27. $35 - 48 = -(48 - 35) = -(13) = -13$

29. $-12 + 5 = -(12 - 5) = -(7) = -7$

31. $-12 - (-5) = -12 + 5 = -7$

33. $-0.05 + 12 = 12.00 - 0.05 = 11.95$

35. $-0.1 - (-0.05) = -0.1 + 0.05$
$$= -0.10 + 0.05 = -0.05$$

37. $\frac{1}{3} - \frac{1}{2} = \frac{1 \cdot 2}{3 \cdot 2} - \frac{1 \cdot 3}{2 \cdot 3} = \frac{2}{6} - \frac{3}{6} = -\frac{1}{6}$

39. $-\frac{1}{3} + \left(-\frac{2}{5}\right) = -\frac{1 \cdot 5}{3 \cdot 5} + \left(-\frac{2 \cdot 3}{5 \cdot 3}\right)$
$$= -\frac{5}{15} + \left(-\frac{6}{15}\right) = -\frac{11}{15}$$

41. $(-3)(5) = -15$

43. $(-8) \div (-2) = 4$

45. $\frac{-20}{-4} = 5$

47. $\left(-\frac{1}{2}\right)\left(-\frac{1}{3}\right) = \frac{1}{6}$

49. $-0.09 \div 0.3 = -0.9 \div 3 = -0.3$

51. $(0.3)(-0.8) = -0.24$

53. $(-5)(-0.2) = 1$

55. $3 + 7(9) = 3 + 63 = 66$

57. $(3 + 4)^2 = 7^2 = 49$

59. $3 + 2 \cdot |5 - 6 \cdot 4| = 3 + 2 \cdot |5 - 24|$
$$= 3 + 2 \cdot |-19| = 3 + 2 \cdot 19 = 3 + 38 = 41$$

61. $(3 - 7) - (4 - 9) = (-4) - (-5)$
$$= -4 + 5 = 1$$

63. $-2 - 4(2 - 3 \cdot 5) = -2 - 4(2 - 15)$
$$= -2 - 4(-13) = -2 + 52 = 50$$

65. $3^2 - (7 + 5)^2 = 9 - 12^2 = 9 - 144 = -135$

67. $\frac{-3 - 5}{2 - (-2)} = \frac{-8}{2 + 2} = \frac{-8}{4} = -2$

69. $\frac{6 + 3}{3} - 5 \cdot 4 + 1 = 3 - 20 + 1 = -16$

71. $b^2 - 4ac = (-2)^2 - 4(-1)(3) = 4 - (-12)$
$$= 4 + 12 = 16$$

73. $(c - b)(c + b) = (3 - (-2))(3 + (-2))$
$$= (3 + 2)(3 - 2) = (5)(1) = 5$$

75. $a^2 + 2ab + b^2 = (-1)^2 + 2(-1)(-2) + (-2)^2$
$$= 1 + 4 + 4 = 9$$

77. $a^3 - b^3 = (-1)^3 - (-2)^3 = -1 - (-8)$
$$= -1 + 8 = 7$$

79. $\dfrac{b+c}{a+b} = \dfrac{-2+3}{-1+(-2)} = \dfrac{1}{-3} = -\dfrac{1}{3}$

81. $|a - b| = |-1 - (-2)| = |-1 + 2|$
$$= |1| = 1$$

83. $(a+b)c = (-1 + (-2))(3) = (-3)(3) = -9$

85. $3(4) - 2 = 10$
$12 - 2 = 10$ Correct.
So 4 is a solution to the equation.

87. $\dfrac{3(-6)}{2} = 9$

$\dfrac{-18}{2} = 9$ Incorrect.

So -6 is not a solution to the equation.

89. $\dfrac{15+3}{2} = 9$

$\dfrac{18}{2} = 9$ Correct.

So 15 is a solution to the equation.

91. $-4 - 3 = 1$

$-7 = 1$ Incorrect.

So 4 is not a solution to the equation.

93. The statement $a(x + y) = ax + ay$ is correct because of the distributive property.

95. The statement $(0.001)(1000) = 1$ is correct because of the inverse property.

97. The statement $0 + y = y$ is correct because of the identity property. Zero is the additive identity.

99. The statement $3 + (2 + x) = (3 + 2) + x$ is correct because of the associative property of addition.

101. The statement $5 \cdot 200 = 200 \cdot 5$ is correct because of the commutative property of multiplication.

103. The statement $-50 + 50 = 0$ is correct because of the inverse property. The numbers -50 and 50 are additive inverses of each other.

105. The statement $12 \cdot 1 = 12$ is correct because of the identity property. One is the multiplicative identity.

107. $3a + 7 - (4a - 5) = 3a + 7 - 4a + 5$
$$= -a + 12$$

109. $2a(3a - 5) + 4a = 6a^2 - 10a + 4a$
$$= 6a^2 - 6a$$

111. $3(t - 2) - 5(3t - 9) = 3t - 6 - 15t + 45$
$$= -12t + 39$$

113. $0.1(a + 0.3) - (a + 0.6)$
$= 0.1a + 0.03 - a - 0.6 = -0.9a - 0.57$

115. $0.05(x - 20) - 0.1(x + 30)$
$= 0.05x - 1 - 0.1x - 3 = -0.05x - 4$

117. $5 - 3x(-5x - 2) + 12x^2$
$= 5 + 15x^2 + 6x + 12x^2$
$= 27x^2 + 6x + 5$

119. $-(a - 2) - 2 - a = -a + 2 - 2 - a = -2a$

121. $x(x + 1) + 3(x - 1) = x^2 + x + 3x - 3$
$$= x^2 + 4x - 3$$

123. $752(-13) + 752(13) = 752(-13 + 13)$
$$= 752(0) = 0$$

125. $|15 - 23| = |-8| = 8$

127. $-6^2 + 3(5) = -36 + 15 = -21$

129. $\dfrac{2}{5} + \dfrac{1}{10} = \dfrac{2 \cdot 2}{5 \cdot 2} + \dfrac{1}{10} = \dfrac{4}{10} + \dfrac{1}{10} = \dfrac{5}{10} = \dfrac{1}{2}$

131. $(0.05) \div (-0.1) = (0.5) \div (-1) = -0.5$

133. $2\left(-\dfrac{1}{2}\right)^2 + \left(-\dfrac{1}{2}\right) - 1 = 2 \cdot \dfrac{1}{4} - \dfrac{1}{2} - 1$
$$= \dfrac{2}{4} - \dfrac{1}{2} - 1 = \dfrac{1}{2} - \dfrac{1}{2} - 1 = -1$$

135. $\dfrac{2x + 4}{2} = \dfrac{1}{2}(2x + 4) = x + 2$

137. The terms in $4 + 2x$ cannot be combined because they are not like terms.

139. $4 \cdot \dfrac{x}{2} = 2 \cdot 2 \cdot \dfrac{x}{2} = 2x$

141. $-4(x - 2) = -4x - (-8) = -4x + 8$

143. $4x + 2x = (4 + 2)x = 6x$

145. $4 \cdot \dfrac{x}{4} = x$

147. $2 \cdot x \cdot 4 = 2 \cdot 4 \cdot x = 8x$

149. $\dfrac{1 \text{ membership}}{0.125 \text{ hours}} + \dfrac{1 \text{ membership}}{0.1 \text{ hours}}$

$= 8 \text{ mbs/hr} + 10 \text{ mbs/hr}$

$= 18 \text{ memberships/hr}$

CHAPTER 1 TEST

1. The only whole numbers listed are 0 and 8.

2. The only integers listed are -3, 0, and 8.

3. The rational numbers listed are -3, $-\frac{1}{4}$, 0, and 8.

4. The irrational numbers are the real numbers that are not rational. So the irrational numbers listed are $-\sqrt{3}$, $\sqrt{5}$, and π.

5. $6 + 3(-9) = 6 + (-27) = -21$

6. $(-2)^2 - 4(-2)(-1) = 4 - 8 = -4$

7. $\dfrac{-3^2 - 9}{3 - 5} = \dfrac{-9 - 9}{-2} = \dfrac{-18}{-2} = 9$

8. $-5 + 6 - 12 + 4 = -17 + 10 = -7$

9. $0.05 - 1 = -(1.00 - 0.05) = -0.95$

10. $(5 - 9)(5 + 9) = (-4)(14) = -56$

11. $(878 + 89) + 11 = 878 + (89 + 11)$

$= 878 + 100 = 978$

12. $6 + |3 - 5(2)| = 6 + |3 - 10|$

$= 6 + |-7| = 6 + 7 = 13$

13. $8 - 3|7 - 10| = 8 - 3|-3| = 8 - 3 \cdot 3$

$= 8 - 9 = -1$

14. $(839 + 974)[3(-4) + 12] = (839 + 974)[0] = 0$

15. $974(7) + 974(3) = 974(10) = 9740$

16. $-\dfrac{2}{3} + \dfrac{3}{8} = -\dfrac{2 \cdot 8}{3 \cdot 8} + \dfrac{3 \cdot 3}{8 \cdot 3} = -\dfrac{16}{24} + \dfrac{9}{24} = -\dfrac{7}{24}$

17. $(-0.05)(400) = -20$

18. $\left(-\dfrac{3}{4}\right)\left(\dfrac{2}{9}\right) = \left(-\dfrac{3}{2 \cdot 2}\right)\left(\dfrac{2}{3 \cdot 3}\right) = -\dfrac{1}{6}$

19. $13 \div \left(-\dfrac{1}{3}\right) = 13 \cdot (-3) = -39$

20. The statement $2(x + 7) = 2x + 14$ is correct because of the distributive property.

21. The statement $48 \cdot 1000 = 1000 \cdot 48$ is correct because of the commutative property of multiplication.

22. The statement $2 + (6 + x) = (2 + 6) + x$ is correct because of the associative property of addition.

23. The statement $-348 + 348 = 0$ is correct because of the inverse property.

24. The statement $1 \cdot (-6) = -6$ is correct because of the identity property.

25. The statement $0 \cdot 388 = 0$ is correct because of the multiplication property of 0.

26. $3x + 30 = 3x + 3 \cdot 10 = 3(x + 10)$

27. $7w - 7 = 7w - 7 \cdot 1 = 7(w - 1)$

28. $6 + 4x + 2x = 6x + 6$

29. $6 + 4(x - 2) = 6 + 4x - 8 = 4x - 2$

30. $5x - (3 - 2x) = 5x - 3 + 2x = 7x - 3$

31. $x + 10 - 0.1(x + 25) = x + 10 - 0.1x - 2.5$

$= 0.9x + 7.5$

32. $2a(4a - 5) - 3a(-2a - 5)$

$= 8a^2 - 10a + 6a^2 + 15a = 14a^2 + 5a$

33. $\dfrac{6x + 12}{6} = \dfrac{1}{6}(6x + 12) = \dfrac{1}{6} \cdot 6x + \dfrac{1}{6} \cdot 12$

$= x + 2$

34. $8 \cdot \dfrac{t}{2} = 4 \cdot 2 \cdot \dfrac{t}{2} = 4t$

35. $(-9xy)(-6xy) = 54x^2y^2$

36. $b^2 - 4ac = 3^2 - 4(-2)(4) = 9 - (-32)$

$= 9 + 32 = 41$

37. $\dfrac{a - b}{b - c} = \dfrac{-2 - 3}{3 - 4} = \dfrac{-5}{-1} = 5$

38. $(a - c)(a + c) = (-2 - 4)(-2 + 4)$

$= (-6)(2) = -12$

39. $3(-2) - 4 = 2$

$-10 = 2$ Incorrect.

So -2 is not a solution to the equation.

40. $\dfrac{13 + 3}{8} = 2$

$\dfrac{16}{8} = 2$ Correct.

So 13 is The solution to the equation is 13.

41. $-(-3) + 5 = 8$

$3 + 5 = 8$ Correct.

So -3 is a solution to the equation.

42. $\dfrac{1 \text{ delivery}}{0.25 \text{ hr}} + \dfrac{1 \text{ delivery}}{0.2 \text{ hr}} = 4 \text{ del/hr} + 5 \text{ del/hr}$

$= 9 \text{ deliveries per hour}$

43. $80.405 + 3.66R - 0.06(A - 30)$

$= 3.66R - 0.06A + 82.205$

$3.66(25) - 0.06(80) + 82.205$

$= 168.905 \text{ cm}$

2.1 WARM-UPS

1. True, because $2 \cdot 5 - 5 = 5$ is correct.

2. True, because 8 satisfies $\frac{x}{2} = 4$ and 8 satisfies $x - 8 = 0$.   **3.** False, because we should multiply by the reciprocal of $\frac{3}{4}$, $\frac{4}{3}$.   **4.** False, because 5 is not a solution to $w + 5 = 0$.

5. False, because the number must be a nonzero real number.

6. False, because we should first add 7 to each side and then divide each side by 3.

7. True, because $\frac{3}{2}$ is the reciprocal of $\frac{2}{3}$.

8. True, because $-3/2$ satisfies both equations.
9. True, because simplifying the left side of $x - (x - 3) = 5x$ gives us $3 = 5x$.
10. True, because $4 - (-4) = -2(-4)$ is correct.

2.1 EXERCISES

1.
$$x - 6 = -5$$
$$x - 6 + 6 = -5 + 6$$
$$x = 1$$
The solution to the equation is 1.

3.
$$x + 13 = -4$$
$$x + 13 - 13 = -4 - 13$$
$$x = -17$$
The solution to the equation is -17.

5.
$$x - \frac{1}{2} = \frac{1}{2}$$
$$x - \frac{1}{2} + \frac{1}{2} = \frac{1}{2} + \frac{1}{2}$$
$$x = 1$$

The solution to the equation is 1.

7.
$$\frac{1}{3} + x = \frac{1}{3}$$
$$\frac{1}{3} + x - \frac{1}{3} = \frac{1}{3} - \frac{1}{3}$$
$$x = 0$$

The solution to the equation is 0.

9.
$$5a = -2 + 4a$$
$$5a - 4a = -2 + 4a - 4a$$
$$a = -2$$
The solution to the equation is -2.

11.
$$-3y = 5 - 4y$$
$$-3y + 4y = 5 - 4y + 4y$$
$$y = 5$$
The solution to the equation is 5.

13.
$$-x + 6 = 5$$
$$-x + 6 - 6 = 5 - 6$$
$$-x = -1$$
$$-(-x) = -(-1)$$
$$x = 1$$
The solution to the equation is 1.

15.
$$-9 - a = -3$$
$$-9 - a + 9 = -3 + 9$$
$$-a = 6$$
$$-(-a) = -6$$
$$a = -6$$
The solution to the equation is -6.

17.
$$2q + 5 = q - 7$$
$$2q + 5 - 5 = q - 7 - 5$$
$$2q = q - 12$$
$$2q - q = q - 12 - q$$
$$q = -12$$
The solution to the equation is -12.

19.
$$-3x + 1 = 5 - 2x$$
$$-3x + 1 - 1 = 5 - 2x - 1$$
$$-3x = 4 - 2x$$
$$-3x + 2x = 4 - 2x + 2x$$
$$-x = 4$$
$$x = -4$$
The solution to the equation is -4.

21.
$$-12 - 5x = -4x + 1$$
$$-12 - 5x + 5x = -4x + 1 + 5x$$
$$-12 = x + 1$$
$$-12 - 1 = x + 1 - 1$$
$$-13 = x$$
The solution to the equation is -13.

23.
$$3x + 0.3 = 2 + 2x$$
$$3x + 0.3 - 0.3 = 2 + 2x - 0.3$$
$$3x = 2x + 1.7$$
$$3x - 2x = 2x + 1.7 - 2x$$
$$x = 1.7$$
The solution to the equation is 1.7.

25.
$$1.2k - 0.6 = 0.2k - 0.5$$
$$1.2k - 0.6 + 0.6 = 0.2k - 0.5 + 0.6$$
$$1.2k = 0.2k + 0.1$$
$$1.2k - 0.2k = 0.2k + 0.1 - 0.2k$$
$$k = 0.1$$
The solution to the equation is 0.1.

27.
$$0.2x - 4 = 0.6 - 0.8x$$
$$0.2x - 4 + 4 = 0.6 - 0.8x + 4$$
$$0.2x = 4.6 - 0.8x$$
$$0.2x + 0.8x = 4.6 - 0.8x + 0.8x$$
$$x = 4.6$$
The solution to the equation is 4.6.

29.
$$\frac{x}{2} = -4$$
$$2 \cdot \frac{x}{2} = 2 \cdot (-4)$$
$$x = -8$$
The solution to the equation is -8.

31.
$$0.03 = \frac{x}{60}$$
$$60(0.03) = 60 \cdot \frac{x}{60}$$
$$1.8 = x$$
The solution to the equation is 1.8.

33.
$$-3x = 15$$
$$\frac{-3x}{-3} = \frac{15}{-3}$$
$$x = -5$$
The solution to the equation is -5.

35.
$$20 = 4x$$
$$\frac{20}{4} = \frac{4x}{4}$$
$$5 = x$$
The solution to the equation is 5.

37.
$$\frac{a}{2} = \frac{1}{3}$$
$$2 \cdot \frac{a}{2} = 2 \cdot \frac{1}{3}$$
$$a = \frac{2}{3}$$
The solution to the equation is $\frac{2}{3}$.

39.
$$\frac{c}{3} = \frac{1}{6}$$
$$3 \cdot \frac{c}{3} = 3 \cdot \frac{1}{6}$$
$$c = \frac{1}{2}$$
The solution to the equation is $\frac{1}{2}$.

41.
$$0.5w = 10$$
$$\frac{0.5x}{0.5} = \frac{10}{0.5}$$
$$x = 20$$
The solution to the equation is 20.

43.
$$-z = \frac{2}{3}$$
$$(-1)(-z) = -1 \cdot \frac{2}{3}$$
$$z = -\frac{2}{3}$$
The solution to the equation is $-\frac{2}{3}$.

45.
$$\frac{2}{3}x = 8$$
$$\frac{3}{2} \cdot \frac{2}{3}x = \frac{3}{2} \cdot 8$$
$$x = 12$$
The solution to the equation is 12.

47.
$$-\frac{3}{5}x = 6$$
$$-\frac{5}{3}\left(-\frac{3}{5}x\right) = -\frac{5}{3} \cdot 6$$
$$x = -10$$
The solution to the equation is -10.

49.
$$\frac{5a}{7} = -10$$
$$\frac{7}{5} \cdot \frac{5a}{7} = \frac{7}{5}(-10)$$
$$a = -14$$
The solution to the equation is -14.

51.
$$-\frac{10}{3} = -\frac{5t}{6}$$
$$-\frac{6}{5}\left(-\frac{10}{3}\right) = -\frac{6}{5}\left(-\frac{5t}{6}\right)$$
$$4 = t$$
The solution to the equation is 4.

53.
$$2m = \frac{1}{2}$$
$$\frac{1}{2} \cdot 2m = \frac{1}{2} \cdot \frac{1}{2}$$
$$m = \frac{1}{4}$$
The solution to the equation is $\frac{1}{4}$.

55.
$$-2u = \frac{4}{5}$$
$$-\frac{1}{2}(-2u) = -\frac{1}{2} \cdot \frac{4}{5}$$
$$u = -\frac{2}{5}$$
The solution to the equation is $-\frac{2}{5}$.

57.
$$2x - 3 = 0$$
$$2x - 3 + 3 = 0 + 3$$
$$2x = 3$$
$$\frac{2x}{2} = \frac{3}{2}$$
The solution to the equation is $\frac{3}{2}$.

59.
$$-2x + 5 = 7$$
$$-2x + 5 - 5 = 7 - 5$$
$$-2x = 2$$
$$\frac{-2x}{-2} = \frac{2}{-2}$$
$$x = -1$$
The solution to the equation is -1.

61.
$$-2x - 5 = 7$$
$$-2x - 5 + 5 = 7 + 5$$
$$-2x = 12$$
$$\frac{-2x}{-2} = \frac{12}{-2}$$
$$x = -6$$
The solution to the equation is −6.

63.
$$-3(k - 6) = 2 - k$$
$$-3k + 18 = 2 - k$$
$$-3k + 18 - 18 = 2 - k - 18$$
$$-3k = -k - 16$$
$$-3k + k = -k - 16 + k$$
$$-2k = -16$$
$$\frac{-2k}{-2} = \frac{-16}{-2}$$
$$k = 8$$
The solution to the equation is 8.

65.
$$2(p + 1) - p = 36$$
$$2p + 2 - p = 36$$
$$p + 2 = 36$$
$$p + 2 - 2 = 36 - 2$$
$$p = 34$$
The solution to the equation is 34.

67.
$$7 - 3(5 - u) = 5(u - 4)$$
$$7 - 15 + 3u = 5u - 20$$
$$-8 + 3u = 5u - 20$$
$$-8 - 2u = -20$$
$$-2u = -12$$
$$u = 6$$
The solution to the equation is 6.

69.
$$4(x + 3) = 12$$
$$4x + 12 = 12$$
$$4x = 0$$
$$\frac{4x}{4} = \frac{0}{4}$$
$$x = 0$$
The solution to the equation is 0.

71.
$$-3x - 1 = 5 - 2x$$
$$-3x - 1 - 5 = 5 - 2x - 5$$
$$-3x - 6 = -2x$$
$$-3x - 6 + 3x = -2x + 3x$$
$$-6 = x$$
The solution to the equation is −6.

73.
$$0.3(x + 30) = 27$$
$$0.3x + 9 = 27$$
$$0.3x = 18$$
$$\frac{0.3x}{0.3} = \frac{18}{0.3}$$
$$x = 60$$
The solution to the equation is 60.

75.
$$\frac{w}{5} - 4 = -6$$
$$\frac{w}{5} - 4 + 4 = -6 + 4$$
$$\frac{w}{5} = -2$$
$$5 \cdot \frac{w}{5} = 5(-2)$$
$$w = -10$$
The solution to the equation is −10.

77.
$$\frac{2}{3}y - 5 = 7$$
$$\frac{2}{3}y - 5 + 5 = 7 + 5$$
$$\frac{2}{3}y = 12$$
$$\frac{3}{2} \cdot \frac{2}{3}y = \frac{3}{2} \cdot 12$$
$$y = 18$$
The solution to the equation is 18.

79.
$$4 - \frac{2n}{5} = 12$$
$$4 - \frac{2n}{5} - 4 = 12 - 4$$
$$-\frac{2n}{5} = 8$$
$$-\frac{5}{2}\left(-\frac{2n}{5}\right) = -\frac{5}{2} \cdot 8$$
$$n = -20$$
The solution to the equation is −20.

81.
$$-\frac{1}{3}p - \frac{1}{2} = \frac{1}{2}$$
$$-\frac{1}{3}p - \frac{1}{2} + \frac{1}{2} = \frac{1}{2} + \frac{1}{2}$$
$$-\frac{1}{3}p = 1$$
$$-3\left(-\frac{1}{3}p\right) = -3 \cdot 1$$
$$p = -3$$
The solution to the equation is −3.

83.
$$3x - 2(x - 4) = 4 - (x - 5)$$
$$3x - 2x + 8 = 4 - x + 5$$
$$x + 8 = -x + 9$$
$$x = -x + 1$$
$$2x = 1$$
$$x = \frac{1}{2}$$
The solution to the equation is $\frac{1}{2}$.

85.
$$3.5x - 23.7 = -38.75$$
$$3.5x - 23.7 + 23.7 = -38.75 + 23.7$$
$$3.5x = -15.05$$
$$\frac{3.5x}{3.5} = \frac{-15.05}{3.5}$$
$$x = -4.3$$

The solution to the equation is -4.3.

87.
$$\frac{9}{5}C + 32 = 68$$
$$\frac{9}{5}C = 36$$
$$\frac{5}{9} \cdot \frac{9}{5}C = \frac{5}{9} \cdot 36$$
$$C = 20$$

The air temperature is $20°C$.

89.
$$2x + 2(x + 3) = 42$$
$$2x + 2x + 6 = 42$$
$$4x = 36$$
$$x = 9$$

The width is 9 feet.

2.2 WARM-UPS

1. True, because if we simplify the left side we get the inconsistent equation $0 = 99$.
2. True, because if we simplify the left side we get $5n = 5n$. **3.** False, because 0 satisfies the equation. **4.** False, because $1 \div 0$ is undefined.
5. False, because $5a + 3 = 0$ is equivalent to $a = -3/5$. **6.** True, because $2t = t$ is equivalent to $t = 0$. **7.** True, because if we simplify the left side we get $0.9w = 0.9w$.
8. False, because if we multiply each side by 100 we get $20x + 3x = 800$. **9.** True, because $x/x = 1$ is satisfied by every real number except 0 and x/x is undefined if $x = 0$.
10. True, because $3(8/3) - 8 = 0$.

2.2 EXERCISES

1.
$$x + x = 2x$$
$$2x = 2x$$
All real numbers satisfy the equation. The equation is an identity.

3.
$$a - 1 = a + 1$$
$$a - 1 - a = a + 1 - a$$
$$-1 = 1$$
The equation has no solution. It is an inconsistent equation.

5.
$$3y + 4y = 12y$$
$$7y = 12y$$
$$7y - 7y = 12y - 7y$$
$$0 = 5y$$
$$\frac{0}{5} = \frac{5y}{5}$$
$$0 = y$$
The solution to the equation is 0. The equation is a conditional equation.

7.
$$-4 + 3(w - 1) = w + 2(w - 2) - 1$$
$$-4 + 3w - 3 = w + 2w - 4 - 1$$
$$3w - 7 = 3w - 5$$
$$-7 = -5$$
The equation has no solution. It is an inconsistent equation.

9.
$$3(m + 1) = 3(m + 3)$$
$$3m + 3 = 3m + 9$$
$$3m + 3 - 3m = 3m + 9 - 3m$$
$$3 = 9$$
The equation has no solution. It is an inconsistent equation.

11.
$$x + x = 2$$
$$2x = 2$$
$$\frac{2x}{2} = \frac{2}{2}$$
$$x = 1$$
The solution to the equation is 1. It is a conditional equation.

13.
$$2 - 3(5 - x) = 3x$$
$$2 - 15 + 3x = 3x$$
$$-13 + 3x = 3x$$
$$-13 + 3x - 3x = 3x - 3x$$
$$-13 = 0$$
The equation has no solution. It is an inconsistent equation.

15.
$$(3 - 3)(5 - z) = 0$$
$$0(5 - z) = 0$$
$$0 = 0$$
All real numbers satisfy the equation. The equation is an identity.

17.
$$\frac{0}{x} = 0$$
The equation is satisfied by every nonzero real number. The equation is an identity.

19.
$$x \cdot x = x^2$$
$$x^2 = x^2$$
All real numbers satisfy the equation. The equation is an identity.

21.
$$\frac{x}{2} + 3 = x - \frac{1}{2}$$
$$2\left(\frac{x}{2} + 3\right) = 2\left(x - \frac{1}{2}\right)$$
$$x + 6 = 2x - 1$$
$$x + 6 + 1 = 2x - 1 + 1$$
$$x + 7 = 2x$$
$$x + 7 - x = 2x - x$$
$$7 = x$$

The solution to the equation is 7.

23.
$$\frac{x}{2} + \frac{x}{3} = 20$$
$$6\left(\frac{x}{2} + \frac{x}{3}\right) = 6 \cdot 20$$
$$3x + 2x = 120$$
$$5x = 120$$
$$\frac{5x}{5} = \frac{120}{5}$$
$$x = 24$$

The solution to the equation is 24.

25.
$$\frac{w}{2} + \frac{w}{4} = 12$$
$$4\left(\frac{w}{2} + \frac{w}{4}\right) = 4 \cdot 12$$
$$2w + w = 48$$
$$3w = 48$$
$$\frac{3w}{3} = \frac{48}{3}$$
$$w = 16$$

The solution to the equation is 16.

27.
$$\frac{3z}{2} - \frac{2z}{3} = -10$$
$$6\left(\frac{3z}{2} - \frac{2z}{3}\right) = 6(-10)$$
$$9z - 4z = -60$$
$$5z = -60$$
$$\frac{5z}{5} = \frac{-60}{5}$$
$$z = -12$$

The solution to the equation is −12.

29.
$$\frac{1}{3}p - 5 = \frac{1}{4}p$$
$$12\left(\frac{1}{3}p - 5\right) = 12 \cdot \frac{1}{4}p$$
$$4p - 60 = 3p$$
$$4p - 60 + 60 = 3p + 60$$
$$4p = 3p + 60$$
$$4p - 3p = 3p + 60 - 3p$$
$$p = 60$$

The solution to the equation is 60.

31.
$$\frac{1}{6}v + 1 = \frac{1}{4}v - 1$$
$$12\left(\frac{1}{6}v + 1\right) = 12\left(\frac{1}{4}v - 1\right)$$
$$2v + 12 = 3v - 12$$
$$2v + 12 + 12 = 3v - 12 + 12$$
$$2v + 24 = 3v$$
$$2v + 24 - 2v = 3v - 2v$$
$$24 = v$$

The solution to the equation is 24.

33.
$$x - 0.2x = 72$$
$$10(x - 0.2x) = 10 \cdot 72$$
$$10x - 2x = 720$$
$$8x = 720$$
$$\frac{8x}{8} = \frac{720}{8}$$
$$x = 90$$

The solution to the equation is 90.

35.
$$0.3x + 1.2 = 0.5x$$
$$10(0.3x + 1.2) = 10(0.5x)$$
$$3x + 12 = 5x$$
$$3x + 12 - 3x = 5x - 3x$$
$$12 = 2x$$
$$\frac{12}{2} = \frac{2x}{2}$$
$$6 = x$$

The solution to the equation is 6.

37.
$$0.02x - 1.56 = 0.8x$$
$$100(0.02x - 1.56) = 100(0.8x)$$
$$2x - 156 = 80x$$
$$2x - 156 - 2x = 80x - 2x$$
$$-156 = 78x$$
$$\frac{-156}{78} = \frac{78x}{78}$$
$$-2 = x$$

The solution to the equation is −2.

39.
$$0.1a - 0.3 = 0.2a - 8.3$$
$$10(0.1a - 0.3) = 10(0.2a - 8.3)$$
$$a - 3 = 2a - 83$$
$$a - 3 + 83 = 2a - 83 + 83$$
$$a + 80 = 2a$$
$$a + 80 - a = 2a - a$$
$$80 = a$$
The solution to the equation is 80.

41.
$$0.05r + 0.4r = 27$$
$$100(0.05r + 0.4r) = 100 \cdot 27$$
$$5r + 40r = 2700$$
$$45r = 2700$$
$$\frac{45r}{45} = \frac{2700}{45}$$
$$r = 60$$
The solution to the equation is 60.

43.
$$0.05y + 0.03(y + 50) = 17.5$$
$$100[0.05y + 0.03(y + 50)] = 100(17.5)$$
$$5y + 3(y + 50) = 1750$$
$$5y + 3y + 150 = 1750$$
$$8y + 150 = 1750$$
$$8y + 150 - 150 = 1750 - 150$$
$$8y = 1600$$
$$\frac{8y}{8} = \frac{1600}{8}$$
$$y = 200$$
The solution to the equation is 200.

45.
$$0.1x + 0.05(x - 300) = 105$$
$$100[0.1x + 0.05(x - 300)] = 100 \cdot 105$$
$$10x + 5(x - 300) = 10500$$
$$10x + 5x - 1500 = 10500$$
$$15x - 1500 = 10500$$
$$15x - 1500 + 1500 = 10500 + 1500$$
$$15x = 12000$$
$$\frac{15x}{15} = \frac{12000}{15}$$
$$x = 800$$
The solution to the equation is 800.

47.
$$2x - 9 = 0$$
$$2x = 9$$
$$x = \frac{9}{2}$$
The solution to the equation is $\frac{9}{2}$.

49.
$$-2x + 6 = 0$$
$$-2x = -6$$
$$x = 3$$
The solution to the equation is 3.

51.
$$\frac{z}{5} + 1 = 6$$
$$\frac{z}{5} = 5$$
$$z = 25$$
The solution to the equation is 25.

53.
$$\frac{c}{2} - 3 = -4$$
$$\frac{c}{2} = -1$$
$$c = -2$$
The solution to the equation is -2.

55.
$$3 = t + 6$$
$$-3 = t$$
The solution to the equation is -3.

57.
$$5 + 2q = 3q$$
$$5 = q$$
The solution to the equation is 5.

59.
$$8x - 1 = 9 + 9x$$
$$-10 = x$$
The solution to the equation is -10.

61.
$$-3x + 1 = -1 - 2x$$
$$2 = x$$
The solution to the equation is 2.

63.
$$3x - 5 = 2x - 9$$
$$3x - 5 + 5 = 2x - 9 + 5$$
$$3x = 2x - 4$$
$$3x - 2x = 2x - 4 - 2x$$
$$x = -4$$
The solution to the equation is -4.

65.
$$x + 2(x + 4) = 3(x + 3) - 1$$
$$3x + 8 = 3x + 8$$
All real numbers satisfy the equation.

67.
$$23 - 5(3 - n) = -4(n - 2) + 9n$$
$$23 - 15 + 5n = -4n + 8 + 9n$$
$$8 + 5n = 5n + 8$$
All real numbers satisfy the equation.

69.
$$0.05x + 30 = 0.4x - 5$$
$$100(0.05x + 30) = 100(0.4x - 5)$$
$$5x + 3000 = 40x - 500$$
$$5x + 3000 + 500 = 40x - 500 + 500$$
$$5x + 3500 = 40x$$
$$5x + 3500 - 5x = 40x - 5x$$
$$3500 = 35x$$
$$\frac{3500}{35} = \frac{35x}{35}$$
$$100 = x$$
The solution to the equation is 100.

71.
$$-\frac{2}{3}a + 1 = 2$$
$$3\left(-\frac{2}{3}a + 1\right) = 3 \cdot 2$$
$$-2a + 3 = 6$$
$$-2a + 3 - 3 = 6 - 3$$
$$-2a = 3$$
$$\frac{-2x}{-2} = \frac{3}{-2}$$
$$x = -\frac{3}{2}$$

The solution to the equation is $-\frac{3}{2}$.

73.
$$\frac{y}{2} + \frac{y}{6} = 20$$
$$6\left(\frac{y}{2} + \frac{y}{6}\right) = 6 \cdot 20$$
$$3y + y = 120$$
$$4y = 120$$
$$\frac{4y}{4} = \frac{120}{4}$$
$$y = 30$$

The solution to the equation is 30.

75.
$$0.09x - 0.2(x + 4) = -1.46$$
$$0.09x - 0.2x - 0.8 = -1.46$$
$$9x - 20x - 80 = -146$$
$$-11x - 80 = -146$$
$$-11x = -66$$
$$x = 6$$

The solution to the equation is 6.

77.
$$436x - 789 = -571$$
$$436x - 789 + 789 = -571 + 789$$
$$436x = 218$$
$$\frac{436x}{436} = \frac{218}{436}$$
$$x = 0.5$$

The solution to the equation is 0.5.

79.
$$\frac{x}{344} + 235 = 292$$
$$\frac{x}{344} + 235 - 235 = 292 - 235$$
$$\frac{x}{344} = 57$$
$$344\left(\frac{x}{344}\right) = 344 \cdot 57$$
$$x = 19608$$

The solution to the equation is 19,608.

81.
$$0.15(38,000) + 0.28(x - 38,000) = 19,700$$
$$5700 + 0.28x - 10,640 = 19,700$$
$$0.28x = 24,640$$
$$x = 88,000$$

The taxable income is \$88,000.

2.3 WARM-UPS

1. False, because $D = R \cdot T$ solved for T is $T = D/R$. **2.** False, because a also appears on the right side of the equation. **3.** False, because $A = L \cdot W$ solved for L is $L = A/W$. **4.** False, because $D = R \cdot T$ solved for R is $R = D/T$. **5.** False, because the perimeter of a rectangle is $P = 2L + 2W$. **6.** True, because for a rectangular solid, $V = LWH$. **7.** True, because $P = 2(L + W)$. **8.** True, because if we add x to each side of $y - x = 5$ we get $y = x + 5$. **9.** False, because if $x = -1$, then $y = -3(-1) + 6 = 3 + 6 = 9$. **10.** True, because $C = \pi D$.

2.3 EXERCISES

1.
$$D = R \cdot T$$
$$\frac{D}{T} = \frac{R \cdot T}{T}$$
$$\frac{D}{T} = R$$
$$R = \frac{D}{T}$$

3.
$$C = \pi D$$
$$\frac{C}{D} = \pi$$
$$\pi = \frac{C}{D}$$

5.
$$I = Prt$$
$$\frac{I}{rt} = \frac{Prt}{rt}$$
$$P = \frac{I}{rt}$$

7.
$$F = \frac{9}{5}C + 32$$
$$F - 32 = \frac{9}{5}C$$
$$\frac{5}{9}(F - 32) = \frac{5}{9} \cdot \frac{9}{5}C$$
$$\frac{5}{9}(F - 32) = C$$
$$C = \frac{5}{9}(F - 32)$$

9.

$$A = \tfrac{1}{2}bh$$

$$2 \cdot A = 2 \cdot \tfrac{1}{2}bh$$

$$2A = bh$$

$$\frac{2A}{b} = \frac{bh}{b}$$

$$\frac{2A}{b} = h$$

$$h = \frac{2A}{b}$$

11.

$$P = 2L + 2W$$
$$P - 2W = 2L + 2W - 2W$$
$$P - 2W = 2L$$

$$\frac{P - 2W}{2} = \frac{2L}{2}$$

$$\frac{P - 2W}{2} = L$$

$$L = \frac{P - 2W}{2}$$

13.

$$A = \tfrac{1}{2}(a + b)$$

$$2A = 2 \cdot \tfrac{1}{2}(a + b)$$

$$2A = a + b$$
$$2A - b = a + b - b$$
$$2A - b = a$$
$$a = 2A - b$$

15.

$$S = P + Prt$$
$$S - P = P + Prt - P$$
$$S - P = Prt$$

$$\frac{S - P}{Pt} = \frac{Prt}{Pt}$$

$$\frac{S - P}{Pt} = r$$

$$r = \frac{S - P}{Pt}$$

17.

$$A = \tfrac{1}{2}h(a + b)$$

$$2A = 2 \cdot \tfrac{1}{2}h(a + b)$$

$$2A = h(a + b)$$
$$2A = ah + bh$$
$$2A - bh = ah + bh - bh$$
$$2A - bh = ah$$

$$\frac{2A - bh}{h} = \frac{ah}{h}$$

$$a = \frac{2A - hb}{h}$$

19.

$$5x + a = 3x + b$$
$$5x = 3x + b - a$$
$$5x - 3x = b - a$$
$$2x = b - a$$

$$x = \frac{b - a}{2}$$

21.

$$4(a + x) - 3(x - a) = 0$$
$$4a + 4x - 3x + 3a = 0$$
$$x + 7a = 0$$
$$x = -7a$$

23.

$$3x - 2(a - 3) = 4x - 6 - a$$
$$3x - 2a + 6 = 4x - 6 - a$$
$$3x - 2a + a + 6 + 6 = 4x$$
$$3x - a + 12 = 4x$$
$$-a + 12 = 4x - 3x$$
$$12 - a = x$$
$$x = 12 - a$$

25.

$$3x + 2ab = 4x - 5ab$$
$$3x + 2ab + 5ab = 4x$$
$$3x + 7ab = 4x$$
$$7ab = 4x - 3x$$
$$7ab = x$$
$$x = 7ab$$

27.

$$x + y = -9$$
$$y = -x - 9$$

29.

$$x + y - 6 = 0$$
$$y = -x + 6$$

31.

$$2x - y = 2$$
$$2x = y + 2$$
$$2x - 2 = y$$
$$y = 2x - 2$$

33.

$$3x - y + 4 = 0$$
$$3x + 4 = y$$
$$y = 3x + 4$$

35.

$$x + 2y = 4$$
$$2y = -x + 4$$

$$\tfrac{1}{2} \cdot 2y = \tfrac{1}{2} \cdot (-x + 4)$$

$$y = -\tfrac{1}{2}x + 2$$

37.

$$2x - 2y = 1$$

$$-2y = -2x + 1$$

$$-\tfrac{1}{2}(-2y) = -\tfrac{1}{2}(-2x + 1)$$

$$y = x - \tfrac{1}{2}$$

39.

$$y + 2 = 3(x - 4)$$
$$y + 2 = 3x - 12$$
$$y = 3x - 12 - 2$$
$$y = 3x - 14$$

41.
$$y - 1 = \tfrac{1}{2}(x - 2)$$
$$y - 1 = \tfrac{1}{2}x - 1$$
$$y = \tfrac{1}{2}x$$

43.
$$\tfrac{1}{2}x - \tfrac{1}{3}y = -2$$
$$-\tfrac{1}{3}y = -\tfrac{1}{2}x - 2$$
$$-3\left(-\tfrac{1}{3}y\right) = -3\left(-\tfrac{1}{2}x - 2\right)$$
$$y = \tfrac{3}{2}x + 6$$

45. Let $x = 2$ in the equation $y = 3x - 4$.
$$y = 3(2) - 4 = 6 - 4 = 2$$

47. Let $x = 2$ in the equation $3x - 2y = -8$.
$$3(2) - 2y = -8$$
$$6 - 2y = -8$$
$$-2y = -14$$
$$y = 7$$

49. Let $x = 2$ in the equation $\tfrac{3x}{2} - \tfrac{5y}{3} = 6$.
$$\frac{3(2)}{2} - \frac{5y}{3} = 6$$
$$3 - \frac{5y}{3} = 6$$
$$9 - 5y = 18$$
$$-5y = 9$$
$$y = -\tfrac{9}{5}$$

51. Let $x = 2$ in the equation $y - 3 = \tfrac{1}{2}(x - 6)$.
$$y - 3 = \tfrac{1}{2}(2 - 6)$$
$$y - 3 = \tfrac{1}{2}(-4)$$
$$y - 3 = -2$$
$$y = 1$$

53. Let $x = 2$ in $y - 4.3 = 0.45(x - 8.6)$.
$$y - 4.3 = 0.45(2 - 8.6)$$
$$y - 4.3 = 0.45(-6.6)$$
$$y - 4.3 = -2.97$$
$$y = -2.97 + 4.3$$
$$y = 1.33$$

55. Use $P = 5000$, $t = 3$, and $I = 600$ in the formula for simple interest, $I = Prt$.
$$600 = 5000r(3)$$
$$600 = 15000r$$
$$\frac{600}{15000} = r$$
$$0.04 = r$$

The interest rate is 4%.

57. Use $I = 500$, $P = 2500$, and $r = 5\%$ in the formula for simple interest $I = Prt$.
$$500 = 2500(0.05)t$$
$$500 = 125t$$
$$\frac{500}{125} = t$$
$$4 = t$$

The time was 4 years.

59. The formula for the area of a rectangle is $A = LW$. Use $A = 28$ and $W = 4$ to find L.
$$28 = 4L$$
$$7 = L$$

The length is 7 yards.

61. Use $P = 600$ and $W = 75$ in the formula for the perimeter of a rectangle, $P = 2L + 2W$.
$$600 = 2L + 2(75)$$
$$600 = 2L + 150$$
$$450 = 2L$$
$$225 = L$$

The length is 225 feet.

63. Use $S = 255$, and $r = 15\%$ in the formula for the sale price, $S = L - rL$.
$$255 = L - 0.15L$$
$$255 = 0.85L$$
$$\frac{255}{0.85} = L$$
$$300 = L$$

The original price is $300.

65. Use $d = 40$ and $b = 200$ in the formula for discount, $d = br$.
$$40 = 200r$$
$$\frac{40}{200} = r$$
$$0.2 = r$$

The rate of discount was 20%.

67. The length of a football field is 100 yards or 300 feet. Use $P = 920$ and $L = 300$ in the formula $P = 2L + 2W$.
$$920 = 2(300) + 2W$$
$$920 = 600 + 2W$$
$$320 = 2W$$
$$160 = W$$

The width is 160 feet.

69. Use $W = 2$, $L = 3$, and $H = 4$ in the formula for the volume of a rectangular solid, $V = LWH$.
$$V = 3 \cdot 2 \cdot 4 = 24$$
The volume is 24 cubic feet.

71. Use $C = 8\pi$ in the formula for the circumference of a circle, $C = 2\pi r$.
$$8\pi = 2\pi r$$
$$\frac{8\pi}{2\pi} = \frac{2\pi r}{2\pi}$$
$$4 = r$$
The radius is 4 inches.

73. Use $A = 16$ and $b = 4$ in the formula for the area of a triangle, $A = \frac{1}{2}bh$.
$$16 = \frac{1}{2} 4h$$
$$16 = 2h$$
$$8 = h$$

The height is 8 feet.

75. Use $A = 200$, $h = 20$, and $b_1 = 8$ in the formula for the area of a trapezoid, $A = \frac{1}{2}h(b_1 + b_2)$.
$$200 = \frac{1}{2} \cdot 20(8 + b_2)$$
$$200 = 10(8 + b_2)$$
$$200 = 80 + 10b_2$$
$$120 = 10b_2$$
$$12 = b_2$$

The length of the upper base is 12 inches.

77. Let $D = 1000$ and $a = 8$ in $d = 0.08aD$:
$$d = 0.08(8)(1000) = 640$$
Child's dosage is 640 milligrams.
Child gets same dosage as an adult at about 13 years of age.

79. $\text{amount} = \dfrac{750 \text{ mg}}{1 \text{ gram}} \times 5 \text{ milliliters}$

$= \dfrac{750 \text{ mg}}{1000 \text{ mg}} \times 5 \text{ ml} = 3.75 \text{ ml}$

2.4 WARM-UPS

1. True, because $x + 6$ is 6 more than x.
2. True, because $a + (10 - a) = 10$.   **3.** True, because $D = RT$.   **4.** False, because $T = D/R$ means that her time is $10/x$ hours.   **5.** True, because the selling price minus the commission is what the owner gets.   **6.** False, because $55{,}000 - 0.10(55{,}000) \neq 50{,}000$.   **7.** False, because if x is odd then $x + 1$ is even.   **8.** False, because n nickels at 5 cents each and d dimes at 10 cents each have total value in cents of $5n + 10d$.   **9.** True, because to get the total bill we add the amount of tax 0.05x to the amount of goods x, to get $x + 0.05x = 1.05x$.   **10.** False, because the perimeter is found by adding twice the length to twice the width.

2.4 EXERCISES

1. The sum of a number and 3 indicates addition and so the algebraic expression for that phrase is $x + 3$.
3. Three less than a number indicates subtraction. So the algebraic expression for the phrase is $x - 3$.
5. The product of a number and 5 indicates multiplication. So the algebraic expression is 5x.
7. Ten percent of a number is found by multiplying the number by 0.10 or 0.1. So the algebraic expression is 0.1x.
9. $x/3$ **11.** $\frac{1}{3}x$

13. x and $x + 15$ **15.** x and $6 - x$

17. x and $-4 - x$ **19.** x and $x + 3$

21. x and 0.05x   **23.** x and 1.30x

25. x and $90 - x$ **27.** x and $120 - x$
29. Since consecutive even integers differ by 2, two consecutive even integers are expressed as x and $x + 2$.
31. Consecutive integers differ by 1. So two consecutive integers are represented as x and $x + 1$.
33. m and $m - 2$ **35.** y, $y + 2$, $y + 4$
37. If we use $R = x$ and $t = 3$ in the formula $D = RT$, we get $D = 3x$. So an expression for the distance is 3x mi.
39. Since the discount is 25% of the original price q, the discount is 0.25q dollars.
41. Use $D = x$ and $R = 20$ in the formula $T = D/R$, to get $T = x/20$. So an expression for the time is $x/20$ hr.
43. Use $D = x - 100$ and $T = 12$ in the formula $R = D/T$, to get $R = (x - 100)/12$. So an expression for the rate is $(x - 100)/12$ m/sec.
45. Since the area of a rectangle is the length times the width, the area is $5x$ m^2.

47. Since the perimeter of a rectangle is twice the length plus twice the width, the perimeter in this case is $2w + 2(w + 3)$ in.

49. If the perimeter of the rectangle is 300, then the total of the length and width is 150. If the length is x, then the width is expressed as $150 - x$ ft.

51. If the width is x and the length is 1 foot longer than twice the width, then the length is expressed as $2x + 1$ ft.

53. If the width is x and the length is 5 meters longer, then the length is $x + 5$. Since the area is length times width for a rectangle, we can express the area as $x(x + 5)$ m^2.

55. The simple interest is given by the formula $I = Prt$. So if P is $x + 1000$, r is 18%, and t is 1 year, then the simple interest is expressed as $0.18(x + 1000)$.

57. To find the price per pound we divide the total price by the number of pounds. So the price per pound for the peaches is expressed as $16.50/x$ dollars per pound.

59. Since the sum of complementary angles is 90°, the degree measure is $90 - x$ degrees.

61. Two numbers that differ by 5 are expressed as x and $x + 5$. If their product is 8, we can write the equation $x(x + 5) = 8$.

63. If x is the selling price, the agent gets $0.07x$. Since Herman receives the selling price less the commission, $x - 0.07x = 84{,}532$.

65. To find a percent of 500 we multiply the rate x by 500: $500x = 100$.

67. The value in dollars of x nickels is $0.05x$. The value in dollars of $x + 2$ dimes is $0.10(x + 2)$. Since we know that the total value is $3.80, we can write the equation $0.05x + 0.10(x + 2) = 3.80$. We could express the total value in cents as $5x + 10(x + 2) = 380$.

69. Sum indicates addition. The sum of a number (x) and 5 is 13 is written as the equation $x + 5 = 13$.

71. Three consecutive integers are represented as x, $x + 1$, and $x + 2$. Since their sum is 42, we can write $x + (x + 1) + (x + 2) = 42$.

73. Two consecutive integers are represented as x and $x + 1$. Since their product is 182, we can write the equation $x(x + 1) = 182$.

75. To find 12% of Harriet's income we multiply Harriet's income (x) by 0.12. Since we know that 12% of her income is $3000, we write the equation $0.12x = 3000$.

77. To find 5% of a number we multiply the number (x) by 0.05. Since we know that 5% of the number is 13, we can write the equation $0.05x = 13$.

79. Since the length is 5 feet longer than the width, we can represent the length and width by x and $x + 5$. Since the area is 126, we can write the equation $x(x + 5) = 126$.

81. The number of cents in n nickels is 5n and the number of cents in $n - 1$ dimes is $10(n - 1)$. Since the total value is 95 cents, we can write the equation $5n + 10(n - 1) = 95$.

83. The measures of the two angles are x and $x - 38$. Since the angles are supplementary, we have $x + x - 38 = 180$.

85. $x(x + 3) = 24$ **87.** $w(w - 4) = 24$

2.5 WARM-UPS

1. False, because before you write an equation you must read the problem and identify the variables.   **2.** True, because we must be certain what the variable represents in order to write an equation about it.   **3.** True, because diagrams help us to understand the situation.

4. False, because consecutive odd integers differ by 2.   **5.** False, because if 5x is larger than $3(x + 20)$ we can write the equation $5x - 2 = 3(x + 20)$.   **6.** True, because $x + 6 - x = 6$ is true for any real number x.

7. True, because $x + 7 - x = 7$ for any real number x.   **8.** True, because complementary angles have a sum of 90°.   **9.** False, because x and $x + 180$ do not have a sum of 180 for all values of x.   **10.** True, because $x = \frac{1}{2}(x + 50)$ or $2x = x + 50$.

2.5 EXERCISES

1. Let x = the first integer, $x + 1$ = the second integer, and $x + 2$ = the third integer. Since their sum is 141, we can write the following equation.

$$x + x + 1 + x + 2 = 141$$
$$3x + 3 = 141$$
$$3x = 138$$
$$x = 46$$
$$x + 1 = 47$$
$$x + 2 = 48$$

The three integers are 46, 47, and 48.

3. Let x = the first odd integer and x + 2 = the second odd integer. Since their sum is 152, we can write the following equation.

$$x + x + 2 = 152$$
$$2x + 2 = 152$$
$$2x = 150$$
$$x = 75$$
$$x + 2 = 77$$

The two consecutive odd integers are 75 and 77.

5. Let x = the first integer, x + 1 = the second integer, x + 2 = the third integer, and x + 3 = the fourth integer. Since their sum is 194, we can write the following equation.

$$x + x + 1 + x + 2 + x + 3 = 194$$
$$4x + 6 = 194$$
$$4x = 188$$
$$x = 47$$
$$x + 1 = 48$$
$$x + 2 = 49$$
$$x + 3 = 50$$

The three integers are 47, 48, 49, and 50.

7. Let x = the width and 2x = the length. Since the perimeter is 150, we can write the following.

$$2(x) + 2(2x) = 150$$
$$6x = 150$$
$$x = 25$$
$$2x = 50$$

The length is 50 m and the width is 25 m.

9. Let x = the width and x + 4 = the length. Since the perimeter is 176, we can write the following equation.

$$2x + 2(x + 4) = 176$$
$$2x + 2x + 8 = 176$$
$$4x + 8 = 176$$
$$4x = 168$$
$$x = 42$$
$$x + 4 = 46$$

The width is 42 inches and the length is 46 inches.

11. Let x = the length of each of the equal sides and x − 5 = the length of the base of the triangle. Since the perimeter is 34 inches, we can write the following equation.

$$x + x + x - 5 = 34$$
$$3x = 39$$
$$x = 13$$

The length of each of the equal sides is 13 inches.

13.
$$2w + 2w + 40 = 180$$
$$4w + 40 = 180$$
$$4w = 140$$
$$w = 35$$

So the angle marked w is 35°.

15. Let x = his speed on the freeway and x − 20 = his speed on the country road. Since D = RT, his distance on the freeway was 4x and his distance on the country road was 5(x − 20). Since his total distance was 485 miles, we can write the following equation.

$$4x + 5(x - 20) = 485$$
$$4x + 5x - 100 = 485$$
$$9x = 585$$
$$x = 65$$

He traveled 65 mph on the freeway.

17. Let x = her speed after dawn and x + 5 = her speed before dawn. Her distance after dawn was 6x and her distance before dawn was 5(x + 5). Since her total distance was 630 miles, we can write the following equation.

$$6x + 5(x + 5) = 630$$
$$6x + 5x + 25 = 630$$
$$11x + 25 = 630$$
$$11x = 605$$
$$x = 55$$

Her speed after dawn was 55 mph.

19. Let x = the time in hours to L.A. and x + 48/60 = the time in hours to Chicago. Since D = RT, we have

$$640x = 512(x + 0.8)$$
$$640x = 512x + 409.6$$
$$128x = 409.6$$
$$x = 3.2$$
$$640x = 2048$$

The trip from L.A. to Chicago was 3.2 hours and the trip from Chicago to L.A. was 4 hours. The distance from Chicago to L.A. is 2048 miles.

21. Let x = the number of points scored by the Raiders and x − 18 = the number scored by the Vikings.

$$x + x - 18 = 46$$
$$2x = 64$$
$$x = 32$$
$$x - 18 = 14$$

The scored was Raiders 32, Vikings 14.

23. Let $x =$ the driving time before lunch and $x - 1 =$ the driving time after lunch. Since $D = RT$,

$$50x + 53(x - 1) = 256$$
$$103x - 53 = 256$$
$$103x = 309$$
$$x = 3$$
$$x - 1 = 2$$

She drove for 3 hours before lunch. The distance from Ardmore to Lawton is 2(53) or 106 miles.

25. Let $x =$ Crawford's age in 1950, $x - 1 =$ John Wayne's age in 1950, and $x - 2 =$ James Stewart's age in 1950.

$$x + x - 1 + x - 2 = 129$$
$$3x = 132$$
$$x = 44$$
$$x - 1 = 43$$
$$x - 2 = 42$$

So Crawford was born in 1906, Wayne in 1907, and Stewart in 1908.

2.6 WARM-UPS

1. True, because selling price is original price minus the discount. **2.** True. **3.** False, because $4,000 \neq 0.10(44,000)$. **4.** True, because $20\%(10) + 30\%(x) = 2 + 0.3x$. **5.** False, the percent of acid in the mixture will be between 10% and 14%. **6.** True, because $x + 0.05x = 1.05x$.

2.6 EXERCISES

1. Let $x =$ the original price of the television and $0.25x =$ the amount of the discount. Since the amount of the discount is $80, we can write the following equation.

$$0.25x = 80$$
$$x = \frac{80}{0.25} = 320$$

The original price was $320.

3. Let $x =$ the original price and $0.20x =$ the amount of the discount. Since the price after the discount was $320, we can write the following equation.

$$x - 0.20x = 320$$
$$0.80x = 320$$

$$x = \frac{320}{0.80} = 400$$

The original price was $400.

5. Let $x =$ the selling price, and $0.10x =$ the real estate commission. The selling price minus the commission is what Kirk receives.

$$x - 0.10x = 72,000$$
$$0.90x = 72,000$$
$$x = 80,000$$

The house should sell for $80,000.

7. Let $x =$ the amount of her sales and $0.07x =$ the amount of sales tax. Since her total receipts were $462.24, we can write the following equation.

$$x + 0.07x = 462.24$$
$$1.07x = 462.24$$
$$x = 432$$

The sales tax was $0.07(432) = \$30.24$.

9. Let $x =$ the amount invested in the 100 fund and $x + 3000 =$ the amount invested in the 101 fund.

$$0.18x + 0.15(x + 3000) = 3750$$
$$0.33x + 450 = 3750$$
$$0.33x = 3300$$
$$x = 10,000$$
$$x + 3000 = 13,000$$

He invested $10,000 in the 100 fund and $13,000 in the 101 fund.

11. Let $x =$ the amount invested at 5% and $25000 - x =$ the amount invested at 4%. His income on the first investment was $0.05x$ and his income from the second investment was $0.04(25000 - x)$. Since his total income was actually $1140, we can write the following equation.

$$0.05x + 0.04(25000 - x) = 1140$$
$$0.05x + 1000 - 0.04x = 1140$$
$$0.01x = 140$$
$$x = 14,000$$
$$25,000 - x = 11,000$$

He invested $14,000 in Fidelity and $11,000 in Price.

13. Let $x =$ the amount of 1% milk. The x gallons of 1% milk are mixed with 30 gallons of 3% milk to obtain $x + 30$ gallons of 2% milk. In the 1% milk there are $0.01x$ gallons of fat. In the 3% milk there are $0.03(30)$ gallons of fat. In the 2% milk there are $0.02(x + 30)$ gallons of fat. We can write an equation expressing the fact that the total of the fat in the two milks that are mixed is equal to the fat in the final mixture.

$$0.01x + 0.03(30) = 0.02(x + 30)$$
$$0.01x + 0.9 = 0.02x + 0.6$$

$$0.9 = 0.01x + 0.6$$
$$0.3 = 0.01x$$
$$100(0.3) = 100(0.01x)$$
$$30 = x$$

Use 30 gallons of 1% milk.

15. Let $x =$ the number of liters of 5% solution and $30 - x =$ the number of liters of 20% solution. The amount of alcohol in the 5% solution is 0.05x. The amount of alcohol in the 20% solution is 0.20(30 − x). The amount of alcohol in the final 10% solution is 0.10(30). We can write an equation expressing the fact that the total of the alcohol in each of the two solutions mixed is equal to the alcohol in the final result.

$$0.05x + 0.20(30 - x) = 0.10(30)$$
$$0.05x + 6 - 0.20x = 3$$
$$6 - 0.15x = 3$$
$$-0.15x = -3$$
$$x = \frac{-3}{-0.15} = 20$$
$$30 - x = 10$$

He should use 20 liters of 5% alcohol and 10 liters of 20% alcohol.

17. Let $x =$ the number of registered voters.

We can write the following equation.

$$0.60x = 33420$$
$$x = \frac{33420}{0.6} = 55,700$$

There are 55,700 registered voters.

19. Let $x =$ the price of the car and $0.08x =$ the amount of sales tax. Since the amount of sales tax was $1200, we can write the following equation.

$$0.08x = 1200$$
$$x = \frac{1200}{0.08} = 15000$$

The price of the car was $15,000.

21. Let $x =$ the percent increase and $8x =$ the amount of increase. Since the actual amount of increase is $6, we can write the following equation.

$$8x = 6$$
$$x = \frac{6}{8} = 0.75$$

The price of the shirts is increased 75%.

23. Let $x =$ the number of students at Jefferson and $x + 400 =$ the number of students in the combined school. The number of black students at Jefferson is 0.60x. The number of black students at Wilson is 0.20(400). The number of black students in the combined school will be 0.44(x + 400). Write an equation expressing the fact that the total of the black students from each school is equal to the number of black students in the combined school.

$$0.60x + 0.20(400) = 0.44(x + 400)$$
$$0.60x + 80 = 0.44x + 176$$
$$0.16x + 80 = 176$$
$$0.16x = 96$$
$$x = 600$$

The number of students at Jefferson is 600.

25. Let $x =$ the number of people in private rooms and $x + 18 =$ the number of people in semiprivate rooms. The revenue from the private rooms is 200x dollars and the revenue from the semiprivate rooms is 150(x + 18) dollars. Write an equation for the total receipts.

$$200x + 150(x + 18) = 17,400$$
$$200x + 150x + 2700 = 17,400$$
$$350x + 2700 = 17,400$$
$$350x = 14700$$
$$x = 42$$
$$x + 18 = 60$$

They have 42 private rooms and 30 semi-private rooms (holding 60 people).

27. Let $x =$ the number of pounds of pistachios. We can write an equation expressing the total cost of the mixture.

$$6.40x + 4.80(20) = 5.40(x + 20)$$
$$6.4x + 96 = 5.4x + 108$$
$$x = 12$$

We should mix 12 pounds of pistachios with 20 pounds of cashews to get a mix that sells for $5.40 per pound.

29. Let $x =$ the number of nickels and $10 - x =$ the number of dimes. The value in cents of the nickels is 5x and the value in cents of the dimes is 10(10 − x). Since she has 80 cents altogether, we can write the following equation.

$$5x + 10(10 - x) = 80$$
$$5x + 100 - 10x = 80$$
$$-5x = -20$$
$$x = 4$$
$$10 - x = 6$$

She used 4 nickels and 6 dimes.

31. Let x = the number of gallons of corn oil.

$$0.14x + 0.07(600) = 0.11(x + 600)$$
$$0.14x + 42 = 0.11x + 66$$
$$0.03x = 24$$
$$x = 800$$

Crisco should us 800 gallons of corn oil.

2.7 WARM-UPS

1. True, because −2 is equal to −2. **2.** True, because −5 < 4 and 4 < 6 are both correct.
3. False, because 0 < −1 is false.
4. True, because any number to the right of 7 on the number line satisfies both 7 < x and x > 7. **5.** False, because the graph of x < −3 includes only points to the left of −3.
6. True, because 5 > 2 is correct. **7.** False, because −2 < −3 is incorrect.
8. False, because 2(4) − 1 < 4 is incorrect.
9. True, because 2(0) − 3 ≤ 5(0) − 3 is equivalent to −3 ≤ −3 and that is correct.
10. False, because 0 satisfies 2x − 1 < x but does not satisfy x < 2x − 1.

2.7 EXERCISES

1. True, because −3 is to the left of 5 on the number line.
3. True, because 4 ≤ 4 is true if either 4 < 4 or 4 = 4 is correct.
5. False, because −6 is to the left of −5 on the number line.
7. True, because −4 < −3 is correct.
9. True, (−3)(4) − 1 < 0 − 3 is equivalent to −13 < −3.
11. True, because −4(5) − 6 ≥ 5(−6) is equivalent to −26 ≥ −30.
13. True, because 7(4) − 12 ≤ 3(9) − 2 is equivalent to 16 ≤ 25.
15. The graph of x ≤ 3 consists of the numbers to the left of 3 including 3 on the number line.

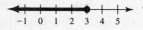

17. The graph of x > −2 consists of the numbers to the right of −2 on the number line.

19. The inequality −1 > x is the same as x < −1.

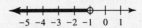

21. The graph of −2 ≤ x is the same as the graph of x ≥ −2, the numbers to the right of and including −2.

23. The graph of x ≥ ½ consists of the numbers to the right of and including ½.

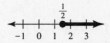

25. The graph of x ≤ 5.3 consists of the numbers to the left of and including 5.3.

27. The graph of −3 < x < 1 consists of the numbers between −3 and 1.

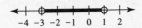

29. The graph of 3 ≤ x ≤ 7 consists of the numbers between 3 and 7, including 3 and 7.

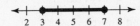

31. The graph of −5 ≤ x < 0 consists of the numbers between −5 and 0, including −5 but not including 0.

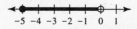

33. The graph of 40 < x ≤ 100 consists of the numbers between 40 and 100, including 100 but not including 40.

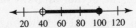

35. The graph shows the numbers to the right of 3. The inequality x > 3 describes this graph.
37. The graph shows the numbers to the left of and including 2. The inequality x ≤ 2 describes this graph.

39. The graph shows the numbers between 0 and 2. The inequality $0 < x < 2$ describes this graph.

41. The graph shows the numbers between -5 and 7, including 7 but not -5. The inequality $-5 < x \leq 7$ describes this graph.

43. The graph shows the numbers to the right of -4. The inequality $x > -4$ describes this graph.

45. Replace x by -9 in $-x > 3$.
$$-(-9) > 3$$
$$9 > 3$$
Since the last inequality is correct, -9 satisfies $-x > 3$.

47. Replace x by -2 in $5 \leq x$.
$$5 \leq -2$$
Since the inequality is incorrect, -2 does not satisfy $5 \leq x$.

49. Replace x by -6 in $2x - 3 > -11$.
$$2(-6) - 3 > -11$$
$$-15 > -11$$
Since the last inequality is incorrect, -6 does not satisfy $2x - 3 > -11$.

51. Replace x by 3 in $-3x + 4 > -7$.
$$-3(3) + 4 > -7$$
$$-5 > -7$$
Since $-5 > -7$ is correct, 3 satisfies $-3x + 4 > -7$.

53. Replace x by 0 in $3x - 7 \leq 5x - 7$.
$$3(0) - 7 \leq 5(0) - 7$$
$$-7 \leq -7$$
Since $-7 \leq -7$ is correct, 0 satisfies $3x - 7 \leq 5x - 7$.

55. Replace x by 2.5 in $-10x + 9 \leq 3(x + 3)$.
$$-10(0) + 9 \leq 3(0 + 3)$$
$$9 \leq 9$$
Since $9 \leq 9$ is correct, 0 satisfies the inequality.

57. Replace x by -7 in $-5 < x < 9$.
$$-5 < -7 < 9$$
Since -7 is not between -5 and 9, -7 does not satisfy $-5 < x < 9$.

59. Replace x by -2 in $-3 \leq 2x + 5 \leq 9$.
$$-3 \leq 2(-2) + 5 \leq 9$$
$$-3 \leq 1 \leq 9$$
Since 1 is between -3 and 9, -2 does satisfy $-3 \leq 2x + 5 \leq 9$.

61. Replace x by -3.4 in $-4.25x - 13.29 < 0.89$.
$$-4.25(-3.4) - 13.29 < 0.89$$
$$1.16 < 0.89$$
Since $1.16 < 0.89$ is incorrect, -3.4 does not satisfy $-4.25x - 13.29 < 0.89$.

63. Only 0 and 5.1 satisfy $x > -5$.

65. Only 5.1 satisfies $x > 5$.

67. Only 5.1 is between 5 and 7.

69. -5.1, 0, and 5.1 satisfy $-6 < -x < 6$.

71. Let $x =$ the sale price of the car and $0.08x =$ the amount of sales tax. The sales tax was more than \$1500 is expressed as $0.08x > 1500$.

73. Let $x =$ his score on the remaining test. The average is found by adding the scores and then dividing by 3. Since the average must be at least 60, we can write $\frac{44 + 72 + x}{3} \geq 60$.

75. Let $x =$ his speed and $8x =$ his daily distance. His distance was between 396 and 453 is expressed as $396 < 8x < 453$.

77. The angle at the base of the ladder is $90 - x$.
So $60 < 90 - x < 70$.

2.8 WARM-UPS

1. True, because dividing each side of $2x > 18$ by 2 gives $x > 9$. **2.** False, because 0 satisfies $x < 5$ but 0 does not satisfy $x - 5 > 0$. **3.** False, because we cannot divide each side by 0. **4.** True, because dividing each side of $-2x \leq 6$ by 2 gives $-x \leq 3$. **5.** False, because x is at most 7 means that x is less than or equal to 7. **6.** True, because at least means greater than or equal to. **7.** False, because if x is not more than 85 then x is less than or equal to 85. **8.** True, because in either inequality x is greater than -9 and x is less than -3. **9.** True, because $1.08x$ gives the truck price plus the sales tax. **10.** False, because at least means greater than or equal to.

2.8 EXERCISES

1. $x > -7$

3. $w \geq 3$

5. $k > 1$

7. $y \leq -8$

9.
$$x + 3 > 0$$
$$x + 3 - 3 > 0 - 3$$
$$x > -3$$

-5 -4 -3 -2 -1 0 1

11. $-3 < w - 1$
$-2 < w$
$w > -2$

(number line: -2 -1 0 1 2 3, open circle at -2, shaded right)

13. $8 > 2b$
$4 > b$
$b < 4$

(number line: -1 0 1 2 3 4 5, open circle at 4, shaded left)

15. $-4z \leq 8$
$z \geq -2$

(number line: -4 -3 -2 -1 0 1 2, closed circle at -2, shaded right)

17. $3y - 2 < 7$
$3y < 9$
$y < 3$

(number line: -1 0 1 2 3 4 5, open circle at 3, shaded left)

19. $3 - 7z \leq 17$
$-7z \leq 14$

$\dfrac{-7z}{-7} \geq \dfrac{14}{-7}$

$z \geq -2$

(number line: -2 -1 0 1 2 3, closed circle at -2, shaded right)

21. $6 > -r + 3$
$r > -3$

(number line: -4 -3 -2 -1 0 1, open circle at -3, shaded right)

23. $5 - 4p > -8 - 3p$
$-p > -13$
$p < 13$

(number line: 9 10 11 12 13 14 15, open circle at 13, shaded left)

25. $-\dfrac{5}{6}q \geq -20$

$-\dfrac{6}{5}\left(-\dfrac{5}{6}\right)q \leq -\dfrac{6}{5}(-20)$

$q \leq 24$

(number line: 20 22 24 26, closed circle at 24, shaded left)

27. $1 - \dfrac{1}{4}t \geq -2$

$-\dfrac{1}{4}t \geq -3$

$-4\left(-\dfrac{1}{4}t\right) \leq -4(-3)$

$t \leq 12$

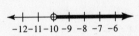
(number line: 8 9 10 11 12 13 14, closed circle at 12, shaded left)

29. $2x + 5 < x - 6$
$x + 5 < -6$
$x < -11$

(number line: -15 -14 -13 -12 -11 -10 -9, open circle at -11, shaded left)

31. $x - 4 < 2(x + 3)$
$x - 4 < 2x + 6$
$-4 < x + 6$
$-10 < x$
$x > -10$

(number line: -12 -11 -10 -9 -8 -7 -6, open circle at -10, shaded right)

33. $0.52x - 35 < 0.45x + 8$
$0.52x < 0.45x + 43$
$0.07x < 43$

$x < \dfrac{43}{0.07}$

$x < 614.3$

(number line: open circle at 614.3, shaded left)

35. $5 < x - 3 < 7$
$5 + 3 < x - 3 + 3 < 7 + 3$
$8 < x < 10$

(number line: 6 7 8 9 10 11 12, open circles at 8 and 10, shaded between)

37. $3 < 2v + 1 < 10$
$2 < 2v < 9$

$1 < v < \dfrac{9}{2}$

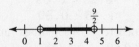
(number line: 0 1 2 3 4 5 6, open circle at 1 and $\frac{9}{2}$, shaded between)

39.
$$-4 \le 5 - k \le 7$$
$$-9 \le -k \le 2$$
$$(-1)(-9) \ge (-1)(-k) \ge (-1)(2)$$
$$9 \ge k \ge -2$$
$$-2 \le k \le 9$$

41.
$$-2 < 7 - 3y \le 22$$
$$-9 < -3y \le 15$$
$$3 > y \ge -5$$
$$-5 \le y < 3$$

43.
$$5 < \frac{2u}{3} - 3 < 17$$
$$8 < \frac{2u}{3} < 20$$
$$24 < 2u < 60$$
$$12 < u < 30$$

45.
$$-7 < \frac{3m + 1}{2} \le 8$$
$$-14 < 3m + 1 \le 16$$
$$-15 < 3m \le 15$$
$$-5 < m \le 5$$

47.
$$0.02 < 0.54 - 0.0048x < 0.05$$
$$-0.52 < -0.0048x < -0.49$$
$$\frac{-0.52}{-0.0048} > \frac{-0.0048x}{-0.0048} > \frac{-0.49}{-0.0048}$$
$$108.3 > x > 102.1$$
$$102.1 < x < 108.3$$

49.
$$\frac{1}{2}x - 1 \le 4 - \frac{1}{3}x$$
$$6\left(\frac{1}{2}x - 1\right) \le 6\left(4 - \frac{1}{3}x\right)$$
$$3x - 6 \le 24 - 2x$$
$$5x \le 30$$
$$x \le 6$$

51.
$$\frac{1}{2}\left(x - \frac{1}{4}\right) > \frac{1}{4}\left(6x - \frac{1}{2}\right)$$
$$\frac{1}{2}x - \frac{1}{8} > \frac{3}{2}x - \frac{1}{8}$$
$$4x - 1 > 12x - 1$$
$$-8x > 0$$
$$x < 0$$

53.
$$\frac{1}{3} < \frac{1}{4}x - \frac{1}{6} < \frac{7}{12}$$
$$4 < 3x - 2 < 7$$
$$6 < 3x < 9$$
$$2 < x < 3$$

55. Let w = the width and w + 4 = the length. The perimeter is $2w + 2(w + 4)$. Since the perimeter is at least 120, we can write the following inequality.
$$2w + 2(w + 4) \ge 120$$
$$4w + 8 \ge 120$$
$$4w \ge 112$$
$$w \ge 28$$

The width must be at least 28 meters.

57. Let x = the price of the car. Since the tax is 0.05x, we can write the following inequality.
$$x + 0.05x + 144 < 9970$$
$$1.05x < 9826$$
$$x < 9358$$

The price of the car must be less than $9358.

59. Let x = the price of the microwave. The cost of the microwave plus the tax is 1.08x. Since she has at most $594, we can write the following inequality.
$$1.08x \le 594$$
$$x \le \frac{594}{1.08}$$
$$x \le 550$$

The price of the microwave must be less than or equal to $550.

61. Let $x =$ Tilak's score on the last test. His average for the three tests is $\frac{44 + 72 + x}{3}$. Since his test average must be at least 60, we can write the following inequality.

$$\frac{44 + 72 + x}{3} \geq 60$$
$$44 + 72 + x \geq 180$$
$$116 + x \geq 180$$
$$x \geq 64$$

He must score at least 64 on the last test to pass the course.

63. Let $x =$ the final exam score. Stacy's semester average is $\frac{1}{3}(48) + \frac{2}{3}x$. Since the semester average must be between 70 and 79 inclusive, we can write the following inequality.

$$70 \leq \frac{1}{3}(48) + \frac{2}{3}x \leq 79$$
$$70 \leq 16 + \frac{2}{3}x \leq 79$$
$$54 \leq \frac{2}{3}x \leq 63$$
$$\frac{3}{2} \cdot 54 \leq \frac{3}{2} \cdot \frac{2}{3}x \leq \frac{3}{2} \cdot 63$$
$$81 \leq x \leq 94.5$$

To get a C, Stacy must score between 81 and 94.5 inclusive on the final exam.

65. Let $x =$ her average speed for a day. Her distance each day was 8x. Since her distance was between 396 and 453 we can write the following inequality.

$$396 < 8x < 453$$
$$49.5 < x < 56.625$$

Her average speed each day was between 49.5 and 56.625 miles per hour.

67. The angles at B are 85° and 95°. So the angle at the light house is $180° - 95° - x$. Since this angle is less than 30°, we have the following inequality.

$$180 - 95 - x < 30$$
$$85 - x < 30$$
$$55 < x$$

So $x > 55°$. Since $x + 95 < 180$, we also have $x < 85°$. So x is between 55° and 85°.

CHAPTER 2 REVIEW

1. $2x - 5 = 9$
$$2x = 14$$
$$x = 7$$
The solution to the equation is 7.

3. $3r - 7 = 0$
$$3r = 7$$
$$r = \frac{7}{3}$$
The solution to the equation is $\frac{7}{3}$.

5. $3 - 4y = 11$
$$-4y = 8$$
$$y = -2$$
The solution to the equation is -2.

7. $2(h - 7) = -14$
$$2h - 14 = -14$$
$$2h = 0$$
$$h = 0$$
The solution to the equation is 0.

9. $3(w - 5) = 6(w + 2) - 3$
$$3w - 15 = 6w + 12 - 3$$
$$-3w = 24$$
$$w = -8$$
The solution to the equation is -8.

11. $-\frac{2}{3}b = 20$
$$-\frac{3}{2}\left(-\frac{2}{3}b\right) = -\frac{3}{2} \cdot 20$$
$$b = -30$$
The solution to the equation is -30.

13. $\frac{1}{3}x = \frac{1}{7}$
$$3 \cdot \frac{1}{3}x = 3 \cdot \frac{1}{7}$$
$$x = \frac{3}{7}$$
The solution to the equation is $\frac{3}{7}$.

15. $0.24c + 1 = 97$
$$0.24c = 96$$
$$c = 400$$
The solution to the equation is 400.

17. $p - 0.1p = 90$
$$0.9p = 90$$
$$p = 100$$
The solution to the equation is 100.

19.
$$2(x-7)-5=5-(3-2x)$$
$$2x-14-5=5-3+2x$$
$$2x-19=2+2x$$
$$-19=2$$

There is no solution to this equation. It is an inconsistent equation.

21.
$$2(w-w)=0$$
$$2(0)=0$$
$$0=0$$

All real numbers satisfy this equation. It is an identity.

23.
$$\frac{3x}{3x}=1$$
$$\frac{x}{x}=1$$

A number divided by itself is 1 except for $0/0$, which is undefined. The solution to the equation is all real numbers except 0. It is an identity.

25.
$$\tfrac{1}{2}a-5=\tfrac{1}{3}a-1$$
$$6\left(\tfrac{1}{2}a-5\right)=6\left(\tfrac{1}{3}a-1\right)$$
$$3a-30=2a-6$$
$$a-30=-6$$
$$a=24$$

The solution to the equation is 24. It is a conditional equation.

27.
$$0.06q+14=0.3q-5.2$$
$$0.06q=0.3q-19.2$$
$$-0.24q=-19.2$$
$$q=80$$

The solution to the equation is 80. It is a conditional equation.

29.
$$0.05(x+100)+0.06x=115$$
$$0.05x+5+0.06x=115$$
$$0.11x=110$$
$$x=1000$$

The solution to the equation is 1000. It is a conditional equation.

31.
$$2x+\tfrac{1}{2}=3x+\tfrac{1}{4}$$
$$8x+2=12x+1$$
$$-4x=-1$$
$$x=\tfrac{1}{4}$$

The solution to the equation is $\tfrac{1}{4}$.

33.
$$\tfrac{x}{2}-\tfrac{3}{4}=\tfrac{x}{6}+\tfrac{1}{8}$$
$$12x-18=4x+3$$
$$8x=21$$
$$x=\tfrac{21}{8}$$

The solution to the equation is $\tfrac{21}{8}$.

35.
$$\tfrac{5}{6}x=-\tfrac{2}{3}$$
$$x=\tfrac{6}{5}\left(-\tfrac{2}{3}\right)=-\tfrac{4}{5}$$

The solution to the equation is $-\tfrac{4}{5}$.

37.
$$-\tfrac{1}{2}(x-10)=\tfrac{3}{4}x$$
$$-\tfrac{1}{2}x+5=\tfrac{3}{4}x$$
$$-2x+20=3x$$
$$-5x=-20$$
$$x=4$$

The solution to the equation is 4.

39.
$$3-4(x-1)+6=-3(x+2)-5$$
$$3-4x+4+6=-3x-6-5$$
$$-4x+13=-3x-11$$
$$-x=-24$$
$$x=24$$

The solution to the equation is 24.

41.
$$5-0.1(x-30)=18+0.05(x+100)$$
$$5-0.1x+3=18+0.05x+5$$
$$-0.1x+8=23+0.05x$$
$$-0.15x=15$$
$$x=-100$$

The solution to the equation is -100.

43.
$$ax+b=0$$
$$ax=-b$$
$$x=-\tfrac{b}{a}$$

45.
$$ax-2=b$$
$$ax=b+2$$
$$x=\tfrac{b+2}{a}$$

47.
$$LWx=V$$
$$x=\tfrac{V}{LW}$$

49.
$$2x-b=5x$$
$$-b=3x$$
$$-\tfrac{b}{3}=x$$
$$x=-\tfrac{b}{3}$$

51. $5x + 2y = 6$

$2y = -5x + 6$

$y = -\frac{5}{2}x + 3$

53. $y - 1 = -\frac{1}{2}(x - 6)$

$y - 1 = -\frac{1}{2}x + 3$

$y = -\frac{1}{2}x + 4$

55. $\frac{1}{2}x + \frac{1}{4}y = 4$

$\frac{1}{4}y = -\frac{1}{2}x + 4$

$4 \cdot \frac{1}{4}y = 4\left(-\frac{1}{2}x + 4\right)$

$y = -2x + 16$

57. Use -3 for x in $y = 3x - 4$.

$y = 3(-3) - 4 = -13$

59. Use -3 for x in $5xy = 6$.

$5(-3)y = 6$

$-15y = 6$

$y = -\frac{6}{15} = -\frac{2}{5}$

61. Use -3 for x in $y - 3 = -2(x - 4)$.

$y - 3 = -2(-3 - 4)$

$y - 3 = 14$

$y = 17$

63. Sum indicates addition. So the sum of a number and 9 is $x + 9$.

65. If two numbers differ by 8, then one number is 8 larger than the other. So x and $x + 8$ are used to represent the numbers. We could also use x and $x - 8$.

67. Sixty-five percent of a number is 0.65 times the number or 0.65x.

69. We use x and $x + 5$ to represent the sides. Since the area of the rectangle is 98, we can write $x(x + 5) = 98$.

71. Let x = Barbara's speed and $x + 10 = $ Jim's speed. In 3 hours Barbara travels 3x miles. In 2 hours Jim travels $2(x + 10)$ miles. Since the distances are the same, we can write the equation $2(x + 10) = 3x$.

73. If x is the first even integer, then $x + 2$ and $x + 4$ represent the second and the third. Since their sum is 88, we can write the equation $x + x + 2 + x + 4 = 88$.

75. Since the sum of the measures of a triangle is 180°, we have $t + 2t + t - 10 = 180$.

77. Let x = the first odd integer, $x + 2 = $ the second odd integer, and $x + 4 = $ the third odd integer. Since their sum is 237, we can write the following equation.

$x + x + 2 + x + 4 = 237$

$3x + 6 = 237$

$3x = 231$

$x = 77$

$x + 2 = 79$

$x + 4 = 81$

The three consecutive odd integers are 77, 79, and 81.

79. Let x = Betty's rate of speed and $x + 15 = $ Lawanda's rate of speed. Since $D = RT$, Betty's distance is 4x and Lawanda's distance is $3(x + 15)$. Since their distances are equal, we can write the following equation.

$4x = 3(x + 15)$

$4x = 3x + 45$

$x = 45$

$x + 15 = 60$

Betty drives 45 mph and Lawanda drives 60 mph.

81. Let x = the husband's income and $x + 6000 = $ Wanda's income. Wanda saves $0.10(x + 6000)$ and her husband saves 0.06x. Since they save $5400 together, we can write the following equation.

$0.10(x + 6000) + 0.06x = 5400$

$0.10x + 600 + 0.06x = 5400$

$0.16x = 4800$

$x = 30,000$

$x + 6000 = 36,000$

Wanda makes $36,000 and her husband makes $30,000 per year.

83. Use 3 for x in $-2x + 5 \leq x - 6$.

$-2(3) + 5 \leq 3 - 6$

$-1 \leq -3$

Since this inequality is incorrect, 3 is not a solution to $-2x + 5 \leq x - 6$.

85. Use -1 for x in $-2 \leq 6 + 4x < 0$.

$-2 \leq 6 + 4(-1) < 0$

$-2 \leq 2 < 0$

Since this last inequality is incorrect, -1 is not a solution to $-2 \leq 6 + 4x < 0$.

87. The graph shows the numbers to the right of 1 on the number line. This graph indicates the solution to $x > 1$.

89. The graph shows the number to the right of and including 2. This graph indicates the solution to $x \geq 2$.

91. The graph shows the numbers between -3 and 3, including -3 but not 3. This graph indicates the solution to $-3 \leq x < 3$.

93. The graph shows the numbers to the left of -1 on the number line. This graph indicates the solution to $x < -1$.

95.
$$x + 2 > 1$$
$$x > -1$$

97.
$$3x - 5 < x + 1$$
$$2x < 6$$
$$x < 3$$

99.
$$-\frac{3}{4}x \geq 3$$
$$-\frac{4}{3}\left(-\frac{3}{4}x\right) \leq -\frac{4}{3} \cdot 3$$
$$x \leq -4$$

101.
$$3 - 2x < 11$$
$$-2x < 8$$
$$x > -4$$

103.
$$-3 < 2x - 1 < 9$$
$$-2 < 2x < 10$$
$$-1 < x < 5$$

105.
$$0 \leq 1 - 2x < 5$$
$$-1 \leq -2x < 4$$
$$\frac{1}{2} \geq x > -2$$
$$-2 < x \leq \frac{1}{2}$$

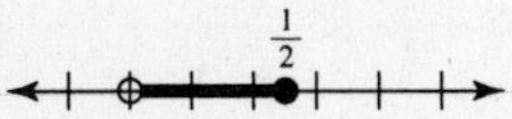

107.
$$-1 \leq \frac{2x - 3}{3} \leq 1$$
$$-3 \leq 2x - 3 \leq 3$$
$$0 \leq 2x \leq 6$$
$$0 \leq x \leq 3$$

109.
$$\frac{1}{3} < \frac{1}{3} + \frac{x}{2} < \frac{5}{6}$$
$$2 < 2 + 3x < 5$$
$$0 < 3x < 3$$
$$0 < x < 1$$

111. A 2-year bond earns 7% interest. In its first year a $10,000 bond earns 0.07(10,000) or $700. A 30-year bond earns 7.6%. In its first year a $10,000 bond earns 0.076(10,000) or $760, which is $60 more than a 2-year bond.

113. Let $x =$ the number of movies at ABC. Since XYZ had 200 movies, the combined store has $x + 200$ movies. The number of children's movies at ABC was 0.60x and the number of children's movies after the merger is $0.40(x + 200)$. Since XYZ had no children's movies, these two amounts of children's movies are equal.

$$0.60x = 0.40(x + 200)$$
$$0.60x = 0.40x + 80$$
$$0.20x = 80$$
$$x = 400$$

So ABC had 400 movies before the merger.

115. Complementary angles have a sum of 90°.

$$x + 2x - 3 = 90$$
$$3x = 93$$
$$x = 31$$

The degree measure is 31°.

117. Let $x =$ the length of the shortest side, $x + 1 =$ the length of the second side, and $2x =$ the length of the third side. Since the perimeter is less than 25 feet we can write the following inequality.

$$x + x + 1 + 2x < 25$$
$$4x + 1 < 25$$
$$4x < 24$$
$$x < 6$$

The shortest side is less than 6 feet in length.

CHAPTER 2 TEST

1.
$$-10x - 6 + 4x = -4x + 8$$
$$-6x - 6 = -4x + 8$$
$$-2x - 6 = 8$$
$$-2x = 14$$
$$x = -7$$

The solution to the equation is -7.

2.
$$5(2x - 3) = x + 3$$
$$10x - 15 = x + 3$$
$$10x = x + 18$$
$$9x = 18$$
$$x = 2$$

The solution to the equation is 2.

3.
$$-\frac{2}{3}x + 1 = 7$$
$$-\frac{2}{3}x = 6$$
$$-\frac{3}{2}\left(-\frac{2}{3}x\right) = -\frac{3}{2} \cdot 6$$
$$x = -9$$

The solution to the equation is -9.

4.
$$x + 0.06x = 742$$
$$1.06x = 742$$
$$x = 700$$

The solution to the equation is 700.

5.
$$2x - 3y = 9$$
$$-3y = -2x + 9$$
$$-\frac{1}{3}(-3y) = -\frac{1}{3}(-2x + 9)$$
$$y = \frac{2}{3}x - 3$$

6.
$$m = aP - w$$
$$m + w = aP$$
$$\frac{m + w}{P} = a$$
$$a = \frac{m + w}{P}$$

7. The graph shows the numbers between -3 and 2, including 2 but not including -3. This graph is the solution to $-3 < x \leq 2$.

8. The graph shows the numbers to the right of 1 on the number line. This graph is the solution to the inequality $x > 1$.

9.
$$4 - 3(w - 5) < -2w$$
$$4 - 3w + 15 < -2w$$
$$19 - 3w < -2w$$
$$19 < w$$
$$w > 19$$

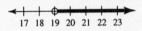

10.
$$1 < \frac{1 - 2x}{3} < 5$$
$$3 < 1 - 2x < 15$$
$$2 < -2x < 14$$
$$-1 > x > -7$$
$$-7 < x < -1$$

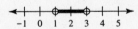

11.
$$1 < 3x - 2 < 7$$
$$3 < 3x < 9$$
$$1 < x < 3$$

12.
$$-\tfrac{2}{3}y < 4$$
$$-\tfrac{3}{2}\left(-\tfrac{2}{3}y\right) > -\tfrac{3}{2}\cdot 4$$
$$y > -6$$

13.
$$2(x+6) = 2x-5$$
$$2x+12 = 2x-5$$
$$12 = -5$$

There is no solution to the equation.

14.
$$x+7x = 8x$$
$$8x = 8x$$

All real numbers satisfy the equation.

15.
$$x-0.03x = 0.97$$
$$0.97x = 0.97$$
$$x = 1$$

The solution to the equation is 1.

16.
$$6x-7 = 0$$
$$6x = 7$$
$$x = \tfrac{7}{6}$$

The solution to the equation is $\tfrac{7}{6}$.

17. Let $x =$ the width and $x+8 =$ the length. Since perimeter is 72, we can write the following equation.
$$2x+2(x+8) = 72$$
$$4x+16 = 72$$
$$4x = 56$$
$$x = 14$$
$$x+8 = 22$$
The width of the rectangle is 14 meters.

18. Use $A = 54$ and $b = 12$ in the formula for the area of a triangle, $A = \tfrac{1}{2}bh$.
$$54 = \tfrac{1}{2}\cdot 12h$$
$$54 = 6h$$
$$9 = h$$
The height is 9 inches.

19. Let $x =$ the number of liters of 20% solution. If she mixes the 20% solution with 50 liters of 60% solution she will obtain $x+50$ liters of 30% solution. The amount of alcohol in the 20% solution is 0.20x. The amount of alcohol in the 60% solution is 0.60(50). The amount of alcohol in the final 30% solution is 0.30(x+50). The alcohol in the final solution is the total of the alcohol in the two solutions.

$$0.20x+0.60(50) = 0.30(x+50)$$
$$0.20x+30 = 0.30x+15$$
$$0.20x+15 = 0.30x$$
$$15 = 0.10x$$
$$150 = x$$

She should use 150 liters of 20% solution.

20. Let $x =$ the original price of the diamonds. His discount is 0.40x. The price he pays is $x-0.40x+250$.
$$x-0.40x+250 \le 1450$$
$$0.60x \le 1200$$
$$x \le 2000$$

The original price of the diamonds can be at most $2000.

21. If $x =$ the degree measure of the smallest angle, then the degree measures of the other two are 2x and 3x.
$$x+2x+3x = 180$$
$$6x = 180$$
$$x = 30$$
The angles are 30°, 60°, and 90°.

Tying It All Together   Chapters 1-2

1. $3x + 5x = (3 + 5)x = 8x$

2. $3x \cdot 5x = 3 \cdot 5 \cdot x \cdot x = 15x^2$

3. $\frac{4x + 2}{2} = \frac{1}{2}(4x + 2) = \frac{1}{2} \cdot 4x + \frac{1}{2} \cdot 2 = 2x + 1$

4. $5 - 4(3 - x) = 5 - 12 + 4x = 4x - 7$

5. $3x + 8 - 5(x - 1) = 3x + 8 - 5x + 5$

$= -2x + 13$

6. $(-6)^2 - 4(-3)2 = 36 - (-24) = 36 + 24 = 60$

7. $3^2 \cdot 2^3 = 9 \cdot 8 = 72$

8. $4(-7) - (-6)(3) = -28 + 18 = -10$

9. $-2x \cdot x \cdot x = -2x^3$

10. $(-1)(-1)(-1)(-1)(-1) = -1$

11. $\quad 3x + 5x = 8$

$8x = 8$

$x = 1$

The solution to the equation is 1.

12. $\quad 3x + 5x = 8x$

$8x = 8x$

All real numbers satisfy this equation.

13. $\quad 3x + 5x = 7x$

$8x = 7x$

$8x - 7x = 7x - 7x$

$x = 0$

The solution to the equation is 0.

14. $\quad 3x + 5 = 8$

$3x = 3$

$x = 1$

The solution to the equation is 1.

15. $\quad 3x + 1 = 7$

$3x = 6$

$x = 2$

The solution to the equation is 2.

16. $\quad 5 - 4(3 - x) = 1$

$5 - 12 + 4x = 1$

$-7 + 4x = 1$

$4x = 8$

$x = 2$

17. $\quad 3x + 8 = 5(x - 1)$

$3x + 8 = 5x - 5$

$3x + 13 = 5x$

$13 = 2x$

$\frac{13}{2} = x$

The solution to the equation is $\frac{13}{2}$.

18. $\quad x - 0.05x = 190$

$0.95x = 190$

$x = \frac{190}{0.95} = 200$

The solution to the equation is 200.

19. $V = C - \frac{C - S}{5}t$

$V = 20{,}000 - \frac{20{,}000 - 4{,}000}{5} \cdot 2 = 13{,}600$

The value after 2 years is $13,600.

$14{,}000 = 20{,}000 - \frac{20{,}000 - S}{5} \cdot 3$

$14{,}000 = 20{,}000 - \frac{60{,}000 - 3S}{5}$

$-6{,}000 = -\frac{60{,}000 - 3S}{5}$

$-30{,}000 = -60{,}000 + 3S$

$30{,}000 = 3S$

$S = 10{,}000$

The scrap value is $10,000.

3.1 WARM-UPS

1. False, because the coefficient of x is −4.
2. False, because the degree is 3, the highest power of the variable x.   3. True, because $x^2 - x$ could be written as $x^2 + (-1)x$.
4. True, because the degree is the highest power of x.   5. False, $x^4 + x$ is a binomial with degree 4.   6. True, because a trinomial has three terms.   7. False, $x^5 - 3x - 8$ is a trinomial with degree 5.   8. True, because the monomials are subtracted correctly.   9. True, because the polynomials are added correctly.
10. False, because the polynomials are not subtracted correctly.

3.1 EXERCISES

1. The coefficient of x^3 is −3 and the coefficient of x^2 is 7.

3. Since x^3 does not appear, it coefficient is 0. The coefficient of x^2 is 6.

5. The coefficient of x^3 is $\frac{1}{3}$ and the coefficient of x^2 is $\frac{7}{2}$.

7. Since −1 has only one term, it is a monomial. Since $-1 = -1x^0$, the degree is 0.

9. Since m^3 has one term only, it is a monomial. The degree is 3, because that is the highest power of m.

11. Since $4x + 7$ has two terms, it is a binomial. The degree is 1, because that is the highest power of x in the polynomial.

13. Since the polynomial has three terms, it is a trinomial. The degree is 10, because that is the highest power of the variable x.

15. Since $x^6 + 1$ has two terms, it is a binomial. The degree is 6, because that is the highest power of x in the binomial.

17. Since the polynomial has 3 terms it is a trinomial. Since the highest power of a is 3, the degree is 3.

19. $2x^2 - 3x + 1 = 2(-1)^2 - 3(-1) + 1$
$$= 2 + 3 + 1 = 6$$
$2x^2 - 3x + 1 = 2(3)^2 - 3(3) + 1 = 18 - 9 + 1 = 10$

21. $-3x^3 - x^2 + 3x - 4$
$$= -3(-1)^3 - (-1)^2 + 3(-1) - 4$$
$$= 3 - 1 - 3 - 4 = -5$$
$-3x^3 - x^2 + 3x - 4 = -3(3)^3 - 3^2 + 3(3) - 4$
$$= -81 - 9 + 9 - 4 = -85$$

23. $x^2 - 6x + 3$
$$= (1.45)^2 - 6(1.45) + 3 = -3.5975$$
$x^2 - 6x + 3$
$$= (-2.36)^2 - 6(-2.36) + 3 = 22.7296$$

25. $1.2x^3 - 4.3x - 2.4$
$$= 1.2(1.45)^3 - 4.3(1.45) - 2.4 = -4.97665$$
$1.2x^3 - 4.3x - 2.4$
$$= 1.2(-2.36)^3 - 4.3(-2.36) - 2.4 = -8.0251072$$

27. $(x - 3) + (3x - 5) = x + 3x - 3 - 5 = 4x - 8$

29. $(q - 3) + (q + 3) = 2q$

31. $(3x + 2) + (x^2 - 4) = x^2 + 3x + 2 - 4$
$$= x^2 + 3x - 2$$

33. $(4x - 1) + (x^3 + 5x - 6)$
$$= x^3 + 4x + 5x - 1 - 6 = x^3 + 9x - 7$$

35. $(a^2 - 3a + 1) + (2a^2 - 4a - 5)$
$$= a^2 + 2a^2 - 3a - 4a + 1 - 5 = 3a^2 - 7a - 4$$

37. $(w^2 - 9w - 3) + (w - 4w^2 + 8)$
$$= w^2 - 4w^2 - 9w + w - 3 + 8 = -3w^2 - 8w + 5$$

39. $(5.76x^2 - 3.14x - 7.09)$
$+ (3.9x^2 + 1.21x + 5.6) = 9.66x^2 - 1.93x - 1.49$

41. $(x - 2) - (5x - 8) = x - 2 - 5x + 8$
$$= -4x + 6$$

43. $(m - 2) - (m + 3) = m - 2 - m - 3 = -5$

45. $(2z^2 - 3z) - (3z^2 - 5z) = 2z^2 - 3z^2 - 3z + 5z$
$$= -z^2 + 2z$$

47. $(w^5 - w^3) - (-w^4 + w^2)$

$= w^5 - w^3 + w^4 - w^2$

$= w^5 + w^4 - w^3 - w^2$

49. $(t^2 - 3t + 4) - (t^2 - 5t - 9)$

$= t^2 - t^2 - 3t + 5t + 4 + 9 = 2t + 13$

51. $(9 - 3y - y^2) - (2 + 5y - y^2)$

$= 9 - 3y - y^2 - 2 - 5y + y^2 = -8y + 7$

53. $(3.55x - 879) - (26.4x - 455.8)$

$= 3.55x - 879 - 26.4x + 455.8$

$= -22.85x - 423.2$

55.
$$\begin{array}{r} 3a - 4 \\ \underline{a + 6} \\ 4a + 2 \end{array}$$

57.
$$\begin{array}{r} 3x + 11 \\ \underline{5x + 7} \\ -2x + 4 \end{array} \qquad 3x - 5x = -2x$$

59.
$$\begin{array}{r} a - b \\ \underline{a + b} \\ 2a \end{array}$$

61.
$$\begin{array}{r} -3m + 1 \\ \underline{2m - 6} \\ -5m + 7 \end{array} \qquad 1 - (-6) = 7$$

63.
$$\begin{array}{r} 2x^2 - x - 3 \\ \underline{2x^2 + x + 4} \\ 4x^2 \qquad + 1 \end{array}$$

65.
$$\begin{array}{r} 3a^3 - 5a^2 \qquad + 7 \\ \underline{2a^3 + 4a^2 - 2a} \\ a^3 - 9a^2 + 2a + 7 \end{array} \qquad \begin{array}{l} -5 - 4 = -9 \\ 0 - (-2) = 2 \end{array}$$

67.
$$\begin{array}{r} x^2 - 3x + 6 \\ \underline{x^2 \qquad - 3} \\ -3x + 9 \end{array} \qquad 6 - (-3) = 9$$

69.
$$\begin{array}{r} y^3 + 4y^2 - 6y - 5 \\ \underline{y^3 + 3y^2 + 2y - 9} \\ 2y^3 + 7y^2 - 4y - 14 \end{array}$$

71. $(2m - 9) + (3m + 4) = 5m - 5$

73. $(9y - 2) - (7y - 3) = 2y + 1$

75. $(2x^2 - x + 3) + (x^2 + 5x - 9) = 3x^2 + 4x - 6$

$3x^2 + 4x - 6 - (x^2 - 3x - 1) = 2x^2 + 7x - 5$

77. $-3m + 3$

79. $-11y - 3$

81. $2x^2 - 6x + 12$

83. $-5z^4 - 8z^3 + 3z^2 + 7$

85. Since profit P is equal to revenue minus cost, we have

$P = x^2 + 400x + 300 - (x^2 + 300x - 200)$

$\quad = 100x + 500$ dollars

If $x = 50$, then

$P = 100(50) + 500 = \$5{,}500$

87. The perimeter is the total of the lengths of the three sides of the triangle.

$x + (3x - 1) + (2x + 4) = 6x + 3$

The perimeter is $6x + 3$ meters.

If $x = 4$, then $6(4) + 3 = 27$ and the perimeter is 27 meters.

89. To find the total distance that she traveled, add the two binomials.

$(2x + 50) + (3x - 10) = 5x + 40$

She traveled a total of $5x + 40$ miles.

If $x = 20$, then $5x + 40 = 140$ miles.

91. $D = (-16t^2 + 7400) - (-16t^2 + 6600)$

$= 800$

If $t = 3$, then $D = 800$ feet.

93. Add the two polynomials to find the total interest.

$0.08(x + 554) + 0.09(x + 335)$

$= 0.08x + 44.32 + 0.09x + 30.15 = 0.17x + 74.47$

His total interest was $0.17x + 74.47$ dollars.

If $x = 1000$, then $0.17(1000) + 74.47 = \$244.47$

95. $655.1 + 9.56w + 1.85h - 4.68a$

$= 655.1 + 9.56(54) + 1.85(147) - 4.68(30)$

$= 1321.39$ calories

3.2 WARM-UPS

1. False, because if x is replaced by -1, the two sides of the equation have different values. A corrected version could be written as $3x^3 \cdot 5x^4 = 15x^7$ for any value of x.

2. False, because if we replace x by 1, the two sides of the equation have different values. A corrected version could be written as $3x^2 \cdot 2x^7 = = 6x^9$ for any value of x.

3. True, because $(3y^3)^2 = 3y^3 \cdot 3y^3 = 9y^6$ for any value of y. **4.** False, because if x is replaced by -1, the equation is incorrect.

40

5. True, because the monomial and the trinomial are correctly multiplied. **6.** True, because $-2(3-x) = -6 - (-2x) = -6 + 2x = 2x - 6$ for any value of x. **7.** True, because $(a+b)(c+d) = (a+b)c + (a+b)d = ac + bc + ad + bd$ for any values of a, b, c, and d. **8.** True, because $-(x-7) = -1(x-7) = -x - (-7) = -x + 7 = 7 - x$ for any value of x. **9.** True, because $-(a-b) = b - a$ for any values of a and b. **10.** False, because if we replace x by 1, then $x+3$ and $x-3$ have values 4 and -2, which are not opposites. The opposite of $x+3$ is $-x-3$ for any number x.

3.2 EXERCISES

1. Multiply the coefficients and add the exponents: $3x^2 \cdot 9x^3 = 27x^5$.

3. Multiply the coefficients and add the exponents: $2a^3 \cdot 7a^8 = 14a^{11}$.

5. $-6x^2 \cdot 5x^2 = -30x^4$

7. $(-9x^{10})(-3x^7) = (-9)(-3)x^{10+7} = 27x^{17}$

9. $-6st \cdot 9st = -54s^2t^2$

11. $3wt \cdot 8w^7t^6 = 3 \cdot 8w^{1+7}t^{1+6} = 24t^7w^8$

13. $(5y)^2 = 5y \cdot 5y = 25y^2$

15. $(2x^3)^2 = 2x^3 \cdot 2x^3 = 4x^6$

17. $4y^2(y^5 - 2y) = 4y^7 - 8y^3$

19. $-3y(6y - 4) = -18y^2 + 12y$

21. $(y^2 - 5y + 6)(-3y)$
$= -3y^3 - (-3)5y^2 + (-3)6y$
$= -3y^3 + 15y^2 - 18y$

23. $-x(y^2 - x^2) = -xy^2 - (-x)x^2 = -xy^2 + x^3$

25. $(3ab^3 - a^2b^2 - 2a^3b)(5a^3)$
$= 15a^4b^3 - 5a^5b^2 - 10a^6b$

27. $-\frac{1}{2}t^2v(4t^3v^2 - 6tv - 4v)$
$= -2t^5v^3 + 3t^3v^2 + 2t^2v^2$

29. $(x+1)(x+2) = (x+1)x + (x+1)2$
$= x^2 + x + 2x + 2 = x^2 + 3x + 2$

31. $(x-3)(x+5) = (x-3)x + (x-3)5$
$= x^2 - 3x + 5x - 15 = x^2 + 2x - 15$

33. $(t-4)(t-9) = (t-4)t - (t-4)9$
$= t^2 - 4t - (9t - 36) = t^2 - 4t - 9t + 36$
$= t^2 - 13t + 36$

35. $(x+1)(x^2 + 2x + 2)$
$= (x+1)x^2 + (x+1)2x + (x+1)2$
$= x^3 + x^2 + 2x^2 + 2x + 2x + 2$
$= x^3 + 3x^2 + 4x + 2$

37. $(3y+2)(2y^2 - y + 3)$
$= (3y+2)2y^2 - (3y+2)y + 3(3y+2)$
$= 6y^3 + 4y^2 - 3y^2 - 2y + 9y + 6$
$= 6y^3 + y^2 + 7y + 6$

39. $(y^2z - 2y^4)(y^2z + 3z^2 - y^4) =$
$(y^2z - 2y^4)y^2z + (y^2z - 2y^4)3z^2 - y^4(y^2z - 2y^4)$
$= y^4z^2 - 2y^6z + 3y^2z^3 - 6y^4z^2 - y^6z + 2y^8$
$= 2y^8 - 3y^6z - 5y^4z^2 + 3y^2z^3$

41.
$$\begin{array}{r} 2a - 3 \\ a + 5 \\ \hline 10a - 15 \\ 2a^2 - 3a \\ \hline 2a^2 + 7a - 15 \end{array}$$

43.
$$\begin{array}{r} 7x + 30 \\ 2x + 5 \\ \hline 35x + 150 \\ 14x^2 + 60x \\ \hline 14x^2 + 95x + 150 \end{array}$$

45.
$$\begin{array}{r} 5x + 2 \\ 4x - 3 \\ \hline -15x - 6 \\ 20x^2 + 8x \\ \hline 20x^2 - 7x - 6 \end{array}$$

47.
$$\begin{array}{r} m - 3n \\ 2a + b \\ \hline mb - 3nb \\ 2am - 6an \\ \hline 2am - 6an + mb - 3nb \end{array}$$

49.

$$x^2 + 3x - 2$$
$$\underline{ x + 6}$$
$$6x^2 + 18x - 12$$
$$\underline{x^3 + 3x^2 - 2x }$$
$$x^3 + 9x^2 + 16x - 12$$

51.

$$2a^3 - 3a^2 + 4$$
$$\underline{ -2a - 3}$$
$$-6a^3 + 9a^2 - 12$$
$$\underline{-4a^4 + 6a^3 - 8a }$$
$$-4a^4 + 9a^2 - 8a - 12$$

53.

$$x - y$$
$$\underline{x + y}$$
$$xy - y^2$$
$$\underline{x^2 - xy }$$
$$x^2 - y^2$$

55.

$$x^2 - xy + y^2$$
$$\underline{ x + y}$$
$$x^2y - xy^2 + y^3$$
$$\underline{x^3 - x^2y + xy^2 }$$
$$x^3 + y^3$$

57. The opposite of $3t - u$ is $-3t + u$ or $u - 3t$.

59. The opposite of $3x + y$ is $-3x - y$.

61. The opposite of $-3a^2 - a + 6$ is $3a^2 + a - 6$.

63. The opposite of $3v^2 + v - 6$ is $-3v^2 - v + 6$.

65. $-3x(2x - 9) = -3x(2x) - (-3x)(9)$

$$= -6x^2 + 27x$$

67. $2 - 3x(2x - 9) = 2 - 6x^2 + 27x$

$$= -6x^2 + 27x + 2$$

69. $(2 - 3x) + (2x - 9) = 2 - 9 - 3x + 2x$

$$= -x - 7$$

71. $(6x^6)^2 = 6x^6 \cdot 6x^6 = 36x^{12}$

73. $3ab^3 \cdot (-2a^2b^7) = -2 \cdot 3aa^2b^3b^7 = -6a^3b^{10}$

75. $(5x + 6)(5x + 6) = (5x + 6)5x + (5x + 6)6$

$$= 25x^2 + 30x + 30x + 36 = 25x^2 + 60x + 36$$

77. $(5x - 6)(5x + 6) = (5x - 6)5x + (5x - 6)6$

$$= 25x^2 - 30x + 30x - 36 = 25x^2 - 36$$

79. $2x^2(3x^5 - 4x^2) = 2x^2 \cdot 3x^5 - 2x^2 \cdot 4x^2$
$$= 6x^7 - 8x^4$$

81. $(m - 1)(m^2 + m + 1)$
$= (m - 1)m^2 + (m - 1)m + (m - 1)1$
$= m^3 - m^2 + m^2 - m + m - 1 = m^3 - 1$

83. $(3x - 2)(x^2 - x - 9)$
$= x^2(3x - 2) - x(3x - 2) - 9(3x - 2)$
$= 3x^3 - 2x^2 - 3x^2 + 2x - 27x + 18$
$= 3x^3 - 5x^2 - 25x + 18$

85. Since area of a rectangle is length times width, the area is $x(x + 4)$ or $x^2 + 4x$ square feet. If $x = 10$, then the area is $10^2 + 4(10)$ or 140 ft^2.

87. For a triangle $A = \frac{1}{2}bh$. So $A = \frac{1}{2}x(2x + 1) = x^2 + \frac{1}{2}x$ square feet. If $x = 5$, then $5^2 + \frac{1}{2} \cdot 5 = 27.5$. The area is 27.5 ft^2.

89. Two numbers that differ by 5 can be represented as x and $x + 5$. Their product is $x^2 + 5x$.

91. The area for a rectangle is length times width. So we find the product:
$(2.3x + 1.2)(3.5x + 5.1)$
$= (2.3x + 1.2)3.5x + (2.3x + 1.2)5.1$
$= 8.05x^2 + 4.2x + 11.73x + 6.12$
$= 8.05x^2 + 15.93x + 6.12$ square meters

93. If $p = 10$, then $40,000 - 1000(10) = 30,000$. At \$10 each 30,000 tickets will be sold. The revenue is $30,000 \cdot 10$, or \$300,000. The total revenue is $p(40,000 - 1000p)$, or $40,000p - 1000p^2$.

95. The values at the end of each year follow: $10x$

$(10x + 10)x = 10x^2 + 10x$

$(10x^2 + 10x + 10)x = 10x^3 + 10x^2 + 10x$

$(10x^3 + 10x^2 + 10x + 10)x$
$$= 10x^4 + 10x^3 + 10x^2 + 10x$$

$(10x^4 + 10x^3 + 10x^2 + 10x + 10)x$
$$= 10x^5 + 10x^4 + 10x^3 + 10x^2 + 10x$$

If $r = 0.10$ then $x = 1.10$:

$10(1.10)^5 + 10(1.10)^4 + 10(1.10)^3 + 10(1.10)^2$

$+ 10(1.10) = 67.16$
So the value of the the investment is \$67.16.

3.3 WARM-UPS

1. False, because if we replace x by 1 we get $(1 + 3)(1 + 2) = 1^2 + 6$, which is incorrect.
2. True, because the binomials are correctly multiplied. 3. True, because the binomials are correctly multiplied. 4. True, because the binomials are correctly multiplied. 5. True, because the binomials are correctly multiplied.
6. False, because $3a^2 \cdot 3a^2 = 9a^4$. 7. True, because the binomials are correctly multiplied.
8. False, because $(-9)(-2) = 18$. 9. False, because $4x - 7x = -3x$. 10. False, because we must be able to multiply binomials quickly and FOIL enables us to do that.

3.3 EXERCISES

1. $(x + 2)(x + 4) = x^2 + 2x + 4x + 8$
$= x^2 + 6x + 8$

3. $(a - 3)(a + 2) = a^2 - 3a + 2a - 6$
$= a^2 - a - 6$

5. $(2x - 1)(x - 2) = 2x^2 - x - 4x + 2$
$= 2x^2 - 5x + 2$

7. $(2a - 3)(a + 1) = 2a^2 - 3a + 2a - 3$
$= 2a^2 - a - 3$

9. $(w - 50)(w - 10) = w^2 - 50w - 10w + 500$
$= w^2 - 60w + 500$

11. $(y - a)(y + 5) = y^2 - ay + 5y - 5a$

13. $(5 - w)(w + m) = 5w - w^2 + 5m - mw$

15. $(2m - 3t)(5m + 3t)$
$= 10m^2 - 15mt + 6mt - 9t^2$
$= 10m^2 - 9mt - 9t^2$

17. $(5a + 2b)(9a + 7b)$
$= 45a^2 + 18ab + 35ab + 14b^2$
$= 45a^2 + 53ab + 14b^2$

19. $(x^2 - 5)(x^2 + 2) = x^4 - 5x^2 + 2x^2 - 10$
$= x^4 - 3x^2 - 10$

21. $(h^3 + 5)(h^3 + 5) = h^6 + 5h^3 + 5h^3 + 25$
$= h^6 + 10h^3 + 25$

23. $(3b^3 + 2)(b^3 + 4) = 3b^6 + 2b^3 + 12b^3 + 8$
$= 3b^6 + 14b^3 + 8$

25. $(y^2 - 3)(y - 2) = y^3 - 2y^2 - 3y + 6$

27. $(3m^3 - n^2)(2m^3 + 3n^2)$
$= 6m^6 + 7m^3n^2 - 3n^4$

29. $(3u^2v - 2)(4u^2v + 6)$
$= 12u^4v^2 - 8u^2v + 18u^2v - 12$
$= 12u^4v^2 + 10u^2v - 12$

31. $(b + 4)(b + 5) = b^2 + 4b + 5b + 20$
$= b^2 + 9b + 20$

33. $(x - 3)(x + 9) = x^2 - 3x + 9x - 27$
$= x^2 + 6x - 27$

35. $(a + 5)(a + 5) = a^2 + 5a + 5a + 25$
$= a^2 + 10a + 25$

37. $(2x - 1)(2x - 1) = 4x^2 - 2x - 2x + 1$
$= 4x^2 - 4x + 1$

39. $(z - 10)(z + 10) = z^2 - 10z + 10z - 100$
$= z^2 - 100$

41. $(a + b)(a + b) = a^2 + ab + ab = b^2$
$= a^2 + 2ab + b^2$

43. $(a - 1)(a - 2) = a^2 - a - 2a + 2$
$= a^2 - 3a + 2$

45. $(2x - 1)(x + 3) = 2x^2 - x + 6x - 3$
$= 2x^2 + 5x - 3$

47. $(5t - 2)(t - 1) = 5t^2 - 2t - 5t + 2$
$= 5t^2 - 7t + 2$

49. $(h - 7)(h - 9) = h^2 - 7h - 9h + 63$
$= h^2 - 16h + 63$

51. $(h + 7w)(h + 7w) = h^2 + 7hw + 7hw + 49w^2$
$= h^2 + 14hw + 49w^2$

53. $(2h^2 - 1)(2h^2 - 1) = 4h^4 - 2h^2 - 2h^2 + 1$
$= 4h^4 - 4h^2 + 1$

55. $\left(2a + \frac{1}{2}\right)\left(4a - \frac{1}{2}\right) = 8a^2 + a - \frac{1}{4}$

57. $\left(\frac{1}{2}x - \frac{1}{3}\right)\left(\frac{1}{4}x + \frac{1}{2}\right) = \frac{1}{8}x^2 + \frac{1}{6}x - \frac{1}{6}$

59. $-2x^4(3x - 1)(2x + 5) = -2x^4(6x^2 + 13x - 5)$
$= -12x^6 - 26x^5 + 10x^4$

61. $(x - 1)(x + 1)(x + 3) = (x - 1)(x^2 + 4x + 3)$
$= x(x^2 + 4x + 3) - 1(x^2 + 4x + 3)$
$= x^3 + 4x^2 + 3x - x^2 - 4x - 3$
$= x^3 + 3x^2 - x - 3$

63. $(3x - 2)(3x + 2)(x + 5) = (9x^2 - 4)(x + 5)$
$= 9x^3 + 45x^2 - 4x - 20$

65. $(x-1)(x+2)-(x+3)(x-4)$
$= (x^2+x-2)-(x^2-x-12)$
$= x^2+x-2-x^2+x+12 = 2x+10$

67. The area of a rectangle is length times width. So the area of the rug is $(x+3)(2x-1) = 2x^2+5x-3$ square feet.

69. The area of a triangle is $\frac{1}{2}$bh. So the area is $\frac{1}{2}(4.57x+3)(2.3x-1.33)$
$= 5.2555x^2 + 0.41095x - 1.995$ square meters.

3.4 WARM-UPS

1. False, because $(2+3)^2 = 5^2 = 25$ and $2^2+3^2 = 4+9 = 13$.

2. True, because the binomial is squared correctly.

3. True, because $(3+5)^2 = 3^2 + 2 \cdot 3 \cdot 5 + 5^2$
$= 9 + 30 + 25$.

4. True, because $(2x+7)^2 = 4x^2 + 2 \cdot 2x \cdot 7 + 7^2$
$= 4x^2 + 28x + 49$.

5. False, because if y is replaced by 1 we get $(1+8)^2 = 1^2 + 64$, which is incorrect.

6. True, because $(a-b)(a+b) = a^2 - b^2$.

7. True, because $(40-1)(40+1) = 40^2 - 1^2$
$= 1600 - 1 = 1599$. **8.** True, because $49 \cdot 51 = (50-1)(50+1) = 2500 - 1 = 2499$.

9. False, because if we replace x by 3 we get $(3-3)^2 = 3^2 - 3(3) + 9$, which is incorrect.

10. False, because $(a+b)^2 = a^2 + 2ab + b^2$.

3.4 EXERCISES

1. $(x+1)^2 = x^2 + 2 \cdot x \cdot 1 + 1^2 = x^2 + 2x + 1$

3. $(y+4)^2 = y^2 + 2 \cdot y \cdot 4 + 4^2 = y^2 + 8y + 16$

5. $(3x+8)^2 = (3x)^2 + 2 \cdot 3x \cdot 8 + 8^2$
$= 9x^2 + 48x + 64$

7. $(s+t)^2 = s^2 + 2st + t^2$

9. $(2x+y)^2 = (2x)^2 + 2 \cdot 2x(y) + (y)^2$
$= 4x^2 + 4xy + y^2$

11. $(2t+3h)^2 = (2t)^2 + 2 \cdot 2t \cdot 3h + (3h)^2$
$= 4t^2 + 12ht + 9h^2$

13. $(a-3)^2 = a^2 - 2 \cdot a \cdot 3 + 3^2 = a^2 - 6a + 9$

15. $(t-1)^2 = (t)^2 - 2 \cdot t \cdot 1 + 1^2$
$= t^2 - 2t + 1$

17. $(3t-2)^2 = (3t)^2 - 2 \cdot 3t \cdot 2 + 2^2$
$= 9t^2 - 12t + 4$

19. $(s-t)^2 = s^2 - 2st + t^2$

21. $(3a-b)^2 = (3a)^2 - 2 \cdot 3a(b) + (b)^2$
$= 9a^2 - 6ab + b^2$

23. $(3z-5y)^2 = (3z)^2 - 2(3z)(5y) + (5y)^2$
$= 9z^2 - 30yz + 25y^2$

25. $(a-5)(a+5) = a^2 - 5^2 = a^2 - 25$

27. $(y-1)(y+1) = y^2 - 1^2 = y^2 - 1$

29. $(3x-8)(3x+8) = (3x)^2 - 8^2$
$= 9x^2 - 64$

31. $(r+s)(r-s) = r^2 - s^2$

33. $(8y-3a)(8y+3a) = 64y^2 - 9a^2$

35. $(5x^2-2)(5x^2+2) = (5x^2)^2 - 2^2 = 25x^4 - 4$

37. $(x+1)^3 = (x+1)(x^2+2x+1)$
$= x^3 + 3x^2 + 3x + 1$

39. $(2a-3)^3 = (2a-3)(4a^2-12a+9)$
$= 2a(4a^2-12a+9) - 3(4a^2-12a+9)$
$= 8a^3 - 36a^2 + 54a - 27$

41. $(a-3)^4 = (a-3)^2(a-3)^2$
$= (a^2-6a+9)(a^2-6a+9)$
$= a^2(a^2-6a+9) - 6a(a^2-6a+9)$
$\qquad + 9(a^2-6a+9)$
$= a^4 - 12a^3 + 54a^2 - 108a + 81$

43. $(a+b)^4 = (a^2+2ab+b^2)(a^2+2ab+b^2)$
$= a^4 + 4a^3b + 6a^2b^2 + 4ab^3 + b^4$

45. $(a-20)(a+20) = a^2 - 20^2 = a^2 - 400$

47. $(x+8)(x+7) = x^2 + 8x + 7x + 56$
$= x^2 + 15x + 56$

49. $(4x-1)(4x+1) = (4x)^2 - 1^2 = 16x^2 - 1$

51. $(9y-1)^2 = 81y^2 - 2 \cdot 9y \cdot 1 + 1^2$
$= 81y^2 - 18y + 1$

53. $(2t-5)(3t+4) = 6t^2 - 15t + 8t - 20$
$= 6t^2 - 7t - 20$

55. $(2t-5)^2 = 4t^2 - 2 \cdot 2t \cdot 5 + 5^2$
$= 4t^2 - 20t + 25$

57. $(2t+5)(2t-5) = (2t)^2 - 5^2 = 4t^2 - 25$

59. $(x^2-1)(x^2+1) = (x^2)^2 - 1^2 = x^4 - 1$

61. $(2y^3-9)^2 = 4y^6 - 36y^3 + 81$

63. $(2x^3+3y^2)^2 = 4x^6 + 12x^3y^2 + 9y^4$

65. $\left(\frac{1}{2}x+\frac{1}{3}\right)^2 = \frac{1}{4}x^2 + \frac{1}{3}x + \frac{1}{9}$

67. $(0.2x-0.1)^2 = 0.04x^2 - 0.04x + 0.01$

69. $(a+b)^3 = (a+b)(a^2+2ab+b^2)$
$= a^3 + 3a^2b + 3ab^2 + b^3$

71. $(1.5x+3.8)^2 = 2.25x^2 + 2(1.5x)(3.8) + (3.8)^2$
$= 2.25x^2 + 11.4x + 14.44$

73. $(3.5t-2.5)(3.5t+2.5) = (3.5t)^2 - (2.5)^2$
$= 12.25t^2 - 6.25$

75. Let $x =$ the length of a side of the square garden. The new length will be $x+5$ feet and the new width will be $x-5$ feet. The old area is x^2 square feet, and the new area is $(x+5)(x-5) = x^2 - 25$ square feet. So the new garden is 25 square feet smaller than the old square garden.

77. The area of a circle is πr^2. The area of a circle with radius $b+1$ is
$3.14(b+1)^2 = 3.14(b^2+2b+1)$
$= 3.14b^2 + 6.28b + 3.14$ square meters

79. $v = k(R-r)(R+r)$
$v = k(R^2 - r^2)$

81. $P(1+r)^2 = P(1+2r+r^2) = P + 2Pr + Pr^2$
If $P = 200$ and $r = 0.10$, then
$200 + 2 \cdot 200 \cdot 0.10 + 200 \cdot (0.10)^2$
$= \$242$

83. $\$14,380.95$

3.5 WARM-UPS

1. False, because $2^{10} \div 2^2 = 1024 \div 4 = 256$ and $2^5 = 32$. By the quotient rule $y^{10} \div y^2 = y^8$ for any nonzero value of y. **2.** False, because if $x = 0$, the left side is $2/7$ and the right side is 2. **3.** True, because $7/7 = 1$. **4.** False, because the quotient is $x^2 + 2$. **5.** True, because $2y(2y-3) = 4y^2 - 6y$. **6.** False, because the quotient times the divisor plus the remainder is equal to the dividend. **7.** True, because $(x+2)(x+1) + 3 = x^2 + 2x + x + 2 + 3$
$= x^2 + 3x + 5$. **8.** True, because of the fact stated in Warm-up number 7. **9.** True, because

of the fact stated in Warm-up number 7. **10.** True, because of the fact that dividend is equal to the divisor times the quotient plus the remainder.

3.5 EXERCISES

1. $9^0 = 1$

3. $(-2x^3)^0 = 1$

5. $2 \cdot 5^0 - 3^0 = 2 \cdot 1 - 1 = 1$

7. $(2x-y)^0 = 1$

9. $\frac{x^8}{x^2} = x^{8-2} = x^6$

11. $\frac{6a^7}{2a^{12}} = \frac{3}{a^{12-7}} = \frac{3}{a^5}$

13. $-12x^5 \div (3x^9) = \frac{-12x^5}{3x^9} = \frac{-4}{x^4}$

15. $-6y^2 \div (6y) = \frac{-6y^2}{6y} = -1y^{2-1} = -y$

17. $\frac{-6x^3y^2}{2x^2y^2} = -3x^{3-2}y^{2-2} = -3x$

19. $\frac{-9x^2y^2}{3x^5y^2} = \frac{-3y^0}{x^3} = \frac{-3}{x^3}$

21. $\frac{3x-6}{3} = \frac{3x}{3} - \frac{6}{3} = x - 2$

23. $\frac{x^5+3x^4-x^3}{x^2} = x^{5-2} + 3x^{4-2} - x^{3-2}$
$= x^3 + 3x^2 - x$

25. $\frac{-8x^2y^2 + 4x^2y - 2xy^2}{-2xy}$

$= \frac{-8x^2y^2}{-2xy} + \frac{4x^2y}{-2xy} - \frac{2xy^2}{-2xy} = 4xy - 2x + y$

27. $(x^2y^3 - 3x^3y^2) \div (x^2y) = \frac{x^2y^3}{x^2y} - \frac{3x^3y^2}{x^2y}$
$= y^2 - 3xy$

29.
$$\begin{array}{r} x+2 \\ x+3\overline{)\,x^2+5x+13} \\ \underline{x^2+3x} \\ 2x+13 \\ \underline{2x+6} \\ 7 \end{array}$$

The quotient is $x+2$ and the remainder is 7.

31.
$$\begin{array}{r} 2 \\ x+5\overline{)\,2x+0} \\ \underline{2x+10} \\ -10 \end{array}$$

The quotient is 2 and the remainder is -10.

45

33.

$$\begin{array}{r} a^2 + 2a + 8 \\ a-2 \overline{\smash{\big)}\ a^3 + 0a^2 + 4a - 3} \\ \underline{a^3 - 2a^2} \\ 2a^2 + 4a \\ \underline{2a^2 - 4a} \\ 8a - 3 \\ \underline{8a - 16} \\ 13 \end{array}$$

The quotient is $a^2 + 2a + 8$ and remainder is 13.

35.

$$\begin{array}{r} x - 4 \\ x+1 \overline{\smash{\big)}\ x^2 - 3x + 0} \\ \underline{x^2 + x} \\ -4x + 0 \\ \underline{-4x - 4} \\ 4 \end{array}$$

The quotient is $x - 4$ and the remainder is 4.

37.

$$\begin{array}{r} h^2 + 3h + 9 \\ h-3 \overline{\smash{\big)}\ h^3 + 0h^2 + 0h - 27} \\ \underline{h^3 - 3h^2} \\ 3h^2 + 0h \\ \underline{3h^2 - 9h} \\ 9h - 27 \\ \underline{9h - 27} \\ 0 \end{array}$$

The quotient is $h^2 + 3h + 9$ and the remainder is 0.

39.

$$\begin{array}{r} 2x - 3 \\ 3x-2 \overline{\smash{\big)}\ 6x^2 - 13x + 7} \\ \underline{6x^2 - 4x} \\ -9x + 7 \\ \underline{-9x + 6} \\ 1 \end{array}$$

The quotient is $2x - 3$ and the remainder is 1.

41.

$$\begin{array}{r} x^2 + 0x + 1 \\ x-1 \overline{\smash{\big)}\ x^3 - x^2 + x - 2} \\ \underline{x^3 - x^2} \\ 0x^2 + x \\ \underline{0x^2 + 0x} \\ x - 2 \\ \underline{x - 1} \\ -1 \end{array}$$

The quotient is $x^2 + 1$ and the remainder is -1.

43.

$$\begin{array}{r} 3 \\ x-5 \overline{\smash{\big)}\ 3x + 0} \\ \underline{3x - 15} \\ 15 \end{array}$$

$$\frac{3x}{x-5} = 3 + \frac{15}{x-5}$$

45.

$$\begin{array}{r} -1 \\ x+3 \overline{\smash{\big)}\ -x + 0} \\ \underline{-x - 3} \\ 3 \end{array}$$

$$\frac{-x}{x+3} = -1 + \frac{3}{x+3}$$

47. $\dfrac{x-1}{x} = \dfrac{x}{x} - \dfrac{1}{x} = 1 - \dfrac{1}{x}$

49. $\dfrac{3x+1}{x} = \dfrac{3x}{x} + \dfrac{1}{x} = 3 + \dfrac{1}{x}$

51.

$$\begin{array}{r} x - 1 \\ x+1 \overline{\smash{\big)}\ x^2 + 0x + 0} \\ \underline{x^2 + x} \\ -x + 0 \\ \underline{-x - 1} \\ 1 \end{array}$$

$$\frac{x^2}{x+1} = x - 1 + \frac{1}{x+1}$$

53.

$$\begin{array}{r} x - 2 \\ x+2 \overline{\smash{\big)}\ x^2 + 0x + 4} \\ \underline{x^2 + 2x} \\ -2x + 4 \\ \underline{-2x - 4} \\ 8 \end{array}$$

$$\frac{x^2+4}{x+2} = x - 2 + \frac{8}{x+2}$$

55.

$$\begin{array}{r} x^2 + 2x + 4 \\ x-2 \overline{\smash{\big)}\ x^3 + 0x^2 + 0x + 0} \\ \underline{x^3 - 2x^2} \\ 2x^2 + 0x \\ \underline{2x^2 - 4x} \\ 4x + 0 \\ \underline{4x - 8} \\ 8 \end{array}$$

$$\frac{x^3}{x-2} = x^2 + 2x + 4 + \frac{8}{x-2}$$

57. $\dfrac{x^3+3}{x} = \dfrac{x^3}{x} + \dfrac{3}{x} = x^2 + \dfrac{3}{x}$

59. $-6a^3b \div (2a^2b) = \dfrac{-6a^3b}{2a^2b} = -3a^{3-2}b^0$

$\qquad\qquad\qquad\qquad\qquad = -3a$

61. $-8w^4t^7 \div (-2w^9t^3) = \dfrac{-8w^4t^7}{-2w^9t^3} = \dfrac{4t^4}{w^5}$

63. $(3a - 12) \div (-3) = \dfrac{3a}{-3} - \dfrac{12}{-3} = -a + 4$

65. $(3x^2 - 9x) \div (3x) = \dfrac{3x^2}{3x} - \dfrac{9x}{3x} = x - 3$

67. $(12x^4 - 4x^3 + 6x^2) \div (-2x^2)$

$= \dfrac{12x^4}{-2x^2} - \dfrac{4x^3}{-2x^2} + \dfrac{6x^2}{-2x^2} = -6x^2 + 2x - 3$

69.
$$
\begin{array}{r}
t + 4 \\
t - 9 \overline{)\ t^2 - 5t - 36} \\
\underline{t^2 - 9t} \\
4t - 36 \\
\underline{4t - 36} \\
0
\end{array}
$$

The quotient is $t + 4$.

71.
$$
\begin{array}{r}
2w + 1 \\
3w - 5 \overline{)\ 6w^2 - 7w - 5} \\
\underline{6w^2 - 10w} \\
3w - 5 \\
\underline{3w - 5} \\
0
\end{array}
$$

The quotient is $2w + 1$.

73.
$$
\begin{array}{r}
4x^2 - 6x + 9 \\
2x + 3 \overline{)\ 8x^3 + 0x^2 + 0x + 27} \\
\underline{8x^3 + 12x^2} \\
-12x^2 + 0x \\
\underline{-12x^2 - 18x} \\
18x + 27 \\
\underline{18x + 27} \\
0
\end{array}
$$

The quotient is $4x^2 - 6x + 9$ and the remainder is 0.

75.
$$
\begin{array}{r}
t^2 - t + 3 \\
t - 2 \overline{)\ t^3 - 3t^2 + 5t - 6} \\
\underline{t^3 - 2t^2} \\
-t^2 + 5t \\
\underline{-t^2 + 2t} \\
3t - 6 \\
\underline{3t - 6} \\
0
\end{array}
$$

The quotient is $t^2 - t + 3$.

77.
$$
\begin{array}{r}
v^2 - 2v + 1 \\
v - 4 \overline{)\ v^3 - 6v^2 + 9v - 4} \\
\underline{v^3 - 4v^2} \\
-2v^2 + 9v \\
\underline{-2v^2 + 8v} \\
v - 4 \\
\underline{v - 4} \\
0
\end{array}
$$

The quotient is $v^2 - 2v + 1$.

79. Since $A = LW$ for a rectangle, we can write $W = A/L$. To find the width divide the area by the length.

$$
\begin{array}{r}
x - 5 \\
x + 6 \overline{)\ x^2 + x - 30} \\
\underline{x^2 + 6x} \\
-5x - 30 \\
\underline{-5x - 30} \\
0
\end{array}
$$

The width is $x - 5$ meters.

3.6 WARM-UPS

1. False, because $-3^0 = -1$.

2. False, because $2^3 \cdot 2^8 = 2^{13}$.

3. False, because $2^3 \cdot 3^2 = 8 \cdot 9 = 72$ and $6^5 \neq 72$.

4. False, because $(2x)^4 = 16x^4$ for any real number x.

5. False, because $(q^3)^5 = q^{15}$.

6. False, because $(-3x^2)^3 = -27x^6$.

7. True. **8.** False, because $a^{12} \div a^4 = a^8$.

9. False, because $\dfrac{6w^4}{3w^9} = \dfrac{2}{w^5}$. **10.** True.

3.6 EXERCISES

1. $2^3 \cdot 5^2 = 8 \cdot 25 = 200$

3. $2^5 \cdot 2^{10} = 2^{15}$

5. $(-3u^8v^7)(-2u^2v) = 6u^{10}v^8$

7. $a^3b^4 \cdot ab^6(ab)^0 = a^4b^{10} \cdot 1 = a^4b^{10}$

9. $\dfrac{-2a^3}{4a^7} = \dfrac{-2}{2 \cdot 2a^{7-3}} = \dfrac{-1}{2a^4}$

11. $\dfrac{2a^5 \cdot 3a^7}{15a^6} = \dfrac{2 \cdot 3a^{12}}{3 \cdot 5a^6} = \dfrac{2a^6}{5}$

13. $(x^2)^3 = x^{2 \cdot 3} = x^6$

15. $2x^2 \cdot (x^2)^5 = 2x^2 \cdot x^{10} = 2x^{12}$

17. $\dfrac{(t^2)^5}{(t^3)^4} = \dfrac{t^{10}}{t^{12}} = \dfrac{1}{t^2}$

19. $\dfrac{3x(x^5)^2}{6x^3(x^2)^4} = \dfrac{3xx^{10}}{6x^3x^8} = \dfrac{3x^{11}}{6x^{11}} = \dfrac{1}{2}$

21. $(xy^2)^3 = x^3(y^2)^3 = x^3y^6$

23. $(-2t^5)^3 = (-2)^3(t^5)^3 = -8t^{15}$

25. $(-2x^2y^5)^3 = (-2)^3(x^2)^3(y^5)^3 = -8x^6y^{15}$

27. $\dfrac{(a^4b^2c^5)^3}{a^3b^4c} = \dfrac{a^{12}b^6c^{15}}{a^3b^4c} = a^9b^2c^{14}$

29. $\left(\dfrac{x^4}{4}\right)^3 = \dfrac{(x^4)^3}{4^3} = \dfrac{x^{12}}{64}$

31. $\left(\dfrac{-2a^2}{b^3}\right)^4 = \dfrac{(-2)^4(a^2)^4}{(b^3)^4} = \dfrac{16a^8}{b^{12}}$

33. $\left(\dfrac{2x^2y}{-4y^2}\right)^3 = \dfrac{2^3(x^2)^3y^3}{(-4)^3(y^2)^3} = \dfrac{8x^6y^3}{-64y^6}$

$= \dfrac{x^6}{-8y^3} = -\dfrac{x^6}{8y^3}$

35. $\left(\dfrac{-6x^2y^4z^9}{3x^6y^4z^3}\right)^2 = \dfrac{(-6)^2(x^2)^2(y^4)^2(z^9)^2}{3^2(x^6)^2(y^4)^2(z^3)^2}$

$= \dfrac{36x^4y^8z^{18}}{9x^{12}y^8z^6} = \dfrac{4z^{12}}{x^8}$

37. $2^3 + 3^3 = 8 + 27 = 35$

39. $(2+3)^3 = 5^3 = 125$

41. $\left(\dfrac{2}{3}\right)^3 = \dfrac{2^3}{3^3} = \dfrac{8}{27}$

43. $5^2 \cdot 2^3 = 25 \cdot 8 = 200$

45. $3x^4 \cdot 5x^7 = 15x^{11}$

47. $(-5x^4)^3 = -125x^{12}$

49. $-3y^5z^{12} \cdot 9yz^7 = -27y^6z^{19}$

51. $\dfrac{-9u^4v^9}{-3u^5v^8} = \dfrac{3v}{u}$

53. $(-xt^2)(-2x^2t)^4 = -xt^2 \cdot 16x^8t^4$

$= -16x^9t^6$

55. $\left(\dfrac{2x^2}{x^4}\right)^3 = \dfrac{2^3(x^2)^3}{(x^4)^3} = \dfrac{8x^6}{x^{12}} = \dfrac{8}{x^6}$

57. $\left(\dfrac{-8a^3b^4}{4c^5}\right)^5 = \dfrac{(-2)^5(a^3)^5(b^4)^5}{(c^5)^5} = \dfrac{-32a^{15}b^{20}}{c^{25}}$

59. $\left(\dfrac{-8x^4y^7}{-16x^5y^6}\right)^5 = \left(\dfrac{y}{2x}\right)^5 = \dfrac{y^5}{32x^5}$

61. By the product rule,

$P(1+r)^{10}(1+r)^5 = P(1+r)^{15}.$

CHAPTER 3 REVIEW

1. $(2w-6)+(3w+4) = 5w-2$

3. $(x^2-2x-5)-(x^2+4x-9)$

$= x^2 - 2x - 5 - x^2 - 4x + 9 = -6x + 4$

5. $(5-3w+w^2)+(w^2-4w-9)$

$= w^2 + w^2 - 3w - 4w - 9 + 5 = 2w^2 - 7w - 4$

7. $(4-3m-m^2)-(m^2-6m+5)$

$= 4 - 3m - m^2 - m^2 + 6m - 5 = -2m^2 + 3m - 1$

9. $5x^2 \cdot (-10x^9) = -50x^{2+9} = -50x^{11}$

11. $(-11a^7)^2 = (-11a^7)(-11a^7) = 121a^{14}$

13. $x - 5(x-3) = x - 5x + 15 = -4x + 15$

15. $5x + 3(x^2-5x+4) = 5x + 3x^2 - 15x + 12$

$= 3x^2 - 10x + 12$

17. $3m^2(5m^3-m+2)$

$= 3m^2 \cdot 5m^3 - 3m^2 \cdot m + 3m^2 \cdot 2$

$= 15m^5 - 3m^3 + 6m^2$

19. $(x-5)(x^2-2x+10)$

$= (x-5)x^2 - (x-5)(2x) + (x-5)(10)$

$= x^3 - 5x^2 - 2x^2 + 10x + 10x - 50$

$= x^3 - 7x^2 + 20x - 50$

21. $(x^2 - 2x + 4)(3x - 2)$

$\quad = (x^2 - 2x + 4)3x - (x^2 - 2x + 4)2$

$\quad = 3x^3 - 6x^2 + 12x - 2x^2 + 4x - 8$

$\quad = 3x^3 - 8x^2 + 16x - 8$

23. $(q - 6)(q + 8) = q^2 - 6q + 8q - 48$

$\quad\quad\quad\quad\quad\quad = q^2 + 2q - 48$

25. $(2t - 3)(t - 9) = 2t^2 - 3t - 18t + 27$

$\quad\quad\quad\quad\quad\quad = 2t^2 - 21t + 27$

27. $(4y - 3)(5y + 2) = 20y^2 - 15y + 8y - 6$

$\quad\quad\quad\quad\quad\quad\quad = 20y^2 - 7y - 6$

29. $(3x^2 + 5)(2x^2 + 1) = 6x^4 + 10x^2 + 3x^2 + 5$

$\quad\quad\quad\quad\quad\quad\quad\quad = 6x^4 + 13x^2 + 5$

31. $(z - 7)(z + 7) = z^2 - 7z + 7z - 49$

$\quad\quad\quad\quad\quad\quad = z^2 - 49$

33. $(y + 7)^2 = y^2 + 2 \cdot y \cdot 7 + 7^2 = y^2 + 14y + 49$

35. $(w - 3)^2 = w^2 - 2 \cdot 3w + 3^2 = w^2 - 6w + 9$

37. $(x^2 - 3)(x^2 + 3) = x^4 - 3x^2 + 3x^2 - 9$

$\quad\quad\quad\quad\quad\quad\quad = x^4 - 9$

39. $(3a + 1)^2 = 9a^2 + 2 \cdot 3a \cdot 1 + 1^2$

$\quad\quad\quad\quad\quad = 9a^2 + 6a + 1$

41. $(4 - y)^2 = 4^2 - 2 \cdot 4 \cdot y + y^2 = y^2 - 8y + 16$

43. $-10x^5 \div (2x^3) = \dfrac{-10x^5}{2x^3} = -5x^2$

45. $\dfrac{6a^5b^7c^6}{-3a^3b^9c^6} = \dfrac{-2a^2}{b^2}$

47. $\dfrac{3x - 9}{-3} = \dfrac{3x}{-3} - \dfrac{9}{-3} = -x + 3$

49. $\dfrac{9x^3 - 6x^2 + 3x}{-3x} = \dfrac{9x^3}{-3x} - \dfrac{6x^2}{-3x} + \dfrac{3x}{-3x}$

$\quad\quad\quad\quad\quad\quad\quad = -3x^2 + 2x - 1$

51. $(a - 1) \div (1 - a) = \dfrac{a - 1}{1 - a} = \dfrac{-1(1 - a)}{(1 - a)} = -1$

53.

$$
\begin{array}{r}
m^3 + 2m^2 + 4m + 8 \\
m - 2 \overline{\smash{)}\ m^4 + 0m^3 + 0m^2 + 0m - 16} \\
\underline{m^4 - 2m^3} \\
2m^3 + 0m^2 \\
\underline{2m^3 - 4m^2} \\
4m^2 + 0m \\
\underline{4m^2 - 8m} \\
8m - 16 \\
\underline{8m - 16} \\
0
\end{array}
$$

The quotient is $m^3 + 2m^2 + 4m + 8$.

55. $\dfrac{3m^3 - 9m^2 + 18m}{3m} = \dfrac{3m^3}{3m} - \dfrac{9m^2}{3m} + \dfrac{18m}{3m}$

$$= m^2 - 3m + 6$$

The quotient is $m^2 - 3m + 6$ and the remainder is 0.

57.

$$
\begin{array}{r}
b - 5 \\
b + 2 \overline{\smash{)}\ b^2 - 3b + 5} \\
\underline{b^2 + 2b} \\
-5b + 5 \\
\underline{-5b - 10} \\
15
\end{array}
$$

The quotient is $b - 5$ and the remainder is 15.

59.

$$
\begin{array}{r}
2x - 1 \\
2x + 1 \overline{\smash{)}\ 4x^2 + 0x - 9} \\
\underline{4x^2 + 2x} \\
-2x - 9 \\
\underline{-2x - 1} \\
-8
\end{array}
$$

The quotient is $2x - 1$ and the remainder is -8.

61.

$$
\begin{array}{r}
x^2 + 2x - 9 \\
x - 1 \overline{\smash{)}\ x^3 + x^2 - 11x + 10} \\
\underline{x^3 - x^2} \\
2x^2 - 11x \\
\underline{2x^2 - 2x} \\
-9x + 10 \\
\underline{-9x + 9} \\
1
\end{array}
$$

The quotient is $x^2 + 2x - 9$ and remainder is 1.

63.

$$x - 3 \overline{)\,\begin{matrix} \,2 \\ 2x + 0 \end{matrix}}$$
$$\begin{matrix} \underline{2x - 6} \\ 6 \end{matrix}$$

$$\frac{2x}{x - 3} = 2 + \frac{6}{x - 3}$$

65.

$$1 - x \overline{)\,\begin{matrix} \,-2 \\ 2x + 0 \end{matrix}}$$
$$\begin{matrix} \underline{2x - 2} \\ 2 \end{matrix}$$

$$\frac{2x}{1 - x} = -2 + \frac{2}{1 - x}$$

67.

$$x + 1 \overline{)\,\begin{matrix} x - 1 \\ x^2 + 0x - 3 \end{matrix}}$$
$$\begin{matrix} \underline{x^2 + x} \\ -x - 3 \\ \underline{-x - 1} \\ -2 \end{matrix}$$

$$\frac{x^2 - 3}{x + 1} = x - 1 - \frac{2}{x + 1}$$

69.

$$x + 1 \overline{)\,\begin{matrix} x - 1 \\ x^2 + 0x + 0 \end{matrix}}$$
$$\begin{matrix} \underline{x^2 + x} \\ -x + 0 \\ \underline{-x - 1} \\ 1 \end{matrix}$$

$$\frac{x^2}{x + 1} = x - 1 + \frac{1}{x + 1}$$

71. $2y^{10} \cdot 3y^{20} = 6y^{30}$

73. $\dfrac{-10b^5 c^3}{2b^5 c^9} = \dfrac{-5}{c^6}$

75. $(b^5)^6 = b^{30}$

77. $(-2x^3 y^2)^3 = -8x^9 y^6$

79. $\left(\dfrac{2a}{b}\right)^3 = \dfrac{8a^3}{b^3}$

81. $\left(\dfrac{-6x^2 y^5}{-3z^6}\right)^3 = \left(\dfrac{2x^2 y^5}{z^6}\right)^3 = \dfrac{8x^6 y^{15}}{z^{18}}$

83. $(x + 3)(x + 7) = x^2 + 3x + 7x + 21$
$$= x^2 + 10x + 21$$

85. $(t - 3y)(t - 4y) = t^2 - 3ty - 4ty + 12y^2$
$$= t^2 - 7ty + 12y^2$$

87. $(2x^3)^0 + (2y)^0 = 1 + 1 = 2$

89. $(-3ht^6)^3 = -27h^3 t^{18}$

91. $(2w + 3)(w - 6) = 2w^2 + 3w - 12w - 18$
$$= 2w^2 - 9w - 18$$

93. $(3u - 5v)(3u + 5v) = 9u^2 - 25v^2$

95. $(3h + 5)^2 = 9h^2 + 30h + 25$

97. $(x + 3)^3 = (x + 3)(x + 3)^2$
$$= (x + 3)(x^2 + 6x + 9)$$
$$= x(x^2 + 6x + 9) + 3(x^2 + 6x + 9)$$
$$= x^3 + 9x^2 + 27x + 27$$

99. $(-7s^2 t)(-2s^3 t^5) = 14s^5 t^6$

101. $\left(\dfrac{k^4 m^2}{2k^2 m^2}\right)^4 = \left(\dfrac{k^2}{2}\right)^4 = \dfrac{k^8}{16}$

103. $(5x^2 - 8x - 8) - (4x^2 + x - 3)$
$$= 5x^2 - 8x - 8 - 4x^2 - x + 3$$
$$= x^2 - 9x - 5$$

105. $(2x^2 - 2x - 3) + (3x^2 + x - 9)$
$$= 5x^2 - x - 12$$

107. $(x + 4)(x^2 - 5x + 1)$
$$= x(x^2 - 5x + 1) + 4(x^2 - 5x + 1)$$
$$= x^3 - 5x^2 + x + 4x^2 - 20x + 4$$
$$= x^3 - x^2 - 19x + 4$$

109.

$$x - 2 \overline{)\,\begin{matrix} x + 6 \\ x^2 + 4x - 12 \end{matrix}}$$
$$\begin{matrix} \underline{x^2 - 2x} \\ 6x - 12 \\ \underline{6x - 12} \\ 0 \end{matrix}$$

The quotient is $x + 6$.

111. If w is the width, then the length is $w + 44$.

Since $P = 2L + 2W$, $P = 2(w + 44) + 2(w)$ or

$P = 4w + 88$ and $A = w(w + 44)$ or

$A = w^2 + 44w.$ If $w = 50,$ then $P = 4(50) + 88 = 288$ ft and $A = 50^2 + 44(50) = 4700$ ft^2.

113. $R = p(600 - 15p) = -15p^2 + 600p.$ If $p = 12,$ then $R = -15 \cdot 12^2 + 600 \cdot 12 = 5040.$ The revenue is \$5040 when the price is \$12 each. The maximum weekly revenue occurs when the price is \$20 each.

CHAPTER 3 TEST

1. $(7x^3 - x^2 - 6) + (5x^2 + 2x - 5)$
$$= 7x^3 + 4x^2 + 2x - 11$$

2. $(x^2 - 3x - 5) - (2x^2 + 6x - 7)$
$$= x^2 - 3x - 5 - 2x^2 - 6x + 7 = -x^2 - 9x + 2$$

3. $-5x^3 \cdot 7x^5 = -35x^{3+5} = -35x^8$

4. $3x^3y(2xy^4)^2 = 3x^3y \cdot 4x^2y^8 = 12x^5y^9$

5. $-4a^6b^5 \div (-2a^5b) = 2a^{6-5}b^{5-1} = 2ab^4$

6. $(x - 2) \div (2 - x) = \dfrac{x-2}{2-x} = \dfrac{-1(2-x)}{2-x} = -1$

7. $\dfrac{6y^3 - 9y^2}{-3y} = \dfrac{6y^3}{-3y} - \dfrac{9y^2}{-3y} = -2y^2 + 3y$

8. $\dfrac{-6a^7b^6c^2}{-2a^3b^8c^2} = \dfrac{3a^{7-3}}{b^{8-6}} = \dfrac{3a^4}{b^2}$

9. $\left(\dfrac{-2a}{b}\right)^5 = \dfrac{-32a^5}{b^5}$

10. $\left(\dfrac{12u^3v^9}{-3u^8v^6}\right)^2 = \dfrac{144u^6v^{18}}{9u^{16}v^{12}} = \dfrac{16v^6}{u^{10}}$

11.

$$\require{enclose}\begin{array}{r}x^2 + x - 1 \\ x - 3 \enclose{longdiv}{x^3 - 2x^2 - 4x + 3} \\ \underline{x^3 - 3x^2} \\ x^2 - 4x \\ \underline{x^2 - 3x} \\ -x + 3 \\ \underline{-x + 3} \\ 0 \end{array}$$

$(x^3 - 2x^2 - 4x + 3) \div (x - 3) = x^2 + x - 1$

12. $3x^2(5x^3 - 7x^2 + 4x - 1)$
$$= 15x^5 - 21x^4 + 12x^3 - 3x^2$$

13. $(x + 5)(x - 2) = x^2 + 5x - 2x - 10$
$$= x^2 + 3x - 10$$

14. $(3a - 7)(2a + 5) = 6a^2 + a - 35$

15. $(a - 7)^2 = a^2 - 2 \cdot a \cdot 7 + 7^2 = a^2 - 14a + 49$

16. $(4x + 3y)^2 = 16x^2 + 2 \cdot 4x \cdot 3y + (3y)^2$
$$= 16x^2 + 24xy + 9y^2$$

17. $(b - 3)(b + 3) = b^2 - 3b + 3b - 9 = b^2 - 9$

18. $(3t^2 - 7)(3t^2 + 7) = 9t^4 - 21t^2 + 21t^2 - 49$
$$= 9t^4 - 49$$

19. $(4x^2 - 3)(x^2 + 2) = 4x^4 - 3x^2 + 8x^2 - 6$
$$= 4x^4 + 5x^2 - 6$$

20. $(x - 2)(x + 3)(x - 4) = (x - 2)(x^2 - x - 12)$
$$= x(x^2 - x - 12) - 2(x^2 - x - 12)$$
$$= x^3 - 3x^2 - 10x + 24$$

21.
$$\begin{array}{r}2 \\ x - 3 \enclose{longdiv}{2x + 0} \\ \underline{2x - 6} \\ 6 \end{array}$$

$\dfrac{2x}{x - 3} = 2 + \dfrac{6}{x - 3}$

22.
$$\begin{array}{r}x - 5 \\ x + 2 \enclose{longdiv}{x^2 - 3x + 5} \\ \underline{x^2 + 2x} \\ -5x + 5 \\ \underline{-5x - 10} \\ 15 \end{array}$$

$\dfrac{x^2 - 3x + 5}{x + 2} = x - 5 + \dfrac{15}{x + 2}$

23.
$$\begin{array}{r}x - 2 \\ x - 3 \enclose{longdiv}{x^2 - 5x + 9} \\ \underline{x^2 - 3x} \\ -2x + 9 \\ \underline{-2x + 6} \\ 3 \end{array}$$

The quotient is $x - 2$ and the remainder is 3.

24. $(x^2 - 3x + 6) - (3x^2 - 4x - 9)$
$$= -2x^2 + x + 15$$

25. The width is x and the length is $x + 4$. The area is $x(x + 4)$ or $x^2 + 4x$ ft^2. The perimeter is $2x + 2(x + 4)$ or $4x + 8$ ft.

If $x = 4$, then the area is $4^2 + 4 \cdot 4$ or 32 ft^2. If $x = 4$, then the perimeter is $4(4) + 8$ or 24 ft.

26. $R = q(3000 - 150q) = -150q^2 + 3000q$

If $q = 8$, then $R = -150 \cdot 8^2 + 3000 \cdot 8 = 14{,}400$.

At \$8 each the weekly revenue is \$14,400.

Tying It All Together   Chapters 1-3

1. $-16 \div (-2) = 8$

2. $(-2)^3 - 1 = -8 - 1 = -9$

3. $(-5)^2 - 3(-5) + 1 = 25 + 15 + 1 = 41$

4. $2^{10} \cdot 2^{15} = 2^{10+15} = 2^{25}$

5. $2^{15} \div 2^{10} = 2^{15-10} = 2^5 = 32$

6. $2^{10} - 2^5 = 1024 - 32 = 992$

7. $3^2 \cdot 4^2 = 9 \cdot 16 = 144$

8. $(172 - 85) \div (85 - 172)$
$$= \frac{172 - 85}{85 - 172} = \frac{-1(85 - 172)}{(85 - 172)} = -1$$

9. $(5 + 3)^2 = 8^2 = 64$

10. $5^2 + 3^2 = 25 + 9 = 34$

11. $(30 - 1)(30 + 1) = 30^2 - 1^2 = 900 - 1 = 899$

12. $(30 + 1)^2 = 30^2 + 2 \cdot 30 \cdot 1 + 1^2$
$$= 900 + 60 + 1 = 961$$

13. $(x + 3)(x + 5) = x^2 + 3x + 5x + 15$
$$= x^2 + 8x + 15$$

14. Because of exercise number 13
$$(x^2 + 8x + 15) \div (x + 5) = x + 3.$$

15. $x + 3(x + 5) = x + 3x + 15 = 4x + 15$

16. $(x^2 + 8x + 15)(x + 5)$
$$= (x^2 + 8x + 15)x + (x^2 + 8x + 15)5$$
$$= x^3 + 8x^2 + 15x + 5x^2 + 40x + 75$$
$$= x^3 + 13x^2 + 55x + 75$$

17. $-5t^3v \cdot 3t^2v^6 = -15t^{3+2}v^{1+6} = -15t^5v^7$

18. $\dfrac{-10t^3v^2}{-2t^2v} = 5t^{3-2}v^{2-1} = 5tv$

19. $\dfrac{-6y^3 + 8y^2}{-2y^2} = \dfrac{-6y^3}{-2y^2} + \dfrac{8y^2}{-2y^2} = 3y - 4$

20. $(y^2 - 3y - 9) - (-3y^2 + 2y - 6)$
$$= y^2 - 3y - 9 + 3y^2 - 2y + 6$$
$$= 4y^2 - 5y - 3$$

21. $2x + 1 = 0$
   $2x = -1$
   $x = -\dfrac{1}{2}$

The solution to the equation is $-\dfrac{1}{2}$.

22. $x - 7 = 0$
   $x = 7$

The solution to the equation is 7.

23. $2x - 3 = 0$
   $2x = 3$
   $x = \dfrac{3}{2}$

The solution to the equation is $\dfrac{3}{2}$.

24. $3x - 7 = 5$
   $3x = 12$
   $x = 4$

The solution to the equation is 4.

25. $8 - 3x = x + 20$
   $8 = 4x + 20$
   $-12 = 4x$
   $-3 = x$

The solution to the equation is -3.

26. $4 - 3(x + 2) = 0$
   $4 - 3x - 6 = 0$
   $-3x - 2 = 0$
   $-3x = 2$
   $x = -\dfrac{2}{3}$

The solution to the equation is $-\dfrac{2}{3}$.

27. The average cost is $\dfrac{2.25n + 100{,}000}{n}$.

If $n = 1000$, $\dfrac{2.25(1000) + 100{,}000}{1000} = \$102.25$.

If $n = 100{,}000$, then the average cost is

$$\dfrac{2.25(100{,}000) + 100{,}000}{100{,}000} = \$3.25$$

If $n = 1{,}000{,}000$, then the average cost is

$$\dfrac{2.25(1{,}000{,}000) + 100{,}000}{1{,}000{,}000} = \$2.35$$

At one million CDs the initial investment is actually only \$0.10 per CD. The initial investment is spread over all of the 1,000,000 CDs.

4.1 WARM-UPS

1. False, because there are infinitely many prime numbers.   2. False, because $32 = 2^5$.

3. False, because $51 = 3 \cdot 17$.

4. True, because $12 = 3 \cdot 4$ and $16 = 4 \cdot 4$.

5. True, because $10 = 2 \cdot 5$ and $21 = 3 \cdot 7$.

6. True, because $x^5y^3 - x^4y^7 = x^4y^3(x - y^4)$.

7. True, because $2x^2y - 6xy^2 = 2xy(x - 3y)$ or $2x^2y - 6xy^2 = -2xy(-x + 3y)$.

8. False, because $8a^3b - 12a^2b = 4a^2b(2a - 3)$.

9. False, because if $x = 0$ then $x - 7 = 7 - x$ becomes $-7 = 7$.

10. True, because of the distributive property.

4.1 EXERCISES

1. $18 = 2 \cdot 9 = 2 \cdot 3 \cdot 3 = 2 \cdot 3^2$

3. $52 = 2 \cdot 26 = 2 \cdot 2 \cdot 13 = 2^2 \cdot 13$

5. $98 = 2 \cdot 49 = 2 \cdot 7 \cdot 7 = 2 \cdot 7^2$

7. $460 = 2 \cdot 230 = 2 \cdot 10 \cdot 23 = 2 \cdot 2 \cdot 5 \cdot 23$
$$= 2^2 \cdot 5 \cdot 23$$

9. $924 = 2 \cdot 462 = 2 \cdot 2 \cdot 231 = 2 \cdot 2 \cdot 3 \cdot 77$
$$= 2 \cdot 2 \cdot 3 \cdot 7 \cdot 11 = 2^2 \cdot 3 \cdot 7 \cdot 11$$

11. Since $8 = 2^3$ and $20 = 2^2 \cdot 5$, the GCF is 2^2 or 4.

13. Since $36 = 2^2 \cdot 3^2$ and $60 = 2^2 \cdot 3 \cdot 5$, the GCF is $2^2 \cdot 3 = 12$.

15. Since $40 = 2^3 \cdot 5$, $48 = 2^4 \cdot 3$, and $88 = 2^3 \cdot 11$, the GCF is $2^3 = 8$.

17. Since $76 = 2^2 \cdot 19$, $84 = 2^2 \cdot 3 \cdot 7$, and $100 = 2^2 \cdot 5^2$, the GCF is $2^2 = 4$.

19. Since $39 = 3 \cdot 13$, $68 = 2^2 \cdot 17$, and $77 = 7 \cdot 11$, the GCF is 1.

21. Since $6x = 2 \cdot 3 \cdot x$ and $8x^3 = 2^3x^3$, the GCF is $2x$.

23. Since $12x^3 = 2^2 \cdot 3x^3$, $4x^2 = 2^2x^2$, and $6x = 2 \cdot 3x$, the GCF is $2x$.

25. The GCF for $3x^2y$ and $2xy^2$ is xy.

27. Since $24a^2bc = 2^3 \cdot 3a^2bc$ and $60ab^2 = 2^2 \cdot 3 \cdot 5ab^2$, the GCF is $2^2 \cdot 3ab = 12ab$.

29. Since $12u^3v^2 = 2^2 \cdot 3u^3v^2$ and $25s^2t^4 = 5^2s^2t^4$, the GCF is 1.

31. Since $18a^3b = 2 \cdot 3^2a^3b$, $30a^2b^2 = 2 \cdot 3 \cdot 5a^2b^2$, and $54ab^3 = 2 \cdot 3^3ab^3$, the GCF is $2 \cdot 3ab = 6ab$.

33. $27x = 9(3x)$

35. $24t^2 = 8t(3t)$

37. $36y^5 = 4y^2(9y^3)$

39. $u^4v^3 = uv(u^3v^2)$

41. $-14m^4n^3 = 2m^4(-7n^3)$

43. $-33x^4y^3z^2 = -3x^3yz(11xy^2z)$

45. The GCF for x^3 and $6x$ is x. So $x^3 - 6x = x(x^2 - 6)$.

47. The GCF for $5ax$ and $5ay$ is $5a$. So $5ax + 5ay = 5a(x + y)$.

49. The GCF for h^5 and h^3 is h^3. So $h^5 - h^3 = h^3(h^2 - 1)$.

51. The GCF for $2k^7m^4$ and $4k^3m^6$ is $2k^3m^4$. So $-2k^7m^4 + 4k^3m^6 = 2k^3m^4(-k^4 + 2m^2)$.

53. The GCF for $2x^3$, $6x^2 = 2 \cdot 3x^2$, and $8x = 2^3x$ is $2x$. So we can factor out $2x$: $2x^3 - 6x^2 + 8x = 2x(x^2 - 3x + 4)$.

55. Since $12x^4t = 2^2 \cdot 3x^4t$, $30x^3t = 2 \cdot 3 \cdot 5x^3t$, and $24x^2t\theta = 2^3 \cdot 3x^2t^2$, the GCF is $2 \cdot 3x^2t = 6x^2t$. So $12x^4t + 30x^3t - 24x^2t^2 = 6x^2t(2x^2 + 5x - 4t)$.

57. The GCF of $(x - 3)a$ and $(x - 3)b$ is $x - 3$. So $(x - 3)a + (x - 3)b = (x - 3)(a + b)$.

59. The GCF of $(y+1)^2a$ and $(y+1)^2b$ is $(y+1)^2$. So $(y+1)^2a+(y+1)^2b$

$= (y+1)^2(a+b)$.

61. $36a^3b^5 - 27a^2b^4 + 18a^2b^9$

$= 9a^2b^4(4ab - 3 + 2b^5)$

63. The GCF of 8x and 8y is 8. However, we can factor out either 8 or −8. So $8x - 8y = 8(x-y)$ or $8x - 8y = -8(-x+y)$.

65. The GCF of 4x and $8x^2$ is 4x. We can factor out either 4x or −4x. So $-4x + 8x^2$

$= 4x(-1 + 2x)$ or $-4x + 8x^2 = -4x(1 - 2x)$.

67. $x - 5 = 1(x - 5)$

$x - 5 = -1(-x + 5)$

69. $4 - 7a = 1(4 - 7a)$

$4 - 7a = -1(-4 + 7a)$

71. $-24a^3 + 16a^2 = 8a^2(-3a + 2)$

$-24a^3 + 16a^2 = -8a^2(3a - 2)$

73. $-12x^2 - 18x = 6x(-2x - 3)$

$-12x^2 - 18x = -6x(2x + 3)$

75. $-2x^3 - 6x^2 + 14x = 2x(-x^2 - 3x + 7)$

$-2x^3 - 6x^2 + 14x = -2x(x^2 + 3x - 7)$

77. $4a^3b - 6a^2b^2 - 4ab^3 = 2ab(2a^2 - 3ab - 2b^2)$

$4a^3b - 6a^2b^2 - 4ab^3 = -2ab(-2a^2 + 3ab + 2b^2)$

79. Note that we can write her distance as $20x + 40 = 20(x + 2)$ and that D = RT. Since her rate was 20 mph, her time is x + 2 hours.

81. $S = 2\pi r^2 + 2\pi rh$

$S = 2\pi r(r + h)$

4.2 WARM-UPS

1. False, because $x^2 + 16$ is a sum of two squares.

2. True, because $x^2 - 8x + 16 = (x - 4)^2$.

3. False, because the first term is $(3x)^2$, the last terms is $(7)^2$, but the middle term is not $2 \cdot 3x \cdot 7$ or 42x.

4. False, because if $x = 1$, then $4 \cdot 1^2 + 4 = 8$ and $(2 \cdot 1 + 2)^2 = 16$.

5. True, because $a^2 - b^2 = (a + b)(a - b)$.

6. True, because any monomial is called prime.

7. False, because $(x + 3)(x + 3) = x^2 + 6x + 9$.

8. False, because $x^2 - 1$ is not prime.

9. True, because $y^2 - 2y + 1$ is a perfect square trinomial and it is correctly factored as $(y - 1)^2$.

10. True, because $2x^2 - 18 = 2(x^2 - 9)$

$= 2(x - 3)(x + 3)$ is the correct factorization.

4.2 EXERCISES

1. The polynomial $a^2 - 4$ is the difference of two squares, $a^2 - 2^2$. So $a^2 - 4 = (a - 2)(a + 2)$.

3. The polynomial $x^2 - 49$ is a difference of two squares $x^2 - 7^2$. So $x^2 - 49 = (x - 7)(x + 7)$.

5. The polynomial $y^2 - 9x^2$ is a difference of two squares, $y^2 - (3x)^2$. So

$y^2 - 9x^2 = (y + 3x)(y - 3x)$.

7. The polynomial $25a^2 - 49b^2$ is a difference of two squares, $(5a)^2 - (7b)^2$. So

$25a^2 - 49b^2 = (5a + 7b)(5a - 7b)$.

9. The polynomial $121m^2 - 1$ is a difference of two squares, $(11m)^2 - 1^2$.

So $121m^2 - 1 = (11m + 1)(11m - 1)$.

11. The polynomial $9w^2 - 25c^2$ is a difference of two squares, $(3w)^2 - (5c)^2$. So

$9w^2 - 25c^2 = (3w - 5c)(3w + 5c)$.

13. Since the first and last terms of $x^2 - 20x + 100$ are perfect squares and $-20x = -2 \cdot x \cdot 10$, the polynomial is a perfect square trinomial.

15. Since $y^2 - 40$ has only two terms and 40 is not a perfect square, this polynomial is neither a perfect square trinomial nor a difference of two squares.

17. Since the first and last terms of $4y^2 + 12y + 9$ are perfect squares and $12y = 2 \cdot 2y \cdot 3$, the polynomial is a perfect square trinomial.

19. The first and last terms of $x^2 - 8x + 64$ are perfect squares, but $8x$ is not equal to $2 \cdot x \cdot 8$. So the polynomial is not a perfect square trinomial. Since it has three terms it cannot be a difference of two squares, so it is neither a perfect square trinomial nor a difference of two squares.

21. Since $9y^2 - 25c^2 = (3y)^2 - (5c)^2$, the polynomial is a difference of two squares.

23. The first and last terms of $9a^2 + 6ab + b^2$ are perfect squares. Since $6ab = 2 \cdot 3a \cdot b$, it is a perfect square trinomial.

25. Since x^2 and 36 are perfect squares and $12x = 2 \cdot x \cdot 6$, $x^2 + 12x + 36$ is a perfect square trinomial and $x^2 + 12x + 36 = (x + 6)^2$.

27. Since a^2 and 4 are perfect squares and $-4a = -2 \cdot a \cdot 2$, $a^2 - 4a + 4$ is a perfect square trinomial and $a^2 - 4a + 4 = (a - 2)^2$.

29. $4w^2 + 4w + 1 = (2w)^2 + 2 \cdot 2w \cdot 1 + 1^2$
$$= (2w + 1)^2$$

31. $16x^2 - 8x + 1 = (4x)^2 - 2 \cdot 4x \cdot 1 + 1^2$
$$= (4x - 1)^2$$

33. $4t^2 + 20t + 25 = (2t)^2 + 2 \cdot 2t \cdot 5 + 5^2$
$$= (2t + 5)^2$$

35. $9w^2 + 42w + 49 = (3w)^2 + 2 \cdot 3w \cdot 7 + 7^2$
$$= (3w + 7)^2$$

37. $n^2 + 2nt + t^2 = (n + t)^2$

39. $5x^2 - 125 = 5(x^2 - 25) = 5(x - 5)(x + 5)$

41. We could factor out 2 or -2. If we factor out -2, then we have a difference of two squares and we can factor again.

$-2x^2 + 18 = -2(x^2 - 9) = -2(x - 3)(x + 3)$

43. $a^3 - ab^2 = a(a^2 - b^2) = a(a - b)(a + b)$

45. $3x^2 + 6x + 3 = 3(x^2 + 2x + 1) = 3(x + 1)^2$

47. $-5y^2 + 50y - 125 = -5(y^2 - 10y + 25)$
$$= -5(y - 5)^2$$

49. $x^3 - 2x^2y + xy^2 = x(x^2 - 2xy + y^2)$
$$= x(x - y)^2$$

51. We could factor out 3 or -3. If we factor out -3, then we get a difference of two squares and we can factor completely.

$-3x^2 + 3y^2 = -3(x^2 - y^2)$

$= -3(x - y)(x + y)$

53. $2ax^2 - 98a = 2a(x^2 - 49) = 2a(x - 7)(x + 7)$

55. $3ab^2 - 18ab + 27a = 3a(b^2 - 6b + 9)$
$$= 3a(b - 3)^2$$

57. $-4m^3 + 24m^2n - 36mn^2$

$= -4m(m^2 - 6mn + 9n^2) = -4m(m - 3n)^2$

59. $bx + by + cx + cy = b(x + y) + c(x + y)$
$$= (b + c)(x + y)$$

61. Note that we must factor out -4 from the last two terms to get the common factor $(x + 1)$.

$x^3 + x^2 - 4x - 4 = x^2(x + 1) - 4(x + 1)$
$$= (x^2 - 4)(x + 1)$$
$$= (x - 2)(x + 2)(x + 1)$$

63. Note that we must factor out $-x$ from the last two terms to get the common factor $(a - b)$.

$3a - 3b - xa + xb = 3(a - b) - x(a - b)$
$$= (3 - x)(a - b)$$

65. Note that the GCF for the last two terms is 1. So we factor out 1.

$a^3 + 3a^2 + a + 3 = a^2(a + 3) + 1(a + 3)$
$$= (a^2 + 1)(a + 3)$$

67. $xa + ay + 3y + 3x = a(x + y) + 3(x + y)$
$$= (a + 3)(x + y)$$

69. $abc + c - 3 - 3ab = c(ab + 1) - 3(1 + ab)$
$$= c(ab + 1) - 3(ab + 1)$$
$$= (c - 3)(ab + 1)$$

71. $x^2a - a + bx^2 - b = a(x^2 - 1) + b(x^2 - 1)$
$$= (a + b)(x^2 - 1)$$
$$= (a + b)(x - 1)(x + 1)$$

73. $y^2 + y + by + b = y(y+1) + b(y+1)$

$\qquad = (y+b)(y+1)$

75. $6a^3y + 24a^2y^2 + 24ay^3 = 6ay(a^2 + 4ay + 4y^2)$

$= 6ay(a+2y)^2$

77. $24a^3y - 6ay^3 = 6ay(4a^2 - y^2)$

$= 6ay(2a-y)(2a+y)$

79. $2a^3y^2 - 6a^2y = 2a^2y(ay-3)$

81. $ab + 2bw - 4aw - 8w^2$

$= b(a+2w) - 4w(a+2w) = (b-4w)(a+2w)$

83. $h = -16t^2 + 6400 = -16(t^2 - 400)$

$h = -16(t-20)(t+20)$

If $t = 2$, then

$h = -16(2-20)(2+20) = -16(-18)(22)$

$= 6336$ feet

85. $V = y^3 - 6y^2 + 9y = y(y^2 - 6y + 9)$

$V = y(y-3)^2$

Since $V = LWH$ and the base is square, the side of the base is $y - 3$ inches.

4.3 WARM-UPS

1. True, because $(x-3)^2 = x^2 - 2 \cdot 3 \cdot x + 3^2$

$= x^2 - 6x + 9$. **2.** True, because $(x+3)^2$

$= x^2 + 2 \cdot 3 \cdot x + 3^2 = x^2 + 6x + 9$.

3. False, because $(x-9)(x-1) = x^2 - 10x + 9$.

4. False, because $(x-8)(x-9) = x^2 - 17x + 72$.

5. True, because $(x+9)(x-1) = x^2 + 9x - x - 9$

$= x^2 + 8x - 9$.

6. False, because $(x+3)^2 = x^2 + 6x + 9$.

7. True, because $(x-y)(x-9y)$

$= x^2 - xy - 9xy + 9y^2 = x^2 - 10xy + 9y^2$.

8. False, because $(x+1)(x+1) = x^2 + 2x + 1$.

9. False, because $(x+5y)(x-4y)$

$= x^2 + xy - 20y^2$.

10. False, because $(x+1)(x+1) = x^2 + 2x + 1$.

4.3 EXERCISES

1. Two numbers that have a product of 3 and a sum of 4 are 3 and 1. So $x^2 + 4x + 3 = (x+3)(x+1)$.

3. Two numbers that have a product of 18 and a sum of 9 are 3 and 6. So $x^2 + 9x + 18 = (x+3)(x+6)$.

5. Two numbers that have a product of 12 and a sum of -7 are -3 and -4. So $a^2 - 7a + 12 = (a-3)(a-4)$.

7. Two numbers that have a product of -6 and a sum of -5 are -6 and 1. So $b^2 - 5b - 6 = (b-6)(b+1)$.

9. Since 2 and 5 have a product of 10 and a sum of 7, we have $y^2 + 7y + 10 = (y+2)(y+5)$.

11. Two numbers with a product of 8 and a sum of -6 are -2 and -4. So we can write $a^2 - 6a + 8 = (a-2)(a-4)$

13. Two numbers that have a product of 16 and a sum of -10 are -8 and -2. So we can write $m^2 - 10m + 16 = (m-8)(m-2)$.

15. Two numbers that have a product of -10 and a sum of 9 are 10 and -1. So we can write $w^2 + 9w - 10 = (w+10)(w-1)$.

17. Two numbers with a product of -8 and a sum of -2 are -4 and 2. So $w^2 - 2w - 8 = (w-4)(w+2)$.

19. The only numbers that have a product of -12 are 1 and -12, -1 and 12, 2 and -6, -2 and 6, 3 and -4, -3 and 4. None of these pairs of numbers have a sum of -2. So the polynomial is a prime polynomial.

21. Two numbers that have a product of -16 and a sum of 15 are 16 and -1. So we can write $m^2 + 15m - 16 = (m+16)(m-1)$.

23. The only numbers that have a product of 12 are 1 and 12, -1 and -12, 2 and 6, -2 and -6, 3 and 4, -3 and -4. None of these pairs of numbers have a sum of -4. So the polynomial is a prime polynomial.

25. $z^2 - 25 = z^2 - 5^2 = (z-5)(z+5)$

27. A sum of two squares is prime.

29. $m^2 + 12m + 20 = (m+2)(m+10)$

31. The only numbers that have a product of 10 are 1 and 10, -1 and -10, 2 and 5, -2 and -5. Since none of these pairs have a sum of -3, the polynomial is prime.

33. $m^2 - 17m - 18 = (m-18)(m+1)$

35. Two numbers that have a product of 24 are 1 and 24, −1 and −24, 2 and 12, −2 and −12, 3 and 8, −3 and −8, 4 and 6, −4 and −6. Since none of these pairs has a sum of −23, the polynomials is prime.

37. Two numbers that have a product of −24 and a sum of 5 are 8 and −3. So $t^2 + 5t - 24 = (t + 8)(t - 3)$.

39. Two numbers that have a product of −24 and a sum of −2 are −6 and 4. So $t^2 - 2t - 24 = (t - 6)(t + 4)$.

41. Two numbers that have a product of −200 and a sum of −10 are −20 and 10. So $t^2 - 10t - 200 = (t - 20)(t + 10)$.

43. Two numbers that have a product of −150 and a sum of −5 are −15 and 10. So $x^2 - 5x - 150 = (x - 15)(x + 10)$.

45. Two numbers that have a product of 30 and a sum of 13 are 3 and 10. So $y^2 + 13y + 30 = (y + 3)(y + 10)$.

47. Two numbers that have a product of −12 and a sum of −4 are −6 and 2. So $x^2 - 4xy - 12y^2 = (x - 6y)(x + 2y)$.

49. Two numbers that have a product of 12 and a sum of −13 are −1 and −12. So $x^2 - 13xy + 12y^2 = (x - 12y)(x - y)$.

51. Two numbers that have a product of −24 and a sum of −5 are −8 and 3. So $x^2 - 5xs - 24s^2 = (x - 8s)(x + 3s)$.

53. $w^2 - 8w = w(w - 8)$

55. $2w^2 - 162 = 2(w^2 - 81)$
$$= 2(w - 9)(w + 9)$$

57. $x^2w^2 + 9x^2 = x^2(w^2 + 9)$

59. $w^2 - 18w + 81 = w^2 - 2 \cdot 9w + 9^2$
$$= (w - 9)^2$$

61. $6w^2 - 12w - 18 = 6(w^2 - 2w - 3)$
$$= 6(w - 3)(w + 1)$$

63. $32x^2 - 2x^4 = 2x^2(16 - x^2)$
$$= 2x^2(4 - x)(4 + x)$$

65. $3w^2 + 27w + 54 = 3(w^2 + 9w + 18)$
$$= 3(w + 3)(w + 6)$$

67. $18w^2 + w^3 + 36w = w^3 + 18w^2 + 36w$
$$= w(w^2 + 18w + 36)$$

69. $8vw^2 + 32vw + 32v = 8v(w^2 + 4w + 4)$
$$= 8v(w + 2)^2$$

71. $6x^3y + 30x^2y^2 + 36xy^3$
$$= 6xy(x^2 + 5xy + 6y^2)$$
$$= 6xy(x + 3y)(x + 2y)$$

73. The area of a rectangle is given by the formula $A = LW$ and the area of this rectangle is $x^2 + 6x + 8 = (x + 4)(x + 2)$. If the width is $x + 2$ feet, then the length must be $x + 4$ feet.

75. $V = x^3 + 8x^2 + 15x = x(x^2 + 8x + 15)$
$$= x(x + 3)(x + 5)$$
Since $V = LWH$, the increases to the length and width are 3 feet and 5 feet.

4.4 WARM-UPS

1. True, because $(2x + 1)(x + 1)$
$$= 2x^2 + x + 2x + 1 = 2x^2 + 3x + 1.$$
2. False, because $(2x + 1)(x + 3)$
$$= 2x^2 + x + 6x + 3 = 2x^2 + 7x + 3.$$
3. True, because $(3x + 1)(x + 3)$
$$= 3x^2 + x + 9x + 3 = 3x^2 + 10x + 3.$$
4. False, because $(3x + 7)(5x + 2)$
$$= 15x^2 + 35x + 6x + 14 = 15x^2 + 41x + 14.$$
5. True, because $(2x - 9)(x + 1)$
$$= 2x^2 - 9x + 2x - 9 = 2x^2 - 7x - 9.$$
6. False, because $(2x + 3)(x - 3)$
$$= 2x^2 + 3x - 6x - 9 = 2x^2 - 3x - 9.$$
7. False, because $(2x - 9)(2x + 1)$
$$= 4x^2 - 18x + 2x - 9 = 4x^2 - 16x - 9.$$
8. False, because $(4x - 1)(2x + 5)$
$$= 8x^2 - 2x + 20x - 5 = 8x^2 + 18x - 5.$$
9. False, because $(5x - 1)(4x + 1)$
$$= 20x^2 - 4x + 5x - 1 = 20x^2 + x - 1.$$
10. True, because $(3x - 1)(4x - 3)$
$$= 12x^2 - 4x - 9x + 3 = 12x^2 - 13x + 3.$$

4.4 EXERCISES

1. Two integers with a product of 20 and a sum of 12 are 2 and 10.

3. Two integers with a product of −12 and a sum of −4 are −6 and 2.

5. Since $ac = 12$, we need two numbers with a product of 12 and a sum of 7. The numbers are 3 and 4.

7. Since $ac = 18$, we need two numbers with a product of 18 and a sum of −11. The numbers are −2 and −9.

9. Since $ac = -12$, we need two numbers with a product of −12 and sum of 1. The numbers are −3 and 4.

11. Since $ac = 2 \cdot 1 = 2$, we need two numbers that have a product of 2 and a sum of 3. The numbers are 2 and 1.

$$2x^2 + 3x + 1 = 2x^2 + 2x + x + 1$$
$$= 2x(x+1) + 1(x+1)$$
$$= (2x+1)(x+1)$$

13. Since $ac = 2 \cdot 4 = 8$, we need two numbers that have a product of 8 and a sum of 9. The numbers are 8 and 1.

$$2x^2 + 9x + 4 = 2x^2 + 8x + x + 4$$
$$= 2x(x+4) + 1(x+4)$$
$$= (2x+1)(x+4)$$

15. Since $ac = 3 \cdot 2 = 6$, we need two numbers that have a product of 6 and a sum of 7. The numbers are 6 and 1.

$$3t^2 + 7t + 2 = 3t^2 + 6t + t + 2$$
$$= 3t(t+2) + 1(t+2)$$
$$= (3t+1)(t+2)$$

17. Since $ac = 2 \cdot (-3) = -6$, we need two numbers that have a product of -6 and a sum of 5. The numbers are 6 and -1.

$$2x^2 + 5x - 3 = 2x^2 + 6x - 1x - 3$$
$$= 2x(x+3) - 1(x+3)$$
$$= (2x-1)(x+3)$$

19. Since $ac = 6 \cdot (-3) = -18$, we need two numbers that have a product of -18 and a sum of 7. The numbers are 9 and -2.

$$6x^2 + 7x - 3 = 6x^2 + 9x - 2x - 3$$
$$= 3x(2x+3) - 1(2x+3)$$
$$= (3x-1)(2x+3)$$

21. Since $ac = 2 \cdot 6 = 12$, we need two numbers that have a product of 12 and a sum of -7. The numbers are -3 and -4.

$$2x^2 - 7x + 6 = 2x^2 - 4x - 3x + 6$$
$$= 2x(x-2) - 3(x-2)$$
$$= (2x-3)(x-2)$$

23. Since $ac = 5 \cdot 6 = 30$, we need two numbers that have a product of 30 and a sum of -13. The numbers are -10 and -3.

$$5b^2 - 13b + 6 = 5b^2 - 10b - 3b + 30$$
$$= 5b(b-2) - 3(b-2)$$
$$= (5b-3)(b-2)$$

25. Since $ac = 4 \cdot (-3) = -12$, we need two numbers that have a product of -12 and a sum of -11. The numbers are -12 and 1.

$$4y^2 - 11y - 3 = 4y^2 - 12y + 1y - 3$$
$$= 4y(y-3) + 1(y-3)$$
$$= (4y+1)(y-3)$$

27. Since $ac = 3 \cdot 1 = 3$, we need two numbers that have a product of 3 and a sum of 2. There is no pair of numbers that has a product of 3 and a sum of 2. So the polynomial is prime.

29. Since $ac = 8 \cdot (-1) = -8$, we need two numbers that have a product of -8 and a sum of -2. The numbers are -4 and 2.

$$8x^2 - 2x - 1 = 8x^2 - 4x + 2x - 1$$
$$= 4x(2x-1) + 1(2x-1)$$
$$= (4x+1)(2x-1)$$

31. Since $ac = 9 \cdot 2 = 18$, we need two numbers that have a product of 18 and a sum of -9. The numbers are -6 and -3.

$$9t^2 - 9t + 2 = 9t^2 - 6t - 3t + 2$$
$$= 3t(3t-2) - 1(3t-2)$$
$$= (3t-1)(3t-2)$$

33. Since $ac = 15 \cdot 2 = 30$, we need two numbers that have a product of 30 and a sum of 13. The numbers are 10 and 3.

$$15x^2 + 13x + 2 = 15x^2 + 10x + 3x + 2$$
$$= 5x(3x+2) + 1(3x+2)$$
$$= (5x+1)(3x+2)$$

35. Since $ac = 15 \cdot 2 = 30$, we need two numbers that have a product of 30 and a sum of -13. The numbers are -10 and -3.

$$15x^2 - 13x + 2 = 15x^2 - 10x - 3x + 2$$
$$= 5x(3x-2) - 1(3x-2)$$
$$= (5x-1)(3x-2)$$

37. $3x^2 + 7x + 2 = (x+2)(3x+1)$

39. $5x^2 + 11x + 2 = (5x+1)(x+2)$

41. $6a^2 - 17a + 5 = (3a - 1)(2a - 5)$

43. $5a^2 + 11a + 2 = (5a + 1)(a + 2)$

45. $4w^2 + 8w + 3 = (2w + 3)(2w + 1)$

47. $15x^2 - x - 2 = (5x - 2)(3x + 1)$

49. $8x^2 - 6x + 1 = (4x - 1)(2x - 1)$

51. $15x^2 - 31x + 2 = (15x - 1)(x - 2)$

53. $2x^2 + 18x - 90 = 2(x^2 + 9x - 45)$

55. $3x^2 + x - 10 = (3x - 5)(x + 2)$

57. $10x^2 - 3xy - y^2 = (5x + y)(2x - y)$

59. $42a^2 - 13ab + b^2 = (6a - b)(7a - b)$

61. $81w^3 - w = w(81w^2 - 1)$
$$= w(9w - 1)(9w + 1)$$

63. $4w^2 + 2w - 30 = 2(2w^2 + w - 15)$
$$= 2(2w - 5)(w + 3)$$

65. $12x^2 + 36x + 27 = 3(4x^2 + 12x + 9)$
$$= 3(2x + 3)^2$$

67. $6w^2 - 11w - 35 = 6w^2 - 21w + 10w - 35$
$$= 3w(2w - 7) + 5(2w - 7)$$
$$= (3w + 5)(2w - 7)$$

69. $3x^2z - 3zx - 18z = 3z(x^2 - x - 6)$
$$= 3z(x - 3)(x + 2)$$

71. $10x^2y^2 + xy^2 - 9y^2 = y^2(10x^2 + x - 9)$
$$= y^2(10x - 9)(x + 1)$$

73. Two numbers that have a product of -15 and a sum of 2 are 5 and -3.
$$a^2 + 2ab - 15b^2 = (a + 5b)(a - 3b)$$

75. $-6t^3 - t^2 + 2t = -t(6t^2 + t - 2)$
$$= -t(3t + 2)(2t - 1)$$

77. $12t^4 - 2t^3 - 4t^2 = 2t^2(6t^2 - t - 2)$
$$= 2t^2(3t - 2)(2t + 1)$$

79. $4x^2y - 8xy^2 + 3y^3 = y(4x^2 - 8xy + 3y^2)$
$$= y(2x - y)(2x - 3y)$$

81. $-4w^2 + 7w - 3 = -1(4w^2 - 7w + 3)$
$$= -1(w - 1)(4w - 3)$$

83. $-12a^3 + 22a^2b - 6ab^2$
$$= -2a(6a^2 - 11ab + 3b^2)$$
$$= -2a(2a - 3b)(3a - b)$$

85. $h = -16t^2 + 40t + 24$
$h = -8(2t^2 - 5t - 3)$
$h = -8(2t + 1)(t - 3)$
If $t = 3$, then $h = -8(2 \cdot 3 + 1)(3 - 3) = 0$.
If $t = 3$ seconds, the height is 0.

4.5 WARM-UPS

1. False, because if $x = 3$, $3^2 - 4 = 5$ and $(3 - 2)^2 = 1$. **2.** False, because although the first term is $(2x)^2$ and the last term is 3^2, the middle term is not $2 \cdot 2x \cdot 3 = 12x$. **3.** True, because $4y^2 + 25$ is a sum of two squares and a sum of any two squares is a prime polynomial. **4.** True, because $(x + y)(3 + a)$
$= (x + y)3 + (x + y)a = 3x + 3y + ax + ay$.
5. False, because $3x^2 + 51 = 3(x^2 + 17)$.
6. True, because if more factoring is necessary it will be easier with the GCF factored out.
7. False, because if $x = 1$, $1^2 + 9 = 10$ and $(1 + 3)^2 = 16$.
8. True, because no pair of integers has a product of -5 and a sum of -3.
9. True, because $y^2 - 5y - my + 5m$
$= y(y - 5) - m(y - 5) = (y - m)(y - 5)$.
10. False, because $x^2 + ax = x(x + a)$ but $-3x + 3a = -3(x - a) = 3(-x + a)$. We cannot produce the factor $x + a$ out of the last two terms.

4.5 EXERCISES

1.
$$\begin{array}{r}
x^2 - x - 6 \\
x + 4 \overline{)\ x^3 + 3x^2 - 10x - 24} \\
\underline{x^3 + 4x^2} \\
-x^2 - 10x \\
\underline{-x^2 - 4x} \\
-6x - 24 \\
\underline{-6x - 24} \\
0
\end{array}$$

$x^3 + 3x^2 - 10x - 24 = (x + 4)(x^2 - x - 6)$
$= (x + 4)(x - 3)(x + 2)$

3.
$$x - 1 \overline{\smash{\big)}\ x^3 + 4x^2 + x - 6} \quad \overset{\textstyle x^2 + 5x + 6}{}$$

$$\underline{x^3 - x^2}$$
$$5x^2 + x$$
$$\underline{5x^2 - 5x}$$
$$6x - 6$$
$$\underline{6x - 6}$$
$$0$$

$$x^3 + 4x^2 + x - 6 = (x-1)(x^2 + 5x + 6)$$
$$= (x-1)(x+3)(x+2)$$

5.
$$x - 2 \overline{\smash{\big)}\ x^3 + 0x^2 + 0x - 8} \quad \overset{\textstyle x^2 + 2x + 4}{}$$

$$\underline{x^3 - 2x^2}$$
$$2x^2 + 0x$$
$$\underline{2x^2 - 4x}$$
$$4x - 8$$
$$\underline{4x - 8}$$
$$0$$

$$x^3 - 8 = (x-2)(x^2 + 2x + 4)$$

There is no pair of integers that has a product of 4 and a sum of 2. So $x^2 + 2x + 4$ is prime.

7.
$$x + 5 \overline{\smash{\big)}\ x^3 + 4x^2 - 3x + 10} \quad \overset{\textstyle x^2 - x + 2}{}$$

$$\underline{x^3 + 5x^2}$$
$$-x^2 - 3x$$
$$\underline{-x^2 - 5x}$$
$$2x + 10$$
$$\underline{2x + 10}$$
$$0$$

$$x^3 + 4x^2 - 3x + 10 = (x+5)(x^2 - x + 2)$$

There is no pair of integers that has a product of 2 and a sum of −1. So $x^2 - x + 2$ is prime.

9.
$$x + 1 \overline{\smash{\big)}\ x^3 + 2x^2 + 2x + 1} \quad \overset{\textstyle x^2 + x + 1}{}$$

$$\underline{x^3 + x^2}$$
$$x^2 + 2x$$
$$\underline{x^2 + x}$$
$$x + 1$$
$$\underline{x + 1}$$
$$0$$

$$x^3 + 2x^2 + 2x + 1 = (x+1)(x^2 + x + 1)$$

There is no pair of integers that has a product of 1 and a sum of 1. So $x^2 + x + 1$ is prime.

11. $m^3 - 1 = (m-1)(m^2 + m + 1)$

13. $x^3 + 8 = (x+2)(x^2 - 2x + 4)$

15. $8w^3 + 1 = (2w)^3 + 1^3$
$$= (2w+1)(4w^2 - 2w + 1)$$

17. $(2t)^3 - 3^3 = (2t-3)(4t^2 + 6t + 9)$

19. $x^3 - y^3 = (x-y)(x^2 + xy + y^2)$

21. $(2t)^3 + y^3 = (2t+y)(4t^2 - 2ty + y^2)$

23. $2x^2 - 18 = 2(x^2 - 9)$
$$= 2(x-3)(x+3)$$

25. $4x^2 + 8x - 60 = 4(x^2 + 2x - 15)$
$$= 4(x+5)(x-3)$$

27. $x^3 + 4x^2 + 4x = x(x^2 + 4x + 4)$
$$= x(x+2)^2$$

29. $5max^2 + 20ma = 5am(x^2 + 4)$

31. $9x^2 + 6x + 1 = (3x)^2 + 2 \cdot 3x \cdot 1 + 1^2$
$$= (3x+1)^2$$

33. $6x^2y + xy - 2y = y(6x^2 + x - 2)$
$$= y(3x+2)(2x-1)$$

35. There is no pair of integers that has a product of −25 and a sum of 10. So the polynomial is prime.

37. $16m^2 - 4m - 2 = 2(8m^2 - 2m - 1)$
$$= 2(4m+1)(2m-1)$$

39. $9a^2 + 24a + 16 = (3a)^2 + 2 \cdot 3a \cdot 4 + 4^2$
$$= (3a+4)^2$$

41. $24x^2 - 26x + 6 = 2(12x^2 - 13x + 3)$
$$= 2(3x-1)(4x-3)$$

43. $3a^2 - 27a = 3a(a-9)$

45. $8 - 2x^2 = 2(4 - x^2) = 2(2-x)(2+x)$

47. $6x^3 - 5x^2 + 12x = x(6x^2 - 5x + 12)$
There is no pair of integers that has a product of 72 ($6 \cdot 12$) and a sum of −5. So $6x^2 - 5x + 12$ is prime.

49. $a^3b - 4ab = ab(a^2 - 4)$
$$= ab(a-2)(a+2)$$

51. $x^3 + 2x^2 - 4x - 8 = x^2(x+2) - 4(x+2)$
$$= (x^2 - 4)(x+2)$$
$$= (x-2)(x+2)(x+2)$$
$$= (x-2)(x+2)^2$$

53. $2w^4 - 16w = 2w(w^3 - 8)$
$$= 2w(w-2)(w^2 + 2w + 4)$$

55. $3a^2w - 18aw + 27w = 3w(a^2 - 6a + 9)$
$$= 3w(a-3)^2$$

57. $5x^2 - 500 = 5(x^2 - 100)$
$$= 5(x-10)(x+10)$$

59. $2m + 2n - wm - wn = 2(m+n) - w(m+n)$
$$= (2-w)(m+n)$$

61. $3x^4 + 3x = 3x(x^3 + 1)$
$$= 3x(x+1)(x^2 - x + 1)$$

63. $4w^2 + 4w - 4 = 4(w^2 + w - 1)$

There is no pair of integers that has a product of -1 and a sum of 1. So $w^2 + w - 1$ is prime.

65. $a^4 + 7a^3 - 30a^2 = a^2(a^2 + 7a - 30)$
$$= a^2(a+10)(a-3)$$

67. $4aw^3 - 12aw^2 + 9aw$
$$= aw(4w^2 - 12w + 9)$$
$$= aw(2w - 3)^2$$

69. $t^2 + 6t + 9 = t^2 + 2 \cdot t \cdot 3 + 3^2$
$$= (t+3)^2$$

71.
$$\begin{array}{r} x^2 + 8x + 15 \\ x+2 \enclose{longdiv}{x^3 + 10x^2 + 31x + 30} \\ \underline{x^3 + 2x^2} \\ 8x^2 + 31x \\ \underline{8x^2 + 16x} \\ 15x + 30 \\ \underline{15x + 30} \\ 0 \end{array}$$

$V = (x+2)(x^2 + 8x + 15)$
$V = (x+2)(x+3)(x+5)$
The new length is $x+5$ centimeters and the new width is $x+3$ centimeters.

4.6 WARM-UPS

1. False, because the zero factor property only works for a product that is equal to zero.

2. False, because the equation $(x-3)(x-3) = 0$ has only one solution, 3.

3. True, because of the zero factor property.

4. True, because the area of a rectangle is the product of the length and the width.

5. True, because $(1-1)(1+4) = 0$ and $(-4-1)(-4+4) = 0$ are both correct.

6. False, because the Pythagorean theorem applies to right triangles only.

7. True, because $P = 2L + 2W$, the length plus the width is equal to one-half of the perimeter.

8. True, because $x(x-1)(x-2) = 0$ has three solutions: 0, 1, and 2.

9. True, because $0(0-2) = 0$ and $2(2-2) = 0$ are both correct.

10. False, because 3 is not a solution: $3(3-2)(3+5) = 0$ is not correct.

4.6 EXERCISES

1. $(x+5)(x+4) = 0$
$\quad x+5 = 0 \quad$ or $\quad x+4 = 0$
$\quad\quad x = -5 \quad$ or $\quad\quad x = -4$
The solutions to the equation are -4 and -5.

3. $(2x+5)(3x-4) = 0$
$\quad 2x+5 = 0 \quad$ or $\quad 3x-4 = 0$
$\quad\quad 2x = -5 \quad$ or $\quad\quad 3x = 4$
$\quad\quad x = -\dfrac{5}{2} \quad$ or $\quad\quad x = \dfrac{4}{3}$
The solutions to the equation are $-\dfrac{5}{2}$ and $\dfrac{4}{3}$.

5. $w^2 - 9w + 14 = 0$
$\quad (w-2)(w-7) = 0$
$\quad w-2 = 0 \quad$ or $\quad w-7 = 0$
$\quad\quad w = 2 \quad$ or $\quad\quad w = 7$
The solutions to the equation are 2 and 7.

7. $m^2 + 7m = 0$
$\quad m(m+7) = 0$
$\quad m = 0 \quad$ or $\quad m+7 = 0$
$\quad\quad\quad\quad\quad\quad\quad m = -7$
The solutions to the equation are 0 and -7.

9.
$$a^2 + a = 20$$
$$a^2 + a - 20 = 0$$
$$(a+5)(a-4) = 0$$
$$a + 5 = 0 \quad \text{or} \quad a - 4 = 0$$
$$a = -5 \quad \text{or} \quad a = 4$$
The solutions to the equation are -5 and 4.

11.
$$2x^2 + 5x = 3$$
$$2x^2 + 5x - 3 = 0$$
$$(2x-1)(x+3) = 0$$
$$2x - 1 = 0 \quad \text{or} \quad x + 3 = 0$$
$$2x = 1 \quad \text{or} \quad x = -3$$
$$x = \frac{1}{2}$$
The solutions to the equation are $\frac{1}{2}$ and -3.

13.
$$(x+2)(x+6) = 12$$
$$x^2 + 8x + 12 = 12$$
$$x^2 + 8x = 0$$
$$x(x+8) = 0$$
$$x = 0 \quad \text{or} \quad x + 8 = 0$$
$$x = -8$$
The solutions to the equation are 0 and -8.

15.
$$(a+3)(2a-1) = 15$$
$$2a^2 + 5a - 3 = 15$$
$$2a^2 + 5a - 18 = 0$$
$$(2a+9)(a-2) = 0$$
$$2a + 9 = 0 \quad \text{or} \quad a - 2 = 0$$
$$2a = -9$$
$$a = -\frac{9}{2} \quad \text{or} \quad a = 2$$
The solutions to the equation are $-\frac{9}{2}$ and 2.

17.
$$2(4 - 5h) = 3h^2$$
$$8 - 10h - 3h^2 = 0$$
$$3h^2 + 10h - 8 = 0$$
$$(3h-2)(h+4) = 0$$
$$3h - 2 = 0 \quad \text{or} \quad h + 4 = 0$$
$$3h = 2$$
$$h = \frac{2}{3} \quad \text{or} \quad h = -4$$
The solutions to the equation are $\frac{2}{3}$ and -4.

19.
$$2x^2 + 50 = 20x$$
$$2x^2 - 20x + 50 = 0$$
$$x^2 - 10x + 25 = 0$$
$$(x-5)^2 = 0$$
$$x - 5 = 0$$
$$x = 5$$
The solution to the equation is 5.

21.
$$4m^2 - 12m + 9 = 0$$
$$(2m-3)^2 = 0$$
$$2m - 3 = 0$$
$$m = \frac{3}{2}$$
The solution to the equation is $\frac{3}{2}$.

23.
$$x^3 - 9x = 0$$
$$x(x^2 - 9) = 0$$
$$x(x-3)(x+3) = 0$$
$$x = 0 \quad \text{or} \quad x - 3 = 0 \quad \text{or} \quad x + 3 = 0$$
$$x = 0 \quad \text{or} \quad x = 3 \quad x = -3$$
The solutions to the equation are 0, -3, and 3.

25.
$$w^3 + 4w^2 - 4w - 16 = 0$$
$$w^2(w+4) - 4(w+4) = 0$$
$$(w^2 - 4)(w+4) = 0$$
$$(w-2)(w+2)(w+4) = 0$$
$$w - 2 = 0 \quad \text{or} \quad w + 2 = 0 \quad \text{or} \quad w + 4 = 0$$
$$w = 2 \quad \text{or} \quad w = -2 \quad \text{or} \quad w = -4$$
The solutions to the equation are -4, -2, and 2.

27.
$$n^3 - 3n^2 - n + 3 = 0$$
$$n^2(n-3) - 1(n-3) = 0$$
$$(n^2 - 1)(n-3) = 0$$
$$(n-1)(n+1)(n-3) = 0$$
$$n - 1 = 0 \quad \text{or} \quad n + 1 = 0 \quad \text{or} \quad n - 3 = 0$$
$$n = 1 \quad \text{or} \quad n = -1 \quad \text{or} \quad n = 3$$
The solutions to the equation are -1, 1, and 3.

29.
$$y^3 - 9y^2 + 20y = 0$$
$$y(y^2 - 9y + 20) = 0$$
$$y(y-4)(y-5) = 0$$
$$y = 0 \quad \text{or} \quad y - 4 = 0 \quad \text{or} \quad y - 5 = 0$$
$$y = 0 \quad \text{or} \quad y = 4 \quad \text{or} \quad y = 5$$
The solutions to the equation are 0, 4, and 5.

31.
$$x^2 - 16 = 0$$
$$(x-4)(x+4) = 0$$
$$x - 4 = 0 \quad \text{or} \quad x + 4 = 0$$
$$x = 4 \quad \text{or} \quad x = -4$$
The solutions to the equation are -4 and 4.

33.
$$x^2 = 9$$
$$x^2 - 9 = 0$$
$$(x-3)(x+3) = 0$$
$$x - 3 = 0 \quad \text{or} \quad x + 3 = 0$$
$$x = 3 \quad \text{or} \quad x = -3$$
The solutions to the equation are -3 and 3.

35.
$$a^3 = a$$
$$a^3 - a = 0$$
$$a(a^2 - 1) = 0$$
$$a(a-1)(a+1) = 0$$
$$a = 0 \quad \text{or} \quad a - 1 = 0 \quad \text{or} \quad a + 1 = 0$$
$$a = 0 \quad \text{or} \quad a = 1 \quad \text{or} \quad a = -1$$
The solutions to the equation are 0, -1, and 1.

37.
$$3x^2 + 15x + 18 = 0$$
$$3(x^2 + 5x + 6) = 0$$
$$3(x + 3)(x + 2) = 0$$
$$x + 3 = 0 \quad \text{or} \quad x + 2 = 0$$
$$x = -3 \quad \text{or} \quad x = -2$$
The solutions to the equation are -3 and -2.

39.
$$z^2 + \frac{11}{2}z = -6$$
$$2z^2 + 11z = -12$$
$$2z^2 + 11z + 12 = 0$$
$$(2z + 3)(z + 4) = 0$$
$$2z + 3 = 0 \quad \text{or} \quad z + 4 = 0$$
$$2z = -3 \quad \text{or} \quad z = -4$$
$$z = -\frac{3}{2}$$
The solutions to the equation are $-\frac{3}{2}$ and -4.

41.
$$(t - 3)(t + 5) = 9$$
$$t^2 + 2t - 15 = 9$$
$$t^2 + 2t - 24 = 0$$
$$(t + 6)(t - 4) = 0$$
$$t + 6 = 0 \quad \text{or} \quad t - 4 = 0$$
$$t = -6 \quad \text{or} \quad t = 4$$
The solutions to the equation are -6 and 4.

43.
$$(x - 2)^2 + x^2 = 10$$
$$x^2 - 4x + 4 + x^2 = 10$$
$$2x^2 - 4x - 6 = 0$$
$$x^2 - 2x - 3 = 0$$
$$(x - 3)(x + 1) = 0$$
$$x - 3 = 0 \quad \text{or} \quad x + 1 = 0$$
$$x = 3 \quad \text{or} \quad x = -1$$
The solutions to the equation are -1 and 3.

45.
$$\frac{1}{16}x^2 + \frac{1}{8}x = \frac{1}{2}$$
$$x^2 + 2x = 8$$
$$x^2 + 2x - 8 = 0$$
$$(x + 4)(x - 2) = 0$$
$$x + 4 = 0 \quad \text{or} \quad x - 2 = 0$$
$$x = -4 \quad \text{or} \quad x = 2$$
The solutions to the equation are -4 and 2.

47.
$$a^3 + 3a^2 - 25a = 75$$
$$a^3 + 3a^2 - 25a - 75 = 0$$
$$a^2(a + 3) - 25(a + 3) = 0$$
$$(a^2 - 25)(a + 3) = 0$$
$$(a - 5)(a + 5)(a + 3) = 0$$
$$a - 5 = 0 \quad \text{or} \quad a + 5 = 0 \quad \text{or} \quad a + 3 = 0$$
$$a = 5 \quad \text{or} \quad a = -5 \quad \text{or} \quad a = -3$$
The solutions to the equation are -5, -3 and 5.

49. If the perimeter is 34 feet, then the sum of the length and width is 17 feet. Let $x =$ the length and $17 - x =$ the width. Since the diagonal is the hypotenuse of a right triangle, we can use the Pythagorean theorem to write the following equation.
$$x^2 + (17 - x)^2 = 13^2$$
$$x^2 + 289 - 34x + x^2 = 169$$
$$2x^2 - 34x + 120 = 0$$
$$2(x^2 - 17x + 60) = 0$$
$$2(x - 12)(x - 5) = 0$$
$$x - 12 = 0 \quad \text{or} \quad x - 5 = 0$$
$$x = 12 \quad \text{or} \quad x = 5$$
$$17 - x = 5 \qquad 17 - x = 12$$
The length is 12 feet and the width is 5 feet.

51. Let $x =$ the width and $2x + 2 =$ the length. Since the diagonal is the hypotenuse of a right triangle, we can use the Pythagorean theorem to write the following equation.
$$x^2 + (2x + 2)^2 = 13^2$$
$$x^2 + 4x^2 + 8x + 4 = 169$$
$$5x^2 + 8x - 165 = 0$$
$$(5x + 33)(x - 5) = 0$$
$$5x + 33 = 0 \quad \text{or} \quad x - 5 = 0$$
$$5x = -33 \quad \text{or} \quad x = 5$$
$$x = -\frac{33}{5} \qquad 2x + 2 = 12$$
Since the width is a positive number, the width is 5 feet and the length is 12 feet.

53. Let $x =$ the first integer and $x + 1 =$ the second integer. The sum of their squares is 13, is expressed with the following equation.
$$x^2 + (x + 1)^2 = 13$$
$$x^2 + x^2 + 2x + 1 = 13$$
$$2x^2 + 2x - 12 = 0$$
$$x^2 + x - 6 = 0$$
$$(x + 3)(x - 2) = 0$$
$$x + 3 = 0 \quad \text{or} \quad x - 2 = 0$$
$$x = -3 \quad \text{or} \quad x = 2$$
$$x + 1 = -2 \qquad x + 1 = 3$$
The consecutive integers are 2 and 3, or -3 and -2.

55. Since the sum of the two numbers is 11, we can let x = one number and $11-x$ = the other number. From the fact that their product is 30 we can write the following equation.

$$x(11-x) = 30$$
$$11x - x^2 = 30$$
$$-x^2 + 11x - 30 = 0$$
$$x^2 - 11x + 30 = 0$$
$$(x-5)(x-6) = 0$$
$$x - 5 = 0 \quad \text{or} \quad x - 6 = 0$$
$$x = 5 \quad \text{or} \quad x = 6$$
$$11 - x = 6 \qquad 11 - x = 5$$

The numbers are 5 and 6.

57.
$$-16t^2 + 10,000 = 0$$
$$t^2 - 625 = 0$$
$$(t-25)(t+25) = 0$$
$$t - 25 = 0 \quad \text{or} \quad t + 25 = 0$$
$$t = 25 \quad \text{or} \quad t = -25$$

It would take 25 sec to reach the ground.

59. To find the time that it takes for the sandbag to reach the ground, we must solve the equation $0 = -16t^2 - 24t + 720$.

$$-16t^2 - 24t + 720 = 0$$
$$2t^2 + 3t - 90 = 0$$
$$(2t+15)(t-6) = 0$$
$$2t + 15 = 0 \quad \text{or} \quad t - 6 = 0$$
$$t = -\frac{15}{2} \quad \text{or} \quad t = 6$$

Since the answer must be positive, the time it takes to reach the ground is 6 seconds.

61. Let x = the length of the base and $2x + 1$ = the height. Since the area of the triangle is 39 square inches, we can write the following equation.

$$\frac{1}{2}x(2x+1) = 39$$
$$x^2 + \frac{1}{2}x = 39$$
$$2x^2 + x = 78$$
$$2x^2 + x - 78 = 0$$
$$(2x+13)(x-6) = 0$$
$$2x + 13 = 0 \quad \text{or} \quad x - 6 = 0$$
$$2x = -13 \quad \text{or} \quad x = 6$$
$$x = -\frac{13}{2} \qquad 2x + 1 = 13$$

Since the base must have a positive length, the length of the base is 6 inches and the height is 13 inches.

63. Let x = the length of a side of the original square garden. The new garden will be $x-5$ feet by $x-8$ feet. Since the area of the new garden is 180 square feet, we can write the following equation.

$$(x-5)(x-8) = 180$$
$$x^2 - 13x + 40 = 180$$
$$x^2 - 13x - 140 = 0$$
$$(x-20)(x+7) = 0$$
$$x - 20 = 0 \quad \text{or} \quad x + 7 = 0$$
$$x = 20 \quad \text{or} \quad x = -7$$

Since the length of a side of the garden must be positive, the original garden was 20 feet by 20 feet.

65. Let x = the distance between Imelda and Gordon and $x + 20$ = the altitude of the kite. The distance between the kite and Imelda is the length of the hypotenuse of a right triangle.

$$x^2 + (x+20)^2 = 100^2$$
$$x^2 + x^2 + 40x + 400 = 10,000$$
$$2x^2 + 40x - 9600 = 0$$
$$x^2 + 20x - 4800 = 0$$
$$(x+80)(x-60) = 0$$
$$x + 80 = 0 \quad \text{or} \quad x - 60 = 0$$
$$x = -80 \quad \text{or} \quad x = 60$$
$$x + 20 = 80$$

Since the altitude of the kite is positive, the altitude is 80 feet.

67. Let x = the width of one square room and $x + 3$ = the width of the other square room. Since the total area of the two rooms is 45 square yards, we can write the following equation.

$$x^2 + (x+3)^2 = 45$$
$$x^2 + x^2 + 6x + 9 = 45$$
$$2x^2 + 6x - 36 = 0$$
$$x^2 + 3x - 18 = 0$$
$$(x+6)(x-3) = 0$$
$$x + 6 = 0 \quad \text{or} \quad x - 3 = 0$$
$$x = -6 \quad \text{or} \quad x = 3$$
$$x + 3 = 6$$

Since the width of each room is a positive number, one room is 3 yards by 3 yards and the other is 6 yards by 6 yards.

69. Let x and $17 - x$ represent the two legs of the right triangle.

$$x^2 + (17 - x)^2 = 13^2$$
$$x^2 + 289 - 34x + x^2 = 169$$
$$2x^2 - 34x + 120 = 0$$
$$x^2 - 17x + 60 = 0$$
$$(x - 5)(x - 12) = 0$$
$$x - 5 = 0 \quad \text{or} \quad x - 12 = 0$$
$$x = 5 \quad \text{or} \qquad x = 12$$

The distance from A to B is 12 miles.

71.
$$16,000(1 + r)^2 = 25000$$
$$16000(1 + 2r + r^2) = 25000$$
$$16000r^2 + 32000r - 9000 = 0$$
$$16r^2 + 32r - 9 = 0$$
$$(4r + 9)(4r - 1) = 0$$
$$4r + 9 = 0 \quad \text{or} \quad 4r - 1 = 0$$
$$r = -\frac{9}{4} \quad \text{or} \qquad r = \frac{1}{4} = 25\%$$

So the average annual return was 25%.

CHAPTER 4 REVIEW

1. $144 = 2 \cdot 72 = 2 \cdot 2 \cdot 36 = 2 \cdot 2 \cdot 2 \cdot 18$
$= 2 \cdot 2 \cdot 2 \cdot 2 \cdot 9 = 2 \cdot 2 \cdot 2 \cdot 2 \cdot 3 \cdot 3 = 2^4 \cdot 3^2$

3. $58 = 2 \cdot 29$

5. $150 = 2 \cdot 75 = 2 \cdot 3 \cdot 25 = 2 \cdot 3 \cdot 5 \cdot 5 = 2 \cdot 3 \cdot 5^2$

7. Since $36 = 2^2 \cdot 3^2$ and $90 = 2 \cdot 3^2 \cdot 5$, the GCF is $2 \cdot 3^2 = 18$.

9. Since the GCF for 8 and 12 is 4, the GCF is 4x.

11. $3x + 6 = 3(x + 2)$

13. $2a - 20 = -2(-a + 10)$

15. $2a - a^2 = a(2 - a)$

17. $6x^2y^2 - 9x^5y = 3x^2y(2y - 3x^3)$

19. $3x^2y - 12xy - 9y^2 = 3y(x^2 - 4x - 3y)$

21. $y^2 - 400 = y^2 - 20^2 = (y - 20)(y + 20)$

23. $w^2 - 8w + 16 = w^2 - 2 \cdot w \cdot 4 + 4^2 = (w - 4)^2$

25. $4y^2 + 20y + 25 = (2y)^2 + 2 \cdot 2y \cdot 5 + 5^2$
$$= (2y + 5)^2$$

27. $r^2 - 4r + 4 = r^2 - 2 \cdot r \cdot 2 + 2^2 = (r - 2)^2$

29. $8t^3 - 24t^2 + 18t = 2t(4t^2 - 12t + 9)$
$$= 2t(2t - 3)^2$$

31. $x^2 + 12xy + 36y^2 = x^2 + 2 \cdot x \cdot 6y + (6y)^2$
$$= (x + 6y)^2$$

33. $x^2 + 5x - xy - 5y = x(x + 5) - y(x + 5)$
$$= (x - y)(x + 5)$$

35. Two numbers that have a product of -24 and a sum of 5 are 8 and -3.

$b^2 + 5b - 24 = (b + 8)(b - 3)$

37. Two numbers that have product of -60 and a sum of -4 are -10 and 6.

$r^2 - 4r - 60 = (r - 10)(r + 6)$

39. Two numbers that have a product of -55 and a sum of -6 are -11 and 5.

$y^2 - 6y - 55 = (y - 11)(y + 5)$

41. $u^2 + 26u + 120 = (u + 20)(u + 6)$

43. $3t^3 + 12t^2 = 3t^2(t + 4)$

45. $5w^3 + 25w^2 + 25w = 5w(w^2 + 5w + 5)$

47. $2a^3b + 3a^2b^2 + ab^3 = ab(2a^2 + 3ab + b^2)$
$= ab(2a + b)(a + b)$

49. $9x^3 - xy^2 = x(9x^2 - y^2) = x(3x - y)(3x + y)$

51. Since $ac = 14 \cdot (-3) = -42$, we need two integers that have a product of -42 and a sum of 1. The numbers are 7 and -6.

$$14t^2 + t - 3 = 14t^2 + 7t - 6t - 3$$
$$= 7t(2t + 1) - 3(2t + 1)$$
$$= (7t - 3)(2t + 1)$$

53. Since $ac = 6 \cdot (-7) = -42$, we need two numbers that have a product of -42 and a sum of -19. The numbers are 2 and -21.

$$6x^2 - 19x - 7 = 6x^2 - 21x + 2x - 7$$
$$= 3x(2x - 7) + 1(2x - 7)$$
$$= (3x + 1)(2x - 7)$$

55. Since $ac = 6 \cdot (-4) = -24$, we need two integers that have a product of -24 and a sum of 5. The numbers are 8 and -3.

$$6p^2 + 5p - 4 = 6p^2 - 3p + 8p - 4$$
$$= 3p(2p - 1) + 4(2p - 1)$$
$$= (3p + 4)(2p - 1)$$

57. $-30p^3 + 8p^2 + 8p = -2p(15p^2 - 4p - 4)$

$$= -2p(5p + 2)(3p - 2)$$

59. $6x^2 - 29xy - 5y^2 = (6x + y)(x - 5y)$

61. $32x^2 + 16xy + 2y^2 = 2(16x^2 + 8xy + y^2)$

$$= 2(4x + y)^2$$

63. $5x^3 + 40x = 5x(x^2 + 8)$

65. Since $ac = 9(-2) = -18$, we need two integers that have a product of -18 and a sum of 3. The integers are 6 and -3.

$$9x^2 + 3x - 2 = 9x^2 + 6x - 3x - 2$$
$$= 3x(3x + 2) - 1(3x + 2)$$
$$= (3x - 1)(3x + 2)$$

67. $x^3 + 2x^2 - x - 2 = x^2(x + 2) - 1(x + 2)$

$$= (x + 2)(x^2 - 1)$$
$$= (x + 2)(x - 1)(x + 1)$$

69. $x^2y - 16xy^2 = xy(x - 16y)$

71. $a^2 + 2a + 1 = (a + 1)^2$

73. $x^3 - x^2 + x - 1 = x^2(x - 1) + 1(x - 1)$

$$= (x^2 + 1)(x - 1)$$

75. $a^2 + ab + 2a + 2b = a(a + b) + 2(a + b)$

$$= (a + 2)(a + b)$$

77. $-2x^2 + 16x - 24 = -2(x^2 - 8x + 12)$

$$= -2(x - 6)(x - 2)$$

79. $m^3 - 10^3 = (m - 10)(m^2 + 10m + 100)$

81.

$$
\begin{array}{r}
x^2 - 2x + 5 \\
x + 2 \overline{\smash{\big)}\ x^3 + 0x^2 + x + 10} \\
\underline{x^3 + 2x^2} \\
-2x^2 + x \\
\underline{-2x^2 - 4x} \\
5x + 10 \\
\underline{5x + 10} \\
0
\end{array}
$$

$x^3 + x + 10 = (x + 2)(x^2 - 2x + 5)$

Since no pair of integers has a product of 5 and a sum of -2, $x^2 - 2x + 5$ is prime.

83.

$$
\begin{array}{r}
x^2 + 2x - 15 \\
x + 4 \overline{\smash{\big)}\ x^3 + 6x^2 - 7x - 60} \\
\underline{x^3 + 4x^2} \\
2x^2 - 7x \\
\underline{2x^2 + 8x} \\
-15x - 60 \\
\underline{-15x - 60} \\
0
\end{array}
$$

$$x^3 + 6x^2 - 7x - 60 = (x + 4)(x^2 + 2x - 15)$$
$$= (x + 4)(x + 5)(x - 3)$$

85.
$$x^3 - 5x^2 = 0$$
$$x^2(x - 5) = 0$$
$$x^2 = 0 \quad \text{or} \quad x - 5 = 0$$
$$x = 0 \quad \text{or} \quad x = 5$$

The solutions to the equation are 0 and 5.

87.
$$(a - 2)(a - 3) = 6$$
$$a^2 - 5a + 6 = 6$$
$$a^2 - 5a = 0$$
$$a(a - 5) = 0$$
$$a = 0 \quad \text{or} \quad a - 5 = 0$$
$$a = 5$$

The solutions to the equation are 0 and 5.

89.
$$2m^2 - 9m - 5 = 0$$
$$(2m + 1)(m - 5) = 0$$
$$2m + 1 = 0 \quad \text{or} \quad m - 5 = 0$$
$$2m = -1 \quad \text{or} \quad m = 5$$
$$m = -\frac{1}{2}$$

The solutions to the equation are $-\frac{1}{2}$ and 5.

91.
$$m^3 + 4m^2 - 9m - 36 = 0$$
$$m^2(m + 4) - 9(m + 4) = 0$$
$$(m^2 - 9)(m + 4) = 0$$
$$(m - 3)(m + 3)(m + 4) = 0$$
$$m - 3 = 0 \quad \text{or} \quad m + 3 = 0 \quad \text{or} \quad m + 4 = 0$$
$$m = 3 \quad \text{or} \quad m = -3 \quad \text{or} \quad m = -4$$

The solutions to the equation are -4, -3, and 3.

93.
$$(x+3)^2 + x^2 = 5$$
$$x^2 + 6x + 9 + x^2 - 5 = 0$$
$$2x^2 + 6x + 4 = 0$$
$$x^2 + 3x + 2 = 0$$
$$(x+2)(x+1) = 0$$
$$x + 2 = 0 \quad \text{or} \quad x + 1 = 0$$
$$x = -2 \quad \text{or} \quad x = -1$$
The solutions to the equation are -2 and -1.

95.
$$p^2 + \frac{1}{4}p - \frac{1}{8} = 0$$
$$8p^2 + 2p - 1 = 0$$
$$(4p - 1)(2p + 1) = 0$$
$$4p - 1 = 0 \quad \text{or} \quad 2p + 1 = 0$$
$$p = \frac{1}{4} \quad \text{or} \quad p = -\frac{1}{2}$$
The solutions to the equation are $-\frac{1}{2}$ and $\frac{1}{4}$.

97. Since the numbers differ by 6, we can let $x =$ one number and $x + 6 =$ the other. Since their squares differ by 96, we can write the following equation.
$$(x+6)^2 - x^2 = 96$$
$$x^2 + 12x + 36 - x^2 = 96$$
$$12x + 36 = 96$$
$$12x = 60$$
$$x = 5$$
$$x + 6 = 11$$
The numbers are 5 and 11.

99. If the perimeter is 28 inches, then the sum of the length and width is 14. Let $x =$ the width and $14 - x =$ the length. The diagonal is the hypotenuse of a right triangle.
$$x^2 + (14 - x)^2 = 10^2$$
$$x^2 + 196 - 28x + x^2 = 100$$
$$2x^2 - 28x + 96 = 0$$
$$x^2 - 14x + 48 = 0$$
$$(x-6)(x-8) = 0$$
$$x - 6 = 0 \quad \text{or} \quad x - 8 = 0$$
$$x = 6 \quad \text{or} \quad x = 8$$
$$14 - x = 8 \qquad 14 - x = 6$$

The notebook is 6 inches by 8 inches.

101.
$$v = kR^2 - kr^2$$
$$v = k(R^2 - r^2)$$
$$v = k(R - r)(R + r)$$

103. Let x be the distance from the bottom of the ladder to the building and $x + 2$ be the distance from the top of the ladder to the ground.
$$x^2 + (x+2)^2 = 10^2$$
$$x^2 + x^2 + 4x + 4 = 100$$
$$2x^2 + 4x - 96 = 0$$
$$x^2 + 2x - 48 = 0$$
$$(x+8)(x-6) = 0$$
$$x + 8 = 0 \quad \text{or} \quad x - 6 = 0$$
$$x = -8 \quad \text{or} \quad x = 6$$
The distance from the bottom of the ladder to the building is 6 feet.

CHAPTER 4 TEST

1. $66 = 6 \cdot 11 = 2 \cdot 3 \cdot 11$

2. $336 = 2 \cdot 168 = 2 \cdot 2 \cdot 84 = 2 \cdot 2 \cdot 2 \cdot 42$
$$= 2 \cdot 2 \cdot 2 \cdot 2 \cdot 21 = 2^4 \cdot 3 \cdot 7$$

3. Since $48 = 2^4 \cdot 3$ and $80 = 2^4 \cdot 5$, the GCF is $2^4 = 16$.

4. Since $42 = 2 \cdot 3 \cdot 7$, $66 = 2 \cdot 3 \cdot 11$, and $78 = 2 \cdot 3 \cdot 13$, the GCF of 42, 66, and 78 is $2 \cdot 3 = 6$.

5. The GCF for $6y^2$ and $15y^3$ is $3y^2$.

6. The GCF for $12a^2b$, $18ab^2$, and $24a^3b^3$ is $6ab$.

7. $5x^2 - 10x = 5x(x - 2)$

8. $6x^2y^2 + 12xy^2 + 12y^2 = 6y^2(x^2 + 2x + 2)$
Since there is no pair of integers that has a product of 2 and a sum of 2, $x^2 + 2x + 2$ is prime.

9. $3a^3b = 3ab^3 = 3ab(a^2 - b^2)$
$$= 3ab(a - b)(a + b)$$

10. Two numbers that have a product of -24 and a sum of 2 are 6 and -4.
$$a^2 + 2a - 24 = (a + 6)(a - 4)$$

11. $4b^2 - 28b + 49 = (2b)^2 - 2 \cdot 2b \cdot 7 + 7^2$

$$= (2b - 7)^2$$

12. $3m^3 + 27m = 3m(m^2 + 9)$

Since $m^2 + 9$ is a sum of two squares, it is prime.

13. $ax - ay + bx - by = a(x - y) + b(x - y)$

$$= (a + b)(x - y)$$

14. $ax - 2a - 5x + 10 = a(x - 2) - 5(x - 2)$

$$= (a - 5)(x - 2)$$

15. Since $ac = 6(-5) = -30$, we need two integers that have a product of -30 and a sum of -7. The integers are -10 and 3.

$6b^2 - 7b - 5 = 6b^2 + 3b - 10b - 5$

$$= 3b(2b + 1) - 5(2b + 1)$$

$$= (3b - 5)(2b + 1)$$

16. $m^2 + 4mn + 4n^2 = m^2 + 2 \cdot m \cdot 2n + (2n)^2$

$$= (m + 2n)^2$$

17. Since $ac = 2(15) = 30$, we need two integers that have a product of 30 and a sum of -13. The integers are -10 and -3.

$2a^2 - 13a + 15 = 2a^2 - 10a - 3a + 15$

$$= 2a(a - 5) - 3(a - 5)$$

$$= (2a - 3)(a - 5)$$

18. $z^3 + 9z^2 + 18z = z(z^2 + 9z + 18)$

$$= z(z + 3)(z + 6)$$

19.

$$
\require{enclose}
\begin{array}{r}
x^2 - 5x + 6 \\
x - 1 \enclose{longdiv}{x^3 - 6x^2 + 11x - 6} \\
\underline{x^3 - x^2} \\
-5x^2 + 11x \\
\underline{-5x^2 + 5x} \\
6x - 6 \\
\underline{6x - 6} \\
0
\end{array}
$$

$x^3 - 6x^2 + 11x - 6 = (x - 1)(x^2 - 5x + 6)$

$$= (x - 1)(x - 2)(x - 3)$$

20.

$$2x^2 + 5x - 12 = 0$$

$$2x^2 + 8x - 3x - 12 = 0$$

$$2x(x + 4) - 3(x + 4) = 0$$

$$(2x - 3)(x + 4) = 0$$

$2x - 3 = 0$ or $x + 4 = 0$

$2x = 3$ or $x = -4$

$x = \dfrac{3}{2}$

The solutions to the equation are $\dfrac{3}{2}$ and -4.

21.

$$3x^3 - 12x = 0$$

$$3x(x^2 - 4) = 0$$

$$3x(x - 2)(x + 2) = 0$$

$3x = 0$ or $x - 2 = 0$ or $x + 2 = 0$

$x = 0$ or $x = 2$ or $x = -2$

The solutions to the equation are 0, -2, and 2.

22. $(2x - 1)(3x + 5) = 5$

$$6x^2 + 7x - 5 = 5$$

$$6x^2 + 7x - 10 = 0$$

$$(6x - 5)(x + 2) = 0$$

$6x - 5 = 0$ or $x + 2 = 0$

$x = \dfrac{5}{6}$ or $x = -2$

The solutions to the equation are -2 and $\dfrac{5}{6}$.

23. Let $x =$ the width and $x + 3 =$ the length. Since the diagonal is the hypotenuse of a right triangle, we can write the following equation.

$$x^2 + (x + 3)^2 = 15^2$$

$$x^2 + x^2 + 6x + 9 = 225$$

$$2x^2 + 6x - 216 = 0$$

$$x^2 + 3x - 108 = 0$$

$$(x - 9)(x + 12) = 0$$

$x - 9 = 0$ or $x + 12 = 0$

$x = 9$ or $x = -12$

$x + 3 = 12$

The length is 12 feet and the width is 9 feet.

24. Let x and $4-x$ represent the numbers.

$$x(4-x) = -32$$
$$-x^2 + 4x + 32 = 0$$
$$x^2 - 4x - 32 = 0$$
$$(x-8)(x+4) = 0$$
$$x - 8 = 0 \quad \text{or} \quad x + 4 = 0$$
$$x = 8 \quad \text{or} \quad x = -4$$

If $x = 8$, then $4 - x = 4 - 8 = -4$. If $x = -4$, then $4 - x = 4 - (-4) = 8$. The numbers are -4 and 8.

Tying It All Together Chapters 1-4

1. $\frac{91-17}{17-91} = \frac{-1(17-91)}{17-91} = -1$

2. $\frac{4-18}{-6-1} = \frac{-14}{-7} = 2$

3. $5 - 2(7-3) = 5 - 2(4) = 5 - 8 = -3$

4. $3^2 - 4(6)(-2) = 9 - (-48) = 9 + 48 = 57$

5. $2^5 - 2^4 = 32 - 16 = 16$

6. $0.07(37) + 0.07(63) = 0.07(37 + 63)$
$$= 0.07(100) = 7$$

7. $x \cdot 2x = 2 \cdot x \cdot x = 2x^2$

8. $x + 2x = (1 + 2)x = 3x$

9. $\frac{6+2x}{2} = \frac{6}{2} + \frac{2x}{2} = 3 + x$

10. $\frac{6 \cdot 2x}{2} = 6x \cdot \frac{2}{2} = 6x \cdot 1 = 6x$

11. $2 \cdot 3y \cdot 4z = 2 \cdot 3 \cdot 4yz = 24yz$

12. $2(3y + 4z) = 2 \cdot 3y + 2 \cdot 4z = 6y + 8z$

13. $2 - (3 - 4z) = 2 - 3 + 4z = 4z - 1$

14. $t^8 \div t^2 = t^{8-2} = t^6$

15. $t^8 \cdot t^2 = t^{8+2} = t^{10}$

16. $\frac{8t^8}{2t^2} = 4t^{8-2} = 4t^6$

17.
$$2x - 5 > 3x + 4$$
$$-5 > x + 4$$
$$-9 > x$$
$$x < -9$$

18.
$$4 - 5x \leq -11$$
$$-5x \leq -15$$
$$\frac{-5x}{-5} \geq \frac{-15}{-5}$$
$$x \geq 3$$

19.
$$-\frac{2}{3}x + 3 < -5$$
$$-2x + 9 < -15$$
$$-2x < -24$$
$$x > 12$$

20.
$$0.05(x - 120) - 24 < 0$$
$$0.05x - 6 - 24 < 0$$
$$0.05x < 30$$
$$x < 600$$

21.
$$2x - 3 = 0$$
$$2x = 3$$
$$x = \frac{3}{2}$$
The solution to the equation is $\frac{3}{2}$.

22.
$$2x + 1 = 0$$
$$2x = -1$$
$$x = -\frac{1}{2}$$
The solution to the equation is $-\frac{1}{2}$.

23.
$$(x-3)(x+5) = 0$$
$$x - 3 = 0 \quad \text{or} \quad x + 5 = 0$$
$$x = 3 \quad \text{or} \quad x = -5$$
The solutions to the equation are 3 and −5.

24. $(2x-3)(2x+1)=0$

$\quad 2x-3=0 \quad$ or $\quad 2x+1=0$

$\quad\quad 2x=3 \quad$ or $\quad\quad 2x=-1$

$\quad\quad x=\frac{3}{2} \quad$ or $\quad\quad x=-\frac{1}{2}$

The solutions to the equation are $\frac{3}{2}$ and $-\frac{1}{2}$.

25. $3x(x-3)=0$

$\quad 3x=0 \quad$ or $\quad x-3=0$

$\quad\quad x=0 \quad$ or $\quad\quad x=3$

The solutions to the equation are 0 and 3.

26. $x^2=x$

$\quad x^2-x=0$

$\quad x(x-1)=0$

$\quad x=0 \quad$ or $\quad x-1=0$

$\quad x=0 \quad$ or $\quad\quad x=1$

The solutions to the equation are 0 and 1.

27. $3x-3x=0$

$\quad\quad 0=0$

All real numbers satisfy the equation.

28. $3x-3x=1$

$\quad\quad 0=1$

The equation has no solution.

29. $0.01x-x+14.9=0.5x$

$\quad x-100x+1490=50x$

$\quad\quad -149x=-1490$

$\quad\quad\quad x=10$

The solution to the equation is 10.

30. $0.05x+0.04(x-40)=2$

$\quad 0.05x+0.04x-1.6=2$

$\quad\quad\quad 0.09x=3.6$

$\quad\quad\quad 9x=360$

$\quad\quad\quad x=40$

The solution to the equation is 40.

31. $2x^2-18=0$

$\quad 2(x^2-9)=0$

$\quad 2(x-3)(x+3)=0$

$\quad x-3=0 \quad$ or $\quad x+3=0$

$\quad\quad x=3 \quad$ or $\quad\quad x=-3$

The solutions to the equation are -3 and 3.

32. $2x^2+7x-15=0$

$\quad (2x-3)(x+5)=0$

$\quad 2x-3=0 \quad$ or $\quad x+5=0$

$\quad\quad 2x=3 \quad$ or $\quad\quad x=-5$

$\quad\quad x=\frac{3}{2}$

The solutions to the equation are -5 and $\frac{3}{2}$.

33. If the perimeter is 69 feet, then the sum of the length and width is 34.5 feet. So if w is the width, then the length is $34.5-w$.

$$w(34.5-w)=283.5$$

$$34.5w-w^2-283.5=0$$

$$w^2-34.5w+283.5=0$$

$$2w^2-69w+567=0$$

Factor by the ac method. Two numbers that have a product of 1134 and a sum of -69 are 42 and 27.

$$2w^2-42w-27w+567=0$$

$$2w(w-21)-27(w-21)=0$$

$$(2w-27)(w-21)=0$$

$$2w-27=0 \quad \text{or} \quad w-21=0$$

$$w=13.5 \quad \text{or} \quad w=21$$

If $w=13.5$, then the length is $34.5-13.5=21$. If $w=21$, then the length is $34.5-21=13.5$. So the length is 21 feet and the width is 13.5 feet.

5.1 WARM-UPS

1. False, because 3003 is an odd number and it does not have a factor of 2.

2. True, because 2, 3, and 5 are prime numbers and $2^3 \cdot 3 \cdot 5 = 8 \cdot 15 = 120$.

3. True, because we cannot get zero in the denominator in the expression $\frac{x-2}{5}$.

4. False, because replacing x by -1 gives us $\frac{x+1}{x-3} = \frac{-1+1}{-1-3} = 0$.

5. False, because the two's cannot be divided out since 2 is not a factor of the numerator.

6. True, because 2 is a factor of both the numerator and denominator.

7. True, because of the quotient rule.

8. False, because the expression is already in lowest terms.

9. False, because $a - b = -1(b - a)$ and their quotient is -1.

10. True, because $\frac{-3x-6}{x+2} = \frac{-3(x+2)}{x+2} = -3$.

5.1 EXERCISES

1. We cannot use numbers that cause the denominator to have a value of 0. So $x + 1 = 0$ if $x = -1$. We cannot use -1 in place of x.

3. The denominator has a value of 0, if $3a - 5 = 0$ or $a = 5/3$. So we cannot use 5/3 in place of a.

5. We cannot use numbers that cause the denominator to have a value of 0.

$$x^2 - 16 = 0$$
$$(x - 4)(x + 4) = 0$$
$$x - 4 = 0 \quad \text{or} \quad x + 4 = 0$$
$$x = 4 \quad \text{or} \quad x = -4$$

We cannot use -4 or 4 in place of x.

7. Since the denominator is 2, the denominator is never zero. So any number can be used in place of p in the expression.

9. $\frac{6}{27} = \frac{2 \cdot 3}{3 \cdot 3 \cdot 3} = \frac{2}{9}$

11. $\frac{42}{90} = \frac{2 \cdot 3 \cdot 7}{2 \cdot 3 \cdot 3 \cdot 5} = \frac{7}{15}$

13. $\frac{36a}{90} = \frac{2 \cdot 2 \cdot 3 \cdot 3a}{2 \cdot 3 \cdot 3 \cdot 5} = \frac{2a}{5}$

15. $\frac{78}{30w} = \frac{2 \cdot 3 \cdot 13}{2 \cdot 3 \cdot 5w} = \frac{13}{5w}$

17. $\frac{6x+2}{6} = \frac{2(3x+1)}{2 \cdot 3} = \frac{3x+1}{3}$

19. $\frac{2x+4y}{6y+3x} = \frac{2(x+2y)}{3(x+2y)} = \frac{2}{3}$

21. $\frac{w^2-49}{w+7} = \frac{(w-7)(w+7)}{w+7} = w - 7$

23. $\frac{a^2-1}{a^2+2a+1} = \frac{(a-1)(a+1)}{(a+1)^2} = \frac{a-1}{a+1}$

25. $\frac{2x^2+4x+2}{4x^2-4} = \frac{2(x+1)^2}{4(x-1)(x+1)} = \frac{x+1}{2x-2}$

27. $\frac{3x^2+18x+27}{21x+63} = \frac{3(x+3)^2}{21(x+3)} = \frac{x+3}{7}$

29. $\frac{x^{10}}{x^7} = x^{10-7} = x^3$

31. $\frac{z^3}{z^8} = \frac{1}{z^{8-3}} = \frac{1}{z^5}$

33. $\frac{4x^7}{-2x^5} = -2x^{7-5} = -2x^2$

35. $\frac{-12m^9n^{18}}{8m^6n^{16}} = \frac{-3 \cdot 4m^{9-6}n^{18-16}}{2 \cdot 4} = \frac{-3m^3n^2}{2}$

37. $\frac{6b^{10}c^4}{-8b^{10}c^7} = \frac{2 \cdot 3b^{10-10}}{-2 \cdot 4c^{7-4}} = \frac{-3}{4c^3}$

39. $\frac{30a^3bc}{18a^7b^{17}} = \frac{6 \cdot 5c}{6 \cdot 3a^{7-3}b^{17-1}} = \frac{5c}{3a^4b^{16}}$

41. $\frac{210}{264} = \frac{2 \cdot 3 \cdot 5 \cdot 7}{2 \cdot 2 \cdot 2 \cdot 3 \cdot 11} = \frac{35}{44}$

43. $\frac{231}{168} = \frac{3 \cdot 7 \cdot 11}{2 \cdot 2 \cdot 2 \cdot 3 \cdot 7} = \frac{11}{8}$

45. $\frac{630x^5}{300x^9} = \frac{30 \cdot 21}{30 \cdot 10x^{9-5}} = \frac{21}{10x^4}$

47. $\frac{924a^{23}}{448a^{19}} = \frac{2 \cdot 2 \cdot 3 \cdot 7 \cdot 11a^{23-19}}{2 \cdot 2 \cdot 2 \cdot 2 \cdot 2 \cdot 2 \cdot 7} = \frac{33a^4}{16}$

49. $\frac{3a-2b}{2b-3a} = \frac{-1(2b-3a)}{2b-2a} = -1$

51. $\frac{h^2-t^2}{t-h} = \frac{(h-t)(h+t)}{-1(h-t)} = \frac{h+t}{-1} = -h - t$

53. $\frac{2g-6h}{9h^2-g^2} = \frac{-2(3h-g)}{(3h-g)(3h+g)} = \frac{-2}{3h+g}$

71

55. $\dfrac{x^2 - x - 6}{9 - x^2} = \dfrac{(x-3)(x+2)}{(3-x)(3+x)} = \dfrac{-1(x+2)}{3+x}$

$= \dfrac{-x-2}{x+3}$

57. $\dfrac{-x-6}{x+6} = \dfrac{-1(x+6)}{x+6} = -1$

59. $\dfrac{-2y - 6y^2}{3 + 9y} = \dfrac{-2y(1+3y)}{3(1+3y)} = \dfrac{-2y}{3}$

61. $\dfrac{-3x - 6}{3x - 6} = \dfrac{-3(x+2)}{3(x-2)} = \dfrac{x+2}{-1(x-2)} = \dfrac{x+2}{2-x}$

63. $\dfrac{-12a - 6}{2a^2 + 7a + 3} = \dfrac{-6(2a+1)}{(2a+1)(a+3)} = \dfrac{-6}{a+3}$

65. $\dfrac{2x^{12}}{4x^8} = \dfrac{2x^{12-8}}{2 \cdot 2} = \dfrac{x^4}{2}$

67. $\dfrac{2x + 4}{4x} = \dfrac{2(x+2)}{2(2x)} = \dfrac{x+2}{2x}$

69. $\dfrac{a - 4}{4 - a} = \dfrac{-1(4-a)}{4-a} = -1$

71. $\dfrac{2c - 4}{4 - c^2} = \dfrac{-2(2-c)}{(2-c)(2+c)} = \dfrac{-2}{c+2}$

73. $\dfrac{x^2 + 4x + 4}{x^2 - 4} = \dfrac{(x+2)^2}{(x-2)(x+2)} = \dfrac{x+2}{x-2}$

75. $\dfrac{-2x - 4}{x^2 + 5x + 6} = \dfrac{-2(x+2)}{(x+2)(x+3)} = \dfrac{-2}{x+3}$

77. $\dfrac{2q^8 + q^7}{2q^6 + q^5} = \dfrac{q^7(2q+1)}{q^5(2q+1)} = q^{7-5} = q^2$

79. $\dfrac{u^2 - 6u - 16}{u^2 - 16u + 64} = \dfrac{(u-8)(u+2)}{(u-8)^2} = \dfrac{u+2}{u-8}$

81. $\dfrac{a^3 - 8}{2a - 4} = \dfrac{(a-2)(a^2 + 2a + 4)}{2(a-2)} = \dfrac{a^2 + 2a + 4}{2}$

83. $\dfrac{y^3 - 2y^2 - 4y + 8}{y^2 - 4y + 4} = \dfrac{(y^2 - 4)(y-2)}{(y-2)^2}$

$= \dfrac{(y-2)(y+2)(y-2)}{(y-2)^2} = y + 2$

85. $\dfrac{2 - 5}{2 + 3} = \dfrac{-3}{5} = -0.6$

$\dfrac{-4 - 5}{-4 + 3} = \dfrac{-9}{-1} = 9$

$\dfrac{-3.02 - 5}{-3.02 + 3} = 401$

$\dfrac{-2.96 - 5}{-2.96 + 3} = -199$

87. Since $D = RT$, we have $T = D/R$. So Sergio's time was $\dfrac{300}{x + 10}$ hours.

89. The cost per pound is found by dividing the total cost by the number of pounds. So if $x + 4$ pounds cost \$4.50, then the cost per pound is $\dfrac{4.50}{x + 4}$ dollars per pound.

91. If she could clean the whole pool in 3 hours, she would be cleaning 1/3 of the pool per hour. So if she cleans the whole pool in x hours, she cleans $\dfrac{1}{x}$ of the pool per hour.

93. a) \$0.75   b) \$0.75, \$0.63, \$0.615

c) It approaches \$0.60

5.2 WARM-UPS

1. True, because $\dfrac{2}{3} \cdot \dfrac{5}{3} = \dfrac{10}{9}$.   **2.** True, because

$\dfrac{x - 7}{3} \cdot \dfrac{6}{7 - x} = \dfrac{-1(7 - x)}{3} \cdot \dfrac{2 \cdot 3}{7 - x} = -2$.

3. True, because of the definition of division of fractions.   **4.** False, because if $x = 1$ we get $3 \div 1 = \dfrac{1}{3} \cdot 1$, which is incorrect.

5. True, because we must factor before we can divide out the common factors.

6. False, because $\dfrac{1}{2} \cdot \dfrac{1}{4} = \dfrac{1}{8}$.

7. False, because $\dfrac{1}{2} \div 3 = \dfrac{1}{2} \cdot \dfrac{1}{3} = \dfrac{1}{6}$.

8. True, because
$\dfrac{839 - 487}{487 - 839} = \dfrac{-1(487 - 839)}{487 - 839} = -1$.

9. True, because $\dfrac{a}{3} \div 3 = \dfrac{a}{3} \cdot \dfrac{1}{3} = \dfrac{a}{9}$.

10. True, because $\dfrac{a}{b} \cdot \dfrac{b}{a} = \dfrac{ab}{ab} = 1$.

5.2 EXERCISES

1. $\dfrac{8}{15}\cdot\dfrac{35}{24}=\dfrac{2\cdot2\cdot2}{3\cdot5}\cdot\dfrac{5\cdot7}{2\cdot2\cdot2\cdot3}=\dfrac{7}{9}$

3. $\dfrac{12}{17}\cdot\dfrac{51}{10}=\dfrac{2\cdot6}{17}\cdot\dfrac{3\cdot17}{2\cdot5}=\dfrac{18}{5}$

5. $24\cdot\dfrac{7}{20}=4\cdot6\cdot\dfrac{7}{4\cdot5}=\dfrac{42}{5}$

7. $\dfrac{5a}{12b}\cdot\dfrac{3ab}{55a}=\dfrac{5a}{2\cdot2\cdot3b}\cdot\dfrac{3ab}{5\cdot11a}=\dfrac{a}{44}$

9. $\dfrac{-2x^6}{7a^5}\cdot\dfrac{21a^2}{6x}=\dfrac{-2x^6}{7a^5}\cdot\dfrac{3\cdot7a^2}{2\cdot3x}=\dfrac{-x^5}{a^3}$

11. $\dfrac{15t^3y^5}{20w^7}\cdot24t^5w^3y^2=\dfrac{3\cdot5t^3y^5}{2\cdot2\cdot5w^7}\cdot2^3\cdot3t^5w^3y^2$

$=\dfrac{18t^8y^7}{w^4}$

13. $\dfrac{3a+3b}{15}\cdot\dfrac{10a}{a^2-b^2}=\dfrac{3(a+b)}{3\cdot5}\cdot\dfrac{2\cdot5a}{(a-b)(a+b)}$

$=\dfrac{2a}{a-b}$

15. $(x^2-6x+9)\cdot\dfrac{3}{x-3}=(x-3)^2\cdot\dfrac{3}{x-3}$

$=3(x-3)=3x-9$

17. $\dfrac{16a+8}{5a^2+5}\cdot\dfrac{2a^2+a-1}{4a^2-1}$

$=\dfrac{8(2a+1)}{5(a^2+1)}\cdot\dfrac{(2a-1)(a+1)}{(2a-1)(2a+1)}$

$=\dfrac{8(a+1)}{5(a^2+1)}=\dfrac{8a+8}{5a^2+5}$

19. $12\div\dfrac{2}{5}=12\cdot\dfrac{5}{2}=30$

21. $\dfrac{5}{7}\div\dfrac{15}{14}=\dfrac{5}{7}\cdot\dfrac{14}{15}=\dfrac{2}{3}$

23. $\dfrac{40}{3}\div12=\dfrac{40}{3}\cdot\dfrac{1}{12}=\dfrac{2^3\cdot5}{3\cdot2^2\cdot3}=\dfrac{10}{9}$

25. $\dfrac{5x^2}{3}\div\dfrac{10x}{21}=\dfrac{5x^2}{3}\cdot\dfrac{3\cdot7}{2\cdot5x}=\dfrac{7x}{2}$

27. $\dfrac{8m^3}{n^4}\div(12mn^2)=\dfrac{2^3m^3}{n^4}\cdot\dfrac{1}{2^2\cdot3mn^2}=\dfrac{2m^2}{3n^6}$

29. $\dfrac{y-6}{2}\div\dfrac{6-y}{6}=\dfrac{y-6}{2}\cdot\dfrac{6}{6-y}$

$=\dfrac{-1(6-y)}{2}\cdot\dfrac{2\cdot3}{6-y}=-3$

31. $\dfrac{x^2+4x+4}{8}\div\dfrac{(x+2)^3}{16}=\dfrac{(x+2)^2}{8}\cdot\dfrac{16}{(x+2)^3}$

$=\dfrac{16}{8(x+2)}=\dfrac{2}{x+2}$

33. $\dfrac{t^2+3t-10}{t^2-25}\cdot\dfrac{1}{4t-8}=\dfrac{(t+5)(t-2)}{(t-5)(t+5)}\cdot\dfrac{1}{4(t-2)}$

$=\dfrac{1}{4(t-5)}=\dfrac{1}{4t-20}$

35. $(2x-5)(x+1)\cdot\dfrac{x-1}{2x-5}=x^2-1$

37. $\dfrac{\dfrac{x-2y}{5}}{\dfrac{1}{10}}=\dfrac{x-2y}{5}\cdot\dfrac{10}{1}=2(x-2y)=2x-4y$

39. $\dfrac{\dfrac{x^2-4}{12}}{\dfrac{x-2}{6}}=\dfrac{x^2-4}{12}\cdot\dfrac{6}{x-2}=\dfrac{(x-2)(x+2)}{2\cdot6}\cdot\dfrac{6}{x-2}$

$=\dfrac{x+2}{2}$

41. $\dfrac{\dfrac{x^2+9}{3}}{5}=\dfrac{x^2+9}{3}\cdot\dfrac{1}{5}=\dfrac{x^2+9}{15}$

43. $\dfrac{\dfrac{x^2-y^2}{x-y}}{9}=(x^2-y^2)\cdot\dfrac{9}{x-y}$

$=(x-y)(x+y)\cdot\dfrac{9}{x-y}=9(x+y)=9x+9y$

45. $\dfrac{x-1}{3}\cdot\dfrac{9}{1-x}=\dfrac{-1(1-x)3\cdot3}{3(1-x)}=-3$

47. $\dfrac{3a+3b}{a}\cdot\dfrac{1}{3}=\dfrac{3(a+b)}{a}\cdot\dfrac{1}{3}=\dfrac{a+b}{a}$

49. $\dfrac{\dfrac{b}{a}}{\dfrac{1}{2}}=\dfrac{b}{a}\cdot\dfrac{2}{1}=\dfrac{2b}{a}$

51. $\dfrac{6y}{3}\cdot\dfrac{1}{2x}=\dfrac{6y}{6x}=\dfrac{y}{x}$

53. $\dfrac{a^3b^4}{-2ab^2}\cdot\dfrac{a^5b^7}{ab}=\dfrac{a^8b^{11}}{-2a^2b^3}=\dfrac{-a^6b^8}{2}$

55. $\dfrac{2mn^4}{6mn^2}\div\dfrac{3m^5n^7}{m^2n^4}=\dfrac{2mn^4}{6mn^2}\cdot\dfrac{m^2n^4}{3m^5n^7}=\dfrac{2m^3n^8}{18m^6n^9}$

$=\dfrac{1}{9m^3n}$

57. $\dfrac{3x^2 + 16x + 5}{x} \cdot \dfrac{x^2}{9x^2 - 1}$

$= \dfrac{(3x + 1)(x + 5)}{x} \cdot \dfrac{x^2}{(3x - 1)(3x + 1)}$

$= \dfrac{(x + 5)x}{3x - 1} = \dfrac{x^2 + 5x}{3x - 1}$

59. $\dfrac{a^2 - 2a - 4}{a^2 - 4} \cdot \dfrac{(a + 2)^3}{2a + 4}$

$= \dfrac{a^2 - 2a - 4}{(a - 2)(a + 2)} \cdot \dfrac{(a + 2)^3}{2(a + 2)}$

$= \dfrac{(a^2 - 2a - 4)(a + 2)}{2(a - 2)} = \dfrac{a^3 + 8}{2a - 4}$

61. $\dfrac{2x^2 + 19x - 10}{x^2 - 100} \cdot \dfrac{2x^2 - 19x - 10}{4x^2 - 1}$

$= \dfrac{(2x - 1)(x + 10)}{(x - 10)(x + 10)} \cdot \dfrac{(2x + 1)(x - 10)}{(2x - 1)(2x + 1)} = 1$

63. $\dfrac{m^2 + 6m + 9}{m^2 - 6m + 9} \cdot \dfrac{m^2 - 9}{m^2 + mk + 3m + 3k}$

$= \dfrac{(m + 3)^2}{(m - 3)^2} \cdot \dfrac{(m - 3)(m + 3)}{(m + 3)(m + k)} = \dfrac{(m + 3)^2}{(m - 3)(m + k)}$

65. $A = LW = x \cdot \dfrac{5}{x} = 5 \text{ m}^2$

5.3 WARM-UPS

1. False, because to get an equivalent fraction with a denominator of 18 we must multiply both numerator and denominator by 6.

2. False, because we must factor each denominator to determine the LCD.

3. True, because $\dfrac{3}{2ab^2} = \dfrac{3 \cdot 5a^2b^2}{2ab^2 \cdot 5a^2b^2} = \dfrac{15a^2b^2}{10a^3b^4}$.

4. True, because $2^5 \cdot 3^2$ is a multiple of both $2^5 \cdot 3$ and $2^4 \cdot 3^2$, and it is the smallest number that is a multiple of both.

5. False, because 30 is a multiple of both 6 and 10.

6. False, because the LCD for $6a^2b$ and $4ab^3$ is $12a^2b^3$.

7. False, because $a^2 + 1$ is not a multiple of $a + 1$.

8. False, because if $x = 1$ then $\dfrac{1}{2} \neq \dfrac{1 + 7}{2 + 7}$.

9. True, because $x^2 - 4 = (x - 2)(x + 2)$.

10. True, because $x = x \cdot \dfrac{3}{3} = \dfrac{3x}{3}$.

5.3 EXERCISES

1. $\dfrac{1}{3} = \dfrac{1 \cdot 9}{3 \cdot 9} = \dfrac{9}{27}$

3. $7 = 7 \cdot \dfrac{2x}{2x} = \dfrac{14x}{2x}$

5. $\dfrac{5}{b} = \dfrac{5 \cdot 3t}{b \cdot 3t} = \dfrac{15t}{3bt}$

7. $\dfrac{-9z}{2aw} = \dfrac{-9z \cdot 4z}{2aw \cdot 4z} = \dfrac{-36z^2}{8awz}$

9. $\dfrac{2}{3a} = \dfrac{2 \cdot 5a^2}{3a \cdot 5a^2} = \dfrac{10a^2}{15a^3}$

11. $\dfrac{4}{5xy^2} = \dfrac{4 \cdot 2xy^3}{5xy^2 \cdot 2xy^3} = \dfrac{8xy^3}{10x^2y^5}$

13. $\dfrac{5}{2x + 2} = \dfrac{5(-4)}{(2x + 2)(-4)} = \dfrac{-20}{-8x - 8}$

15. $\dfrac{8a}{5b^2 - 5b} = \dfrac{8a(-4b)}{(5b^2 - 5b)(-4b)} = \dfrac{-32ab}{20b^2 - 20b^3}$

17. $\dfrac{3}{x + 2} = \dfrac{3(x - 2)}{(x + 2)(x - 2)} = \dfrac{3x - 6}{x^2 - 4}$

19. $\dfrac{3x}{x + 1} = \dfrac{3x(x + 1)}{(x + 1)(x + 1)} = \dfrac{3x^2 + 3x}{x^2 + 2x + 1}$

21. $\dfrac{y - 6}{y - 4} = \dfrac{(y - 6)(y + 5)}{(y - 4)(y + 5)} = \dfrac{y^2 - y - 30}{y^2 + y - 20}$

23. Since $12 = 2^2 \cdot 3$ and $16 = 2^4$, the LCD is $2^4 \cdot 3 = 48$.

25. Since $12 = 2^2 \cdot 3$, $18 = 2 \cdot 3^2$, and $20 = 2^2 \cdot 5$, the LCD is $2^2 \cdot 3^2 \cdot 5 = 180$.

27. Since $6a^2 = 2 \cdot 3a^2$ and $15a = 3 \cdot 5a$, the LCD is $2 \cdot 3 \cdot 5a^2 = 30a^2$.

29. The LCD for $2a^4b$, $3ab^6$ and $4a^3b^2$ is $12a^4b^6$.

31. Since $x^2 - 16 = (x-4)(x+4)$ and $x^2 + 8x + 16 = (x+4)^2$, the LCD is $(x-4)(x+4)^2$.

33. The LCD for x, $x+2$, and $x-2$ is $x(x+2)(x-2)$.

35. Since $x^2 - 4x = x(x-4)$, $x^2 - 16 = (x-4)(x+4)$, and $2x = 2x$, the LCD is $2x(x-4)(x+4)$.

37. The LCD for 6 and 8 is 24:
$$\frac{1}{6} = \frac{1 \cdot 4}{6 \cdot 4} = \frac{4}{24}, \quad \frac{3}{8} = \frac{3 \cdot 3}{8 \cdot 3} = \frac{9}{24}$$

39. Since $84 = 2^2 \cdot 3 \cdot 7$ and $63 = 3^2 \cdot 7$, the LCD is $2^2 \cdot 3^2 \cdot 7ab = 252ab$.
$$\frac{3}{84a} = \frac{3 \cdot 3b}{84a \cdot 3b} = \frac{9b}{252ab}, \quad \frac{5}{63b} = \frac{5 \cdot 4a}{63b \cdot 4a} = \frac{20a}{252ab}$$

41. The LCD for $3x^2$ and $2x^5$ is $3 \cdot 2x^5 = 6x^5$.
$$\frac{1}{3x^2} = \frac{1 \cdot 2x^3}{3x^2 \cdot 2x^3} = \frac{2x^3}{6x^5}, \quad \frac{3}{2x^5} = \frac{3 \cdot 3}{2x^5 \cdot 3} = \frac{9}{6x^5}$$

43. Since $9y^5z = 3 \cdot 3y^5z$, $12x^3 = 2^2 \cdot 3x^3$, and $6x^2y = 2 \cdot 3x^2y$, the LCD is $36x^3y^5z$.
$$\frac{x}{9y^5z} = \frac{x \cdot 4x^3}{9y^5z \cdot 4x^3} = \frac{4x^4}{36x^3y^5z},$$
$$\frac{y}{12x^3} = \frac{y \cdot 3y^5z}{12x^3 \cdot 3y^5z} = \frac{3y^6z}{36x^3y^5z}$$
$$\frac{1}{6x^2y} = \frac{1 \cdot 6xy^4z}{6x^2y \cdot 6xy^4z} = \frac{6xy^4z}{36x^3y^5z}$$

45. The LCD for $x-3$ and $x+2$ is $(x-3)(x+2)$.
$$\frac{2x}{x-3} = \frac{2x(x+2)}{(x-3)(x+2)} = \frac{2x^2+4x}{(x-3)(x+2)}$$
$$\frac{5x}{x+2} = \frac{5x(x-3)}{(x+2)(x-3)} = \frac{5x^2-15x}{(x-3)(x+2)}$$

47. Since $6 - a = -1(a-6)$, the LCD is $a-6$. The expression $\frac{4}{a-6}$ already has the required denominator.
$$\frac{5}{6-a} = \frac{5(-1)}{(6-a)(-1)} = \frac{-5}{a-6}$$

49. Since $x^2 - 9 = (x-3)(x+3)$ and $x^2 - 6x + 9 = (x-3)^2$, the LCD is $(x+3)(x-3)^2$.
$$\frac{x}{x^2-9} = \frac{x(x-3)}{(x-3)(x+3)(x-3)} = \frac{x^2-3x}{(x-3)^2(x+3)}$$
$$\frac{5x}{x^2-6x+9} = \frac{5x(x+3)}{(x-3)^2(x+3)} = \frac{5x^2+15x}{(x-3)^2(x+3)}$$

51. Since $w^2 - 2w - 15 = (w-5)(w+3)$ and $w^2 - 4w - 5 = (w-5)(w+1)$, the LCD is $(w-5)(w+1)(w+3)$.
$$\frac{w+2}{(w-5)(w+3)} = \frac{(w+2)(w+1)}{(w-5)(w+3)(w+1)}$$
$$= \frac{w^2+3w+2}{(w-5)(w+3)(w+1)}$$
$$\frac{-2w}{(w-5)(w+1)} = \frac{-2w(w+3)}{(w-5)(w+1)(w+3)}$$
$$= \frac{-2w^2-6w}{(w-5)(w+1)(w+3)}$$

53. $\dfrac{-5}{6(x-2)} = \dfrac{-5(x+2)}{6(x-2)(x+2)} = \dfrac{-5x-10}{6(x-2)(x+2)}$
$$\frac{x}{(x-2)(x+2)} = \frac{6x}{6(x-2)(x+2)}$$
$$\frac{3}{2(x+2)} = \frac{3 \cdot 3(x-2)}{2(x+2) \cdot 3(x-2)} = \frac{9x-18}{6(x-2)(x+2)}$$

55. $\dfrac{2}{(2q+1)(q-3)} = \dfrac{2(q+4)}{(2q+1)(q-3)(q+4)}$
$$= \frac{2q+8}{(2q+1)(q-3)(q+4)}$$
$$\frac{3}{(2q+1)(q+4)} = \frac{3(q-3)}{(2q+1)(q+4)(q-3)}$$
$$= \frac{3q-9}{(2q+1)(q-3)(q+4)}$$
$$\frac{4}{(q+4)(q-3)} = \frac{4(2q+1)}{(q+4)(q-3)(2q+1)}$$
$$= \frac{8q+4}{(2q+1)(q-3)(q+4)}$$

5.4 WARM-UPS

1. False, because $\frac{1}{2}+\frac{1}{3}=\frac{3}{6}+\frac{2}{6}=\frac{5}{6}$.

2. True, because $\frac{7}{12}-\frac{1}{12}=\frac{6}{12}=\frac{1}{2}$.

3. True, because $\frac{3}{5}+\frac{4}{3}=\frac{3\cdot 3+4\cdot 5}{3\cdot 5}=\frac{29}{15}$.

4. True, because $\frac{4}{5}-\frac{5}{7}=\frac{4\cdot 7-5\cdot 5}{5\cdot 7}=\frac{3}{35}$.

5. True, because $\frac{5}{20}+\frac{3}{4}=\frac{1}{4}+\frac{3}{4}=\frac{4}{4}=1$.

6. False, because if $x=2$ the equation becomes $\frac{2}{2}+1=\frac{3}{2}$, which is incorrect.

7. True, because $1+\frac{1}{a}=\frac{a}{a}+\frac{1}{a}=\frac{a+1}{a}$.

8. False, because if $a=0$ we get $0-\frac{1}{4}=\frac{3}{4}\cdot 0$, which is incorrect.

9. True, because $\frac{a}{2}+\frac{b}{3}=\frac{3a}{6}+\frac{2b}{6}=\frac{3a+2b}{6}$.

10. False, because the LCD for x and $x-1$ is $x(x-1)=x^2-x$.

5.4 EXERCISES

1. $\frac{1}{10}+\frac{1}{10}=\frac{2}{10}=\frac{2\cdot 1}{2\cdot 5}=\frac{1}{5}$

3. $\frac{7}{8}-\frac{1}{8}=\frac{6}{8}=\frac{3\cdot 2}{4\cdot 2}=\frac{3}{4}$

5. $\frac{1}{6}-\frac{5}{6}=\frac{-4}{6}=\frac{-2\cdot 2}{3\cdot 2}=-\frac{2}{3}$

7. $-\frac{7}{8}+\frac{1}{8}=\frac{-6}{8}=\frac{-3\cdot 2}{4\cdot 2}=-\frac{3}{4}$

9. $\frac{1}{3}+\frac{2}{9}=\frac{1\cdot 3}{3\cdot 3}+\frac{2}{9}=\frac{3}{9}+\frac{2}{9}=\frac{5}{9}$

11. The LCD for $16=2^4$ and $18=2\cdot 3^2$ is $2^4\cdot 3^2=144$.

$\frac{7}{16}+\frac{5}{18}=\frac{7\cdot 9}{16\cdot 9}+\frac{5\cdot 8}{18\cdot 8}=\frac{63}{144}+\frac{40}{144}=\frac{103}{144}$

13. The LCD for $8=2^3$ and $10=2\cdot 5$ is $2^3\cdot 5=40$.

$\frac{1}{8}-\frac{9}{10}=\frac{1\cdot 5}{8\cdot 5}-\frac{9\cdot 4}{10\cdot 4}=\frac{5}{40}-\frac{36}{40}=-\frac{31}{40}$

15. The LCD for $6=2\cdot 3$ and $8=2^3$ is $2^3\cdot 3=24$.

$-\frac{1}{6}-\left(-\frac{3}{8}\right)=-\frac{1}{6}+\frac{3}{8}=-\frac{1\cdot 4}{6\cdot 4}+\frac{3\cdot 3}{8\cdot 3}=-\frac{4}{24}+\frac{9}{24}$
$$=\frac{5}{24}$$

17. $\frac{3}{2w}+\frac{7}{2w}=\frac{10}{2w}=\frac{2\cdot 5}{2w}=\frac{5}{w}$

19. $\frac{3a}{a+5}+\frac{15}{a+5}=\frac{3a+15}{a+5}=\frac{3(a+5)}{a+5}=3$

21. $\frac{q-1}{q-4}-\frac{3q-9}{q-4}=\frac{q-1-3q+9}{q-4}$

$=\frac{-2q+8}{q-4}=\frac{-2(q-4)}{q-4}=-2$

23. $\frac{4h-3}{h(h+1)}-\frac{h-6}{h(h+1)}=\frac{4h-3-h+6}{h(h+1)}$

$=\frac{3h+3}{h(h+1)}=\frac{3(h+1)}{h(h+1)}=\frac{3}{h}$

25. $\frac{x^2-x-5}{(x+1)(x+2)}+\frac{1-2x}{(x+1)(x+2)}$

$=\frac{x^2-3x-4}{(x+1)(x+2)}=\frac{(x-4)(x+1)}{(x+1)(x+2)}=\frac{x-4}{x+2}$

27. The LCD for $2a$ and $5a$ is $2\cdot 5a=10a$.

$\frac{3}{2a}+\frac{1}{5a}=\frac{3\cdot 5}{2a\cdot 5}+\frac{1\cdot 2}{5a\cdot 2}=\frac{15}{10a}+\frac{2}{10a}=\frac{17}{10a}$

29. $\frac{w-3}{9}-\frac{w-4}{12}=\frac{(w-3)4}{9\cdot 4}-\frac{(w-4)3}{12\cdot 3}$

$=\frac{4w-12}{36}-\frac{3w-12}{36}=\frac{w}{36}$

31. $\frac{b^2}{4a}-c=\frac{b^2}{4a}-c\cdot\frac{4a}{4a}=\frac{b^2}{4a}-\frac{4ac}{4a}=\frac{b^2-4ac}{4a}$

33. $\frac{2}{wz^2}+\frac{3}{w^2z}=\frac{2\cdot w}{wz^2\cdot w}+\frac{3\cdot z}{w^2z\cdot z}=\frac{2w}{w^2z^2}+\frac{3z}{w^2z^2}$

$$=\frac{2w+3z}{w^2z^2}$$

35. $\frac{2}{x+1}-\frac{3}{x}=\frac{2(x)}{(x+1)(x)}-\frac{3(x+1)}{x(x+1)}$

$=\frac{2x}{x(x+1)}-\frac{3x+3}{x(x+1)}=\frac{2x-3x-3}{x(x+1)}=\frac{-x-3}{x(x+1)}$

37. $\dfrac{2}{a-b}+\dfrac{1}{a+b}$

$=\dfrac{2(a+b)}{(a-b)(a+b)}+\dfrac{1(a-b)}{(a+b)(a-b)}$

$=\dfrac{2a+2b}{(a-b)(a+b)}+\dfrac{a-b}{(a+b)(a-b)}=\dfrac{3a+b}{(a-b)(a+b)}$

39. $\dfrac{3}{x^2+x}-\dfrac{4}{5x+5}=\dfrac{3}{x(x+1)}-\dfrac{4}{5(x+1)}$

$=\dfrac{3\cdot 5}{x(x+1)\cdot 5}-\dfrac{4\cdot x}{5(x+1)\cdot x}$

$=\dfrac{15}{5x(x+1)}-\dfrac{4x}{5x(x+1)}=\dfrac{15-4x}{5x(x+1)}$

41. $\dfrac{2a}{a^2-9}+\dfrac{a}{a-3}=\dfrac{2a}{(a-3)(a+3)}+\dfrac{a}{a-3}$

$\qquad =\dfrac{2a}{(a-3)(a+3)}+\dfrac{a(a+3)}{(a-3)(a+3)}$

$\qquad =\dfrac{2a+a^2+3a}{(a-3)(a+3)}=\dfrac{a^2+5a}{(a-3)(a+3)}$

43. $\dfrac{4}{a-b}+\dfrac{4}{b-a}=\dfrac{4}{a-b}+\dfrac{4(-1)}{(b-a)(-1)}$

$=\dfrac{4}{a-b}+\dfrac{-4}{a-b}=\dfrac{0}{a-b}=0$

45. $\dfrac{3}{2a-2}-\dfrac{2}{1-a}=\dfrac{3}{2(a-1)}-\dfrac{2}{1-a}$

$\qquad =\dfrac{3}{2(a-1)}-\dfrac{2(-2)}{(1-a)(-2)}$

$\qquad =\dfrac{3}{2a-2}-\dfrac{-4}{2a-2}=\dfrac{7}{2a-2}$

47. $\dfrac{1}{x^2-4}-\dfrac{3}{x^2-3x-10}$

$=\dfrac{1}{(x-2)(x+2)}-\dfrac{3}{(x-5)(x+2)}$

$=\dfrac{1(x-5)}{(x-2)(x+2)(x-5)}-\dfrac{3(x-2)}{(x-5)(x+2)(x-2)}$

$=\dfrac{x-5}{(x-5)(x+2)(x-2)}-\dfrac{3x-6}{(x-5)(x+2)(x-2)}$

$=\dfrac{x-5-3x+6}{(x-5)(x+2)(x-2)}=\dfrac{-2x+1}{(x-5)(x+2)(x-2)}$

49. $\dfrac{3}{x^2+x-2}+\dfrac{4}{x^2+2x-3}$

$=\dfrac{3}{(x+2)(x-1)}+\dfrac{4}{(x+3)(x-1)}$

$=\dfrac{3(x+3)}{(x+2)(x-1)(x+3)}+\dfrac{4(x+2)}{(x+3)(x-1)(x+2)}$

$=\dfrac{3x+9+4x+8}{(x+2)(x-1)(x+3)}=\dfrac{7x+17}{(x+2)(x-1)(x+3)}$

51. $\dfrac{2}{x}-\dfrac{1}{x-1}+\dfrac{1}{x+2}$

$=\dfrac{2(x-1)(x+2)}{x(x-1)(x+2)}-\dfrac{1(x)(x+2)}{(x-1)(x)(x+2)}+\dfrac{1(x)(x-1)}{(x+2)(x)(x-1)}$

$=\dfrac{2x^2+2x-4}{x(x-1)(x+2)}-\dfrac{x^2+2x}{x(x-1)(x+2)}+\dfrac{x^2-x}{x(x-1)(x+2)}$

$=\dfrac{2x^2+2x-4-x^2-2x+x^2-x}{x(x-1)(x+2)}$

$=\dfrac{2x^2-x-4}{x(x-1)(x+2)}$

53. $\dfrac{5}{3(a-3)}-\dfrac{3}{2a}+\dfrac{4}{a(a-3)}$

$=\dfrac{5\cdot 2a}{3(a-3)\cdot 2a}-\dfrac{3(3)(a-3)}{2a(3)(a-3)}+\dfrac{4\cdot 6}{a(a-3)6}$

$=\dfrac{10a-9a+27+24}{6a(a-3)}=\dfrac{a+51}{6a(a-3)}$

55. $P=2\left(\dfrac{3}{x}\right)+2\left(\dfrac{5}{2x}\right)=\dfrac{6}{x}+\dfrac{5}{x}=\dfrac{11}{x}$ feet

57. Time before $=\dfrac{120}{x}$ hr, time after $=\dfrac{195}{x+5}$ hr.

Total time is

$\dfrac{120}{x}+\dfrac{195}{x+5}=\dfrac{120(x+5)+195x}{x(x+5)}=\dfrac{315x+600}{x(x+5)}$

If $x=60$, then total time is 5 hours

59. Work completed by Kent $=\dfrac{2}{x}$ job, work completed by Keith $=\dfrac{2}{x+3}$ job.

Work completed in 2 days together is

$\dfrac{2}{x}+\dfrac{2}{x+3}=\dfrac{2(x+3)}{x(x+3)}+\dfrac{2x}{(x+3)x}=\dfrac{4x+6}{x(x+3)}$ job.

If $x=6$, then $\dfrac{4\cdot 6+6}{6(6+3)}=\dfrac{30}{54}=\dfrac{5}{9}$ job.

5.5 WARM-UPS

1. False, because the LCD for 4, x, 6, and x^2 is $12x^2$.

2. True, because $2b - 2a = -2(a - b)$.

3. False, because a complex fraction has fractions in its numerator, denominator, or both.

4. False, because $3 - a = -1(a - 3)$.

5. False, because there is no largest common denominator.

6. False, because the LCD for the denominators 2, 3, 4, and 5 is 60.

7. False, because the LCD of the denominators b and a is ab.

8. True, because this complex fraction is simplified correctly and $-3/2$ is the only value that causes the denominator to be zero.

9. True, because $-3/2$ is the only value that causes the denominator to be zero.

10. True, because $\frac{1}{2} + \frac{1}{3} = \frac{5}{6}$ and $1 + \frac{1}{2} = \frac{3}{2}$.

5.5 EXERCISES

1. $\dfrac{\frac{1}{2} + \frac{1}{3}}{\frac{1}{4} - \frac{1}{2}} = \dfrac{\left(\frac{1}{2} + \frac{1}{3}\right)(12)}{\left(\frac{1}{4} - \frac{1}{2}\right)(12)} = \dfrac{6 + 4}{3 - 6} = \dfrac{10}{-3} = -\dfrac{10}{3}$

3. $\dfrac{\frac{2}{5} + \frac{5}{6} - \frac{1}{2}}{\frac{1}{2} - \frac{1}{3} + \frac{1}{15}} = \dfrac{\frac{2}{5} \cdot 30 + \frac{5}{6} \cdot 30 - \frac{1}{2} \cdot 30}{\frac{1}{2} \cdot 30 - \frac{1}{3} \cdot 30 + \frac{1}{15} \cdot 30}$

$= \dfrac{12 + 25 - 15}{15 - 10 + 2} = \dfrac{22}{7}$

5. $\dfrac{3 + \frac{1}{2}}{5 - \frac{3}{4}} = \dfrac{\left(3 + \frac{1}{2}\right)4}{\left(5 - \frac{3}{4}\right)4} = \dfrac{12 + 2}{20 - 3} = \dfrac{14}{17}$

7. $\dfrac{1 - \frac{1}{6} + \frac{2}{3}}{1 + \frac{1}{15} - \frac{3}{10}} = \dfrac{\left(1 - \frac{1}{6} + \frac{2}{3}\right)30}{\left(1 + \frac{1}{15} - \frac{3}{10}\right)30} = \dfrac{30 - 5 + 20}{30 + 2 - 9}$

$= \dfrac{45}{23}$

9. $\dfrac{\frac{1}{a} + \frac{3}{b}}{\frac{1}{b} - \frac{3}{a}} = \dfrac{\frac{1}{a} \cdot ab + \frac{3}{b} \cdot ab}{\frac{1}{b} \cdot ab - \frac{3}{a} \cdot ab} = \dfrac{b + 3a}{a - 3b} = \dfrac{3a + b}{a - 3b}$

11. $\dfrac{5 - \frac{3}{a}}{3 + \frac{1}{a}} = \dfrac{\left(5 - \frac{3}{a}\right)(a)}{\left(3 + \frac{1}{a}\right)(a)} = \dfrac{5a - 3}{3a + 1}$

13. $\dfrac{\frac{1}{2} - \frac{2}{x}}{3 - \frac{1}{x^2}} = \dfrac{\frac{1}{2} \cdot 2x^2 - \frac{2}{x} \cdot 2x^2}{3 \cdot 2x^2 - \frac{1}{x^2} \cdot 2x^2} = \dfrac{x^2 - 4x}{6x^2 - 2}$

15. $\dfrac{\frac{3}{2b} + \frac{1}{b}}{\frac{3}{4} - \frac{1}{b^2}} = \dfrac{\frac{3}{2b} \cdot 4b^2 + \frac{1}{b} \cdot 4b^2}{\frac{3}{4} \cdot 4b^2 - \frac{1}{b^2} \cdot 4b^2} = \dfrac{6b + 4b}{3b^2 - 4}$

$= \dfrac{10b}{3b^2 - 4}$

17. $\dfrac{1 - \frac{3}{y+1}}{3 + \frac{1}{y+1}} = \dfrac{1(y+1) - \frac{3}{y+1}(y+1)}{3(y+1) + \frac{1}{y+1}(y+1)}$

$= \dfrac{y + 1 - 3}{3y + 3 + 1} = \dfrac{y - 2}{3y + 4}$

19. $\dfrac{x + \frac{4}{x-2}}{x - \frac{x+1}{x-2}} = \dfrac{x(x-2) + \frac{4}{x-2}(x-2)}{x(x-2) - \frac{x+1}{x-2}(x-2)}$

$= \dfrac{x(x-2) + 4}{x(x-2) - (x+1)} = \dfrac{x^2 - 2x + 4}{x^2 - 3x - 1}$

21. $\dfrac{\frac{1}{3-x} - 5}{\frac{1}{x-3} - 2} = \dfrac{\frac{1}{3-x}(x-3) - 5(x-3)}{\frac{1}{x-3}(x-3) - 2(x-3)}$

$= \dfrac{-1 - 5x + 15}{1 - 2x + 6} = \dfrac{14 - 5x}{7 - 2x} = \dfrac{5x - 14}{2x - 7}$

23. $\dfrac{1 - \frac{5}{a-1}}{3 - \frac{2}{1-a}} = \dfrac{1(a-1) - \frac{5}{a-1}(a-1)}{3(a-1) - \frac{2}{1-a}(a-1)}$

$= \dfrac{a - 1 - 5}{3a - 3 - 2(-1)} = \dfrac{a - 6}{3a - 1}$

25. $\dfrac{\frac{1}{m-3} - \frac{4}{m}}{\frac{3}{m-3} + \frac{1}{m}} = \dfrac{\frac{1}{m-3}m(m-3) - \frac{4}{m}m(m-3)}{\frac{3}{m-3}m(m-3) + \frac{1}{m}m(m-3)}$

$= \dfrac{m - 4(m-3)}{3m + m - 3} = \dfrac{-3m + 12}{4m - 3}$

27. $\dfrac{\dfrac{2}{w-1}-\dfrac{3}{w+1}}{\dfrac{4}{w+1}+\dfrac{5}{w-1}}$

$=\dfrac{\dfrac{2}{w-1}(w+1)(w-1)-\dfrac{3}{w+1}(w+1)(w-1)}{\dfrac{4}{w+1}(w+1)(w-1)+\dfrac{5}{w-1}(w+1)(w-1)}$

$=\dfrac{2(w+1)-3(w-1)}{4(w-1)+5(w+1)}=\dfrac{-w+5}{9w+1}$

29. $\dfrac{\dfrac{1}{a-b}-\dfrac{1}{a+b}}{\dfrac{1}{b-a}+\dfrac{1}{b+a}}$

$=\dfrac{\dfrac{1}{a-b}(a-b)(a+b)-\dfrac{1}{a+b}(a-b)(a+b)}{\dfrac{1}{b-a}(a-b)(a+b)+\dfrac{1}{b+a}(a-b)(a+b)}$

$=\dfrac{a+b-(a-b)}{-1(a+b)+a-b}=\dfrac{a+b-a+b}{-a-b+a-b}$

$=\dfrac{2b}{-2b}=-1$

31. $\dfrac{2x-9}{6}\cdot\dfrac{9}{2x-3}=\dfrac{2x-9}{2\cdot 3}\cdot\dfrac{3\cdot 3}{2x-3}=\dfrac{6x-27}{4x-6}$

33. $\dfrac{2(x-2y)}{xy^2}\cdot\dfrac{x^3y}{3(x-2y)}=\dfrac{2x^2}{3y}$

35. $\dfrac{(a+6)(a-4)}{a+1}\cdot\dfrac{(a+1)^2}{(a-4)(a+3)}=\dfrac{(a+6)(a+1)}{a+3}$

$=\dfrac{a^2+7a+6}{a+3}$

37. $\dfrac{\dfrac{x}{x+1}}{\dfrac{1}{x^2-1}-\dfrac{1}{x-1}}$

$=\dfrac{\dfrac{x}{x+1}\cdot(x-1)(x+1)}{\dfrac{1}{x^2-1}(x-1)(x+1)-\dfrac{1}{x-1}(x-1)(x+1)}$

$=\dfrac{x(x-1)}{1-1(x+1)}=\dfrac{x(x-1)}{-x}=-1(x-1)=1-x$

39. Let x = the number of males and x = the number of females.

$\dfrac{\text{Calculus students}}{\text{Mathematics students}}=\dfrac{\dfrac{1}{3}x+\dfrac{1}{5}x}{\dfrac{5}{6}x+\dfrac{3}{4}x}=\dfrac{\left(\dfrac{1}{3}x+\dfrac{1}{5}x\right)60}{\left(\dfrac{5}{6}x+\dfrac{3}{4}x\right)60}$

$=\dfrac{20x+12x}{50x+45x}=\dfrac{32x}{95x}=\dfrac{32}{95}$

If x = the number of males and 2x = the number of females.

$\dfrac{\text{Calculus students}}{\text{Mathematics students}}=\dfrac{\dfrac{1}{3}x+\dfrac{1}{5}2x}{\dfrac{5}{6}x+\dfrac{3}{4}2x}=\dfrac{\left(\dfrac{1}{3}x+\dfrac{1}{5}2x\right)60}{\left(\dfrac{5}{6}x+\dfrac{3}{4}2x\right)60}$

$=\dfrac{20x+24x}{50x+90x}=\dfrac{44x}{140x}=\dfrac{11}{35}$

5.6 WARM-UPS

1. False, because the first step is to multiply each side by the LCD.
2. False, because we should not divide each side of an equation by a variable.
3. False, an extraneous solution could be rational.
4. False, because the first step is to multiply each side by the LCD of the denominators in the equation.
5. False, because if x = 2 then the denominator x − 2 has a value of zero.
6. True, because $3x^2-6x$ is the LCD.
7. True, because 1/4 + 1/2 = 3/4.
8. True, 4 + 2x = 3x is obtained by multiplying each side by 4.
9. True, because x^2-1 is the LCD for the denominators x − 1 and x + 1.
10. True, because −1 and 1 cause the denominators to have a value of zero.

5.6 EXERCISES

1. $\dfrac{x}{3}-5=\dfrac{x}{2}-7$

$6\left(\dfrac{x}{3}-5\right)=6\left(\dfrac{x}{2}-7\right)$

$2x-30=3x-42$
$-30=x-42$
$12=x$

The solution to the equation is 12.

3. $\dfrac{y}{5}-\dfrac{2}{3}=\dfrac{y}{6}+\dfrac{1}{3}$

$30\left(\dfrac{y}{5}-\dfrac{2}{3}\right)=30\left(\dfrac{y}{6}+\dfrac{1}{3}\right)$

$6y-20=5y+10$
$y-20=10$
$y=30$

The solution to the equation is 30.

5. $12 \cdot \frac{3}{4} - 12 \cdot \frac{t-4}{3} = 12 \cdot \frac{t}{12}$

$$9 - 4t + 16 = t$$
$$-5t = -25$$
$$t = 5$$

The solution to the equation is 5.

7. $30 \cdot \frac{1}{5} - 30 \cdot \frac{w+10}{15} = 30 \cdot \frac{1}{10} - 30 \cdot \frac{w+1}{6}$

$$6 - 2w - 20 = 3 - 5w - 5$$
$$3w = 12$$
$$w = 4$$

The solution to the equation is 4.

9. $\frac{1}{x} + \frac{1}{2} = \frac{3}{4}$

$$4x\left(\frac{1}{x} + \frac{1}{2}\right) = 4x\left(\frac{3}{4}\right)$$
$$4 + 2x = 3x$$
$$4 = x$$

The solution to the equation is 4.

11. $\frac{2}{3x} + \frac{1}{2x} = \frac{7}{24}$

$$24x\frac{2}{3x} + 24x\frac{1}{2x} = 24x\frac{7}{24}$$
$$16 + 12 = 7x$$
$$28 = 7x$$
$$4 = x$$

The solution to the equation is 4.

13. $2a \cdot \frac{1}{2} + 2a \cdot \frac{a-2}{a} = 2a \cdot \frac{a+2}{2a}$

$$a + 2a - 4 = a + 2$$
$$2a = 6$$
$$a = 3$$

The solution to the equation is 3.

15. $6k \cdot \frac{1}{3} - 6k \cdot \frac{k+3}{6k} = 6k \cdot \frac{1}{3k} - 6k \cdot \frac{k-1}{2k}$

$$2k - k - 3 = 2 - 3k + 3$$
$$4k = 8$$
$$k = 2$$

The solution to the equation is 2.

17. $\frac{x}{2} = \frac{5}{x+3}$

$$2(x+3)\frac{x}{2} = 2(x+3)\frac{5}{x+3}$$
$$x^2 + 3x = 10$$
$$x^2 + 3x - 10 = 0$$
$$(x+5)(x-2) = 0$$
$$x + 5 = 0 \quad \text{or} \quad x - 2 = 0$$
$$x = -5 \quad \text{or} \quad x = 2$$

The solutions to the equation are −5 and 2.

19. $\frac{2}{x+1} = \frac{1}{x} + \frac{1}{6}$

$$6x(x+1)\frac{2}{x+1} = 6x(x+1)\frac{1}{x} + 6x(x+1)\frac{1}{6}$$
$$12x = 6x + 6 + x^2 + x$$
$$0 = x^2 - 5x + 6$$
$$0 = (x-2)(x-3)$$
$$x - 2 = 0 \quad \text{or} \quad x - 3 = 0$$
$$x = 2 \quad \text{or} \quad x = 3$$

The solutions to the equation are 2 and 3.

21. $(a-2)(a+2)\left(\frac{a-1}{a^2-4} + \frac{1}{a-2}\right)$

$$= (a-2)(a+2)\frac{a+4}{a+2}$$
$$a - 1 + a + 2 = (a-2)(a+4)$$
$$2a + 1 = a^2 + 2a - 8$$
$$0 = a^2 - 9$$
$$(a-3)(a+3) = 0$$
$$a - 3 = 0 \quad \text{or} \quad a + 3 = 0$$
$$a = 3 \quad \text{or} \quad a = -3$$

The solution to the equation is −3 and 3.

23. $\frac{1}{x-1} + \frac{2}{x} = \frac{x}{x-1}$

$$x(x-1)\frac{1}{x-1} + x(x-1)\frac{2}{x} = x(x-1)\frac{x}{x-1}$$
$$x + 2x - 2 = x^2$$
$$0 = x^2 - 3x + 2$$
$$0 = (x-2)(x-1)$$
$$x - 2 = 0 \quad \text{or} \quad x - 1 = 0$$
$$x = 2 \quad \text{or} \quad x = 1$$

Since the denominator $x-1$ has a value of 0 for $x = 1$, 1 is not a solution to the equation. The solution to the equation is 2.

25. $\frac{5}{x+2} + \frac{2}{x-3} = \frac{x-1}{x-3}$

$$(x+2)(x-3)\frac{5}{x+2} + (x+2)(x-3)\frac{2}{x-3}$$
$$= (x+2)(x-3)\frac{x-1}{x-3}$$
$$5x - 15 + 2x + 4 = (x+2)(x-1)$$
$$7x - 11 = x^2 + x - 2$$
$$0 = x^2 - 6x + 9$$
$$0 = (x-3)^2$$
$$x - 3 = 0$$
$$x = 3$$

Since the value of the denominator $x-3$ is 0 if $x = 3$, there is no solution to this equation.

27.
$$1 + \frac{3y}{y-2} = \frac{6}{y-2}$$
$$(y-2)1 + (y-2)\frac{3y}{y-2} = (y-2)\frac{6}{y-2}$$
$$y - 2 + 3y = 6$$
$$4y = 8$$
$$y = 2$$

Since the value of the denominator $y - 2$ is 0 if $y = 2$, there is no solution to this equation.

29. $(z+1)(z+2)\left(\frac{z}{z+1} - \frac{1}{z+2}\right)$
$$= (z+1)(z+2)\frac{2z+5}{z^2+3z+2}$$
$$z^2 + 2z - z - 1 = 2z + 5$$
$$z^2 - z - 6 = 0$$
$$(z-3)(z+2) = 0$$
$$z - 3 = 0 \quad \text{or} \quad z + 2 = 0$$
$$z = 3 \quad \text{or} \quad z = -2$$

Since the value of $z + 2$ is 0 if $z = -2$, the only solution to the equation is 3.

31.
$$\frac{a}{4} = \frac{5}{2}$$
$$4 \cdot \frac{a}{4} = 4 \cdot \frac{5}{2}$$
$$a = 10$$

The solution to the equation is 10.

33.
$$\frac{w}{6} = \frac{3w}{11}$$
$$66 \cdot \frac{w}{6} = 66 \cdot \frac{3w}{11}$$
$$11w = 18w$$
$$0 = 7w$$
$$0 = w$$

The solution to the equation is 0.

35.
$$\frac{5}{x} = \frac{x}{5}$$
$$5x \cdot \frac{5}{x} = 5x \cdot \frac{x}{5}$$
$$25 = x^2$$
$$0 = x^2 - 25$$
$$0 = (x-5)(x+5)$$
$$x - 5 = 0 \quad \text{or} \quad x + 5 = 0$$
$$x = 5 \quad \text{or} \quad x = -5$$

The solutions to the equation are -5 and 5.

37.
$$\frac{x-3}{5} = \frac{x-3}{x}$$
$$5x \cdot \frac{x-3}{5} = 5x \cdot \frac{x-3}{x}$$
$$x^2 - 3x = 5x - 15$$
$$x^2 - 8x + 15 = 0$$

$$(x-3)(x-5) = 0$$
$$x - 3 = 0 \quad \text{or} \quad x - 5 = 0$$
$$x = 3 \quad \text{or} \quad x = 5$$

The solutions to the equation are 3 and 5.

39.
$$\frac{1}{x+2} = \frac{x}{x+2}$$
$$(x+2)\frac{1}{x+2} = (x+2)\frac{x}{x+2}$$
$$1 = x$$

The solution to the equation is 1.

41.
$$\frac{1}{2x-4} + \frac{1}{x-2} = \frac{3}{2}$$
$$2(x-2)\frac{1}{2x-4} + 2(x-2)\frac{1}{x-2} = 2(x-2)\frac{3}{2}$$
$$1 + 2 = 3x - 6$$
$$9 = 3x$$
$$3 = x$$

The solution to the equation is 3.

43.
$$\frac{3}{a^2-a-6} = \frac{2}{a^2-4}$$
$$\frac{3}{(a-3)(a+2)} = \frac{2}{(a-2)(a+2)}$$
$$(a-3)(a+2)(a-2)\frac{3}{(a-3)(a+2)}$$
$$= (a-3)(a+2)(a-2)\frac{2}{(a-2)(a+2)}$$
$$(a-2)3 = (a-3)2$$
$$3a - 6 = 2a - 6$$
$$a = 0$$

The solution to the equation is 0.

45.
$$\frac{4}{c-2} - \frac{1}{2-c} = \frac{25}{c+6}$$
$$(c-2)(c+6)\frac{4}{c-2} - (c-2)(c+6)\frac{1}{2-c}$$
$$= (c-2)(c+6)\frac{25}{c+6}$$
$$4(c+6) - (c+6)(-1) = (c-2)25$$
$$4c + 24 + c + 6 = 25c - 50$$
$$5c + 30 = 25c - 50$$
$$5c + 80 = 25c$$
$$80 = 20c$$
$$4 = c$$

The solution to the equation is 4.

47. $\dfrac{1}{x^2-9}+\dfrac{3}{x+3}=\dfrac{4}{x-3}$

$(x+3)(x-3)\dfrac{1}{x^2-9}+(x+3)(x-3)\dfrac{3}{x+3}$

$\qquad\qquad =(x+3)(x-3)\dfrac{4}{x-3}$

$\qquad 1+3(x-3)=4(x+3)$
$\qquad 1+3x-9=4x+12$
$\qquad\qquad -20=x$

The solution to the equation is -20.

49. $\dfrac{3}{2x+4}-\dfrac{1}{x+2}=\dfrac{1}{3x+1}$

$2(x+2)(3x+1)\dfrac{3}{2x+4}-2(x+2)(3x+1)\dfrac{1}{x+2}$

$\qquad\qquad =2(x+2)(3x+1)\dfrac{1}{3x+1}$

$(3x+1)3-2(3x+1)=2(x+2)$
$\qquad 9x+3-6x-2=2x+4$
$\qquad\qquad 3x+1=2x+4$
$\qquad\qquad\qquad x=3$

The solution to the equation is 3.

51. $6(t+1)\dfrac{2t-1}{3(t+1)}+6(t+1)\dfrac{3t-1}{6(t+1)}$

$\qquad\qquad\qquad =6(t+1)\dfrac{t}{t+1}$

$\quad 4t-2+3t-1=6t$

$\qquad\qquad t=3$

The solution to the equation is 3.

53. $\qquad \dfrac{1}{50}=\dfrac{1}{600}+\dfrac{1}{i}$

$\quad 600\,i\dfrac{1}{50}=600\,i\dfrac{1}{600}+600\,i\dfrac{1}{i}$

$\qquad 12i=i+600$

$\qquad 11i=600$

$\qquad i=\dfrac{600}{11}=54\dfrac{6}{11}$ mm

5.7 WARM-UPS

1. True, because the fraction 40/30 reduces to 4/3. **2.** False, because 2 yards = 6 feet and the ratio should be expressed as 3 to 6 or 1 to 2. **3.** False, because the ratio of men to women being 3 to 2 means there would be 30 men to 20 women. **4.** True, because if we multiply the numerator and denominator of 1.5/2 by 2, we get the ratio 3/4. **5.** True, because of the definition of proportion. **6.** True, because of the extremes-means property. **7.** False, because the correct application of the extremes-means property gives us $3x=10$. **8.** False, because the ratio should be given with the same units. One foot to 3 feet is a ratio of 1 to 3. **9.** False, because the number who prefer aspirin is 30 and the number who do not is 70, giving a ratio of 3 to 7. **10.** True, because of the extremes-means property.

5.7 EXERCISES

1. $\dfrac{2.5}{3.5}=\dfrac{(2.5)2}{(3.5)2}=\dfrac{5}{7}$

3. $\dfrac{0.32}{0.6}=\dfrac{0.32(100)}{0.6(100)}=\dfrac{32}{60}=\dfrac{8\cdot4}{15\cdot4}=\dfrac{8}{15}$

5. $\dfrac{35}{10}=\dfrac{7\cdot5}{2\cdot5}=\dfrac{7}{2}$

7. $\dfrac{4.5}{7}=\dfrac{(4.5)(2)}{(7)(2)}=\dfrac{9}{14}$

9. $\dfrac{\frac{1}{2}}{\frac{1}{5}}=\dfrac{\left(\frac{1}{2}\right)10}{\left(\frac{1}{5}\right)10}=\dfrac{5}{2}$

11. $\dfrac{5}{\frac{1}{3}}=\dfrac{5(3)}{\frac{1}{3}(3)}=\dfrac{15}{1}$

13. The ratio of men to women is 12/8 or 3/2.
15. The ratio of smokers to nonsmokers is 72/128, or 9/16.
17. The ratio of violence to kindness is 1240/40 or 31/1.
19. The ratio of rise to run is 8 to 12 or 2 to 3.

21. $\qquad \dfrac{4}{x}=\dfrac{2}{3}$

$\qquad 2x=12$

$\qquad\ x=6$

The solution to the proportion is 6.

23. $\qquad \dfrac{a}{2}=\dfrac{-1}{5}$

$\qquad 5a=-2$

$\qquad\ a=-\dfrac{2}{5}$

The solution to the proportion is $-\dfrac{2}{5}$.

25.
$$-\frac{5}{9} = \frac{3}{x}$$
$$-5x = 27$$
$$x = -\frac{27}{5}$$

The solution to the proportion is $-\frac{27}{5}$.

27.
$$\frac{10}{x} = \frac{34}{x+12}$$
$$34x = 10x + 120$$
$$24x = 120$$
$$x = 5$$

The solution to the proportion is 5.

29.
$$\frac{a}{a+1} = \frac{a+3}{a}$$
$$a^2 = (a+1)(a+3)$$
$$a^2 = a^2 + 4a + 3$$
$$0 = 4a + 3$$
$$-4a = 3$$
$$a = -\frac{3}{4}$$

The solution to the proportion is $-\frac{3}{4}$.

31.
$$\frac{m-1}{m-2} = \frac{m-3}{m+4}$$
$$(m-1)(m+4) = (m-2)(m-3)$$
$$m^2 + 3m - 4 = m^2 - 5m + 6$$
$$3m - 4 = -5m + 6$$
$$8m - 4 = 6$$
$$8m = 10$$
$$m = \frac{10}{8} = \frac{5}{4}$$

The solution to the proportion is $\frac{5}{4}$.

33. Let $x =$ the number of reruns. Since the ratio of new shows to reruns is 2 to 27, we can write the following proportion.
$$\frac{8}{x} = \frac{2}{27}$$
$$2x = 216$$
$$x = 108$$

There were 108 reruns.

35. Let $x =$ the number of votes for the incumbent. Since 220 out of 500 voters said they would vote for the incumbent, we can write the following proportion.
$$\frac{220}{500} = \frac{x}{400,000}$$
$$500x = 88,000,000$$
$$x = 176,000$$

The expected number of votes for the incumbent is 176,000.

37. Let $x =$ the number of points scored by the Tigers and $x + 34 =$ the number scored by the Lions.
$$\frac{x+34}{x} = \frac{5}{3}$$
$$3x + 102 = 5x$$
$$102 = 2x$$
$$51 = x$$
$$x + 34 = 85$$

The score was Lions 85, Tigers 51.

39. Let $x =$ the number of luxury cars and $x + 20 =$ the number of sports cars. Since the ratio of sports cars to luxury cars sold is 3 to 2, we can write the following equation.
$$\frac{x+20}{x} = \frac{3}{2}$$
$$3x = 2x + 40$$
$$x = 40$$
$$x + 20 = 60$$

There were 40 luxury cars and 60 sports cars sold.

41. Let $x =$ the number of inches in 7 feet. We can write the following proportion.
$$\frac{12}{1} = \frac{x}{7}$$
$$x = 84$$

There are 84 inches in 7 feet.

43. Let $x =$ the number of minutes in 0.25 hour.
$$\frac{x \text{ min}}{0.25 \text{ hr}} = \frac{60 \text{ min}}{1 \text{ hr}}$$
$$x = (0.25)(60) = 15 \text{ minutes}$$

45. Let $x =$ the number of miles traveled in 7 hours. Since he travels 230 miles in 3 hours, we can write the following proportion.
$$\frac{230}{3} = \frac{x}{7}$$
$$3x = 1610$$
$$x = \frac{1610}{3} = 536.7$$

He travels 536.7 miles in 7 hours.

47. Let $x =$ the force of a 280 lb player.
$$\frac{x}{280} = \frac{980}{70}$$
$$x = 280 \cdot \frac{980}{70} = 3920 \text{ pounds}$$

49. Let x = the number of trout in Trout Lake.

$$\frac{200}{x} = \frac{5}{150}$$

$$5x = 30000$$
$$x = 6,000$$

So there are 6,000 trout in the lake.

51. From the graph we get that 40% of the students graduate and 60% do not. So the ratio of graduates to nongraduates is 4 to 6 or 2 to 3. Let x = the number of graduates in the 1987 group and x + 200 = the number of nongraduates.

$$\frac{x}{x+200} = \frac{2}{3}$$

$$3x = 2x + 400$$
$$x = 400$$
$$x + 200 = 600$$

So there were 1000 students who started NSC in 1987.

53. k = 1 + 0 + 0.26 + 0 − 0.29 = 0.97
If B = 4200 units and k = 0.97, we have

$$\frac{A}{4200} = 0.97$$
$$A = 4200(0.97) = 4074$$

5.8 WARM-UPS

1. True, because we can obtain the second formula from the first by multiplying each side by m and then dividing by t. **2.** True, because 2mn is the LCD of the denominators 2, m, and n. **3.** False, her average speed is $\frac{300}{x}$ mph. **4.** True, because 20 hard bargains divided by x hours is 20/x hard bargains per hour. **5.** True, because if he paints the whole house in y days then he paints 1/y of the house per day. **6.** False, because if 1/x is the smaller of the two quantities then we must add 1 to 1/x to get an equation. **7.** False, because if a = m/b then b = a/m. **8.** True, because if we divide each side of D = RT by R, we get T = D/R. **9.** False, because if the equation is solved for P then P must not appear on the other side of the equation. **10.** True, because the only way to get R isolated is to factor 3R + yR as R(3 + y).

5.8 EXERCISES

1.
$$\frac{y-1}{x-3} = 2$$

$$(x-3)\frac{y-1}{x-3} = (x-3)2$$
$$y - 1 = 2x - 6$$
$$y = 2x - 5$$

3.
$$\frac{y-1}{x+6} = -\frac{1}{2}$$

$$(x+6)\frac{y-1}{x+6} = -\frac{1}{2}(x+6)$$

$$y - 1 = -\frac{1}{2}x - 3$$

$$y = -\frac{1}{2}x - 2$$

5.
$$\frac{y+a}{x-b} = m$$

$$(x-b)\frac{y+a}{x-b} = m(x-b)$$

$$y + a = mx - mb$$
$$y = mx - mb - a$$

7.
$$\frac{y-1}{x+4} = -\frac{1}{3}$$

$$(x+4)\frac{y-1}{x+4} = -\frac{1}{3}(x+4)$$

$$y - 1 = -\frac{1}{3}x - \frac{4}{3}$$

$$y = -\frac{1}{3}x - \frac{1}{3}$$

9.
$$A = \frac{B}{C}$$

$$C \cdot A = C \cdot \frac{B}{C}$$

$$AC = B$$

$$C = \frac{B}{A}$$

11.
$$\frac{1}{a} + m = \frac{1}{p}$$

$$ap \cdot \frac{1}{a} + ap \cdot m = ap \cdot \frac{1}{p}$$

$$p + apm = a$$
$$p(1 + am) = a$$

$$p = \frac{a}{1 + am}$$

13.
$$F = k\frac{m_1 m_2}{r^2}$$

$$r^2 \cdot F = k\frac{m_1 m_2}{r^2} \cdot r^2$$

$$Fr^2 = km_1m_2$$

$$\frac{Fr^2}{km_2} = \frac{km_1m_2}{km_2}$$

$$m_1 = \frac{r^2F}{km_2}$$

15. $$\frac{1}{a} + \frac{1}{b} = \frac{1}{f}$$

$$abf \cdot \frac{1}{a} + abf \cdot \frac{1}{b} = abf \cdot \frac{1}{f}$$

$$bf + af = ab$$
$$bf = ab - af$$
$$bf = a(b - f)$$

$$\frac{bf}{b - f} = a$$

$$a = \frac{bf}{b - f}$$

17. $$S = \frac{a}{1 - r}$$

$$(1 - r)S = (1 - r)\frac{a}{1 - r}$$

$$S - rS = a$$
$$S = a + rS$$
$$S - a = rS$$

$$\frac{S - a}{S} = r$$

$$r = \frac{S - a}{S}$$

19. $$\frac{P_1V_1}{T_1} = \frac{P_2V_2}{T_2}$$

$$T_1T_2 \cdot \frac{P_1V_1}{T_1} = T_1T_2 \cdot \frac{P_2V_2}{T_2}$$

$$T_2P_1V_1 = T_1P_2V_2$$

$$\frac{P_1V_1T_2}{T_1V_2} = P_2$$

21. $$V = \frac{4}{3}\pi r^2 h$$

$$3V = 3 \cdot \frac{4}{3}\pi r^2 h$$

$$3V = 4\pi r^2 h$$

$$\frac{3V}{4\pi r^2} = h$$

23. Use $A = 12$ and $B = 5$ in the formula $A = \frac{B}{C}$.

$$12 = \frac{5}{C}$$

$$12C = 5$$

$$C = \frac{5}{12}$$

25. Use $p = 6$ and $m = 4$ in the formula $\frac{1}{a} + m = \frac{1}{p}$.

$$\frac{1}{a} + 4 = \frac{1}{6}$$

$$6a \cdot \frac{1}{a} + 6a \cdot 4 = 6a \cdot \frac{1}{6}$$

$$6 + 24a = a$$

$$23a = -6$$

$$a = -\frac{6}{23}$$

27. Use $F = 32$, $r = 4$, $m_1 = 2$, and $m_2 = 6$ in the formula $F = k\frac{m_1m_2}{r^2}$.

$$32 = k\frac{2 \cdot 6}{4^2}$$

$$32 = k\frac{12}{16}$$

$$32 = k\frac{3}{4}$$

$$k = \frac{4}{3} \cdot 32 = \frac{128}{3}$$

29. Use $f = 3$ and $a = 2$ in the formula $\frac{1}{a} + \frac{1}{b} = \frac{1}{f}$.

$$\frac{1}{2} + \frac{1}{b} = \frac{1}{3}$$

$$6b \cdot \frac{1}{2} + 6b \cdot \frac{1}{b} = 6b \cdot \frac{1}{3}$$

$$3b + 6 = 2b$$

$$b = -6$$

31. Use $S = 3/2$ and $r = 1/5$ in the formula $S = \frac{a}{1 - r}$.

$$\frac{3}{2} = \frac{a}{1 - \frac{1}{5}}$$

$$\frac{3}{2} = \frac{a}{\frac{4}{5}}$$

$$a = \frac{3}{2} \cdot \frac{4}{5} = \frac{12}{10} = \frac{6}{5}$$

33. Let $x =$ Frank's rate and $x + 1 =$ Marcie's rate. Since $T = D/R$, Marcie's time is $8/(x + 1)$ and Frank's time is $6/x$. Since their times are equal, we can write the following equation.

$$\frac{6}{x} = \frac{8}{x + 1}$$

$$6(x + 1) = 8x$$
$$6x + 6 = 8x$$
$$6 = 2x$$
$$3 = x$$
$$4 = x + 1$$

Marcie walks 4 mph and Frank walks 3 mph.

35. Let $x =$ Bob's rate and $x - 5 =$ Pat's rate. Since $T = D/R$, Bob's time is $75/x$ and Pat's time is $70/(x - 5)$. Since Pat takes $1/2$ of an hour longer than Bob, we can write the following equation.

$$\frac{75}{x} + \frac{1}{2} = \frac{70}{x - 5}$$

$$2x(x - 5)\frac{75}{x} + 2x(x - 5)\frac{1}{2} = 2x(x - 5)\frac{70}{x - 5}$$

$$150(x - 5) + x(x - 5) = 140x$$
$$150x - 750 + x^2 - 5x = 140x$$
$$x^2 + 5x - 750 = 0$$
$$(x - 25)(x + 30) = 0$$
$$x - 25 = 0 \quad \text{or} \quad x + 30 = 0$$
$$x = 25 \quad \text{or} \quad x = -30$$
$$x - 5 = 20$$

Bob's rate is 25 mph and Pat's rate is 20 mph.

37. Let $x =$ his walking rate and $x + 5 =$ his running rate. Since $T = D/R$, his time walking was $6/x$ and his time running was $8/(x + 5)$. Since his total time was 2 hours, we can write the following equation.

$$\frac{6}{x} + \frac{8}{x + 5} = 2$$

$$x(x + 5)\frac{6}{x} + x(x + 5)\frac{8}{x + 5} = 2x(x + 5)$$

$$6x + 30 + 8x = 2x^2 + 10x$$
$$-2x^2 + 4x + 30 = 0$$
$$x^2 - 2x - 15 = 0$$
$$(x - 5)(x + 3) = 0$$
$$x - 5 = 0 \quad \text{or} \quad x + 3 = 0$$
$$x = 5 \quad \text{or} \quad x = -3$$

His rate walking was 5 mph.

39. Let $x =$ the number of hours for Red to paint the fence by himself. Red paints $1/x$ of the fence per hour and Kiyoshi paints $1/3$ of the fence per hour. Together they paint $1/2$ of the fence per hour. We can write the following equation.

$$\frac{1}{x} + \frac{1}{3} = \frac{1}{2}$$

$$6x \cdot \frac{1}{x} + 6x \cdot \frac{1}{3} = 6x \cdot \frac{1}{2}$$

$$6 + 2x = 3x$$
$$6 = x$$

It would take Red 6 hours to paint the fence by himself.

41. Let $x =$ the number of hours for the dog and the boy working together. The dog destroys $1/2$ of the garden per hour and the boy destroys the whole garden in 1 hour. Since together they destroy $1/x$ of the garden per hour, we can write the following equation.

$$\frac{1}{2} + 1 = \frac{1}{x}$$

$$\frac{3}{2} = \frac{1}{x}$$

$$3x = 2$$

$$x = \frac{2}{3}$$

Working together they can destroy the whole garden in $2/3$ of an hour or 40 minutes.

43. Let $x =$ their time together. Edgar does $1/2$ of the job per hour and Ellen does $1/8$ of the job per hour. Since together they do $1/x$ of the job per hour, we can write the following equation.

$$\frac{1}{2} + \frac{1}{8} = \frac{1}{x}$$

$$8x \cdot \frac{1}{2} + 8x \cdot \frac{1}{8} = 8x \cdot \frac{1}{x}$$

$$4x + x = 8$$
$$5x = 8$$

$$x = \frac{8}{5} \text{ hours}$$

Together they can do the job in 1 hour 36 minutes.

45. Let x = the number of pounds of apples and 18 − x = the number of pounds of bananas. The price per pound for the apples was 9/x and the price per pound for the bananas was 2.40/(18 − x). Since the price per pound for the apples was 3 times the price per pound for the bananas, we can write the following equation.

$$\frac{9}{x} = 3 \cdot \frac{2.40}{18-x}$$

$$\frac{9}{x} = \frac{7.20}{18-x}$$

$$9(18-x) = 7.20x$$
$$162 - 9x = 7.2x$$
$$162 = 16.2x$$
$$10 = x$$
$$8 = 18 - x$$

She bought 8 pounds of bananas and 10 pounds of apples.

47. Let x = the number of gallons used in the small truck and 110 − x = the number of gallons used in the large truck. The small truck got 600/x miles per gallon and the large truck got 800/(110 − x) miles per gallon. Since the small truck got twice as many miles per gallon as the large truck, we can write the following equation.

$$\frac{600}{x} = 2 \cdot \frac{800}{110-x}$$

$$\frac{600}{x} = \frac{1600}{110-x}$$

$$1600x = 66000 - 600x$$
$$2200x = 66000$$
$$x = 30$$
$$110 - x = 80$$

The large truck used 80 gallons of gasoline.

CHAPTER 5 REVIEW

1. $\frac{24}{28} = \frac{2 \cdot 2 \cdot 2 \cdot 3}{2 \cdot 2 \cdot 7} = \frac{6}{7}$

3. $\frac{2a^3c^3}{8a^5c} = \frac{c^{3-1}}{4a^{5-3}} = \frac{c^2}{4a^2}$

5. $\frac{6w-9}{9w-12} = \frac{3(2w-3)}{3(3w-4)} = \frac{2w-3}{3w-4}$

7. $\frac{x^2-1}{3-3x} = \frac{(x-1)(x+1)}{-3(x-1)} = -\frac{x+1}{3}$

9. $\frac{1}{6k} \cdot 3k^2 = \frac{1}{2}k^{2-1} = \frac{1}{2}k$

11. $\frac{2xy}{3} \div y^2 = \frac{2xy}{3} \cdot \frac{1}{y^2} = \frac{2x}{3y}$

13. $\frac{a^2-9}{a-2} \cdot \frac{a^2-4}{a+3}$

$$= \frac{(a-3)(a+3)}{a-2} \cdot \frac{(a-2)(a+2)}{a+3}$$

$$= (a-3)(a+2) = a^2 - a - 6$$

15. $\frac{w-2}{3w} \div \frac{4w-8}{6w} = \frac{w-2}{3w} \cdot \frac{6w}{4w-8}$

$$= \frac{w-2}{3w} \cdot \frac{2 \cdot 3w}{4(w-2)} = \frac{2}{4} = \frac{1}{2}$$

17. Since $36 = 2^2 \cdot 3^2$ and $54 = 2 \cdot 3^3$, the LCD is $2^2 \cdot 3^3$ or 108.

19. Since $6ab^3 = 2 \cdot 3ab^3$ and $8a^7b^2 = 2^3a^7b^2$, the LCD is $2^3 \cdot 3a^7b^3 = 24a^7b^3$.

21. Since $4x = 2^2 \cdot x$ and $6x - 6 = 2 \cdot 3(x-1)$, the LCD is $2^2 \cdot 3x(x-1) = 12x(x-1)$.

23. Since $x^2 - 4 = (x-2)(x+2)$ and $x^2 - x - 2 = (x-2)(x+1)$, the LCD is
$$(x+1)(x-2)(x+2).$$

25. $\frac{5}{12} = \frac{5 \cdot 3}{12 \cdot 3} = \frac{15}{36}$

27. $\frac{2}{3xy} = \frac{2(5x)}{3xy(5x)} = \frac{10x}{15x^2y}$

29. $\frac{5}{y-6} = \frac{5(-2)}{(y-6)(-2)} = \frac{-10}{12-2y}$

31. $\frac{x}{x-1} = \frac{x(x+1)}{(x-1)(x+1)} = \frac{x^2+x}{x^2-1}$

33. $\frac{5}{36} + \frac{9}{28} = \frac{5}{2^23^2} + \frac{9}{2^27}$

$$= \frac{5 \cdot 7}{2^23^2 \cdot 7} + \frac{9 \cdot 3^2}{2^27 \cdot 3^2} = \frac{35}{252} + \frac{81}{252} = \frac{116}{252} = \frac{29}{63}$$

35. $3 - \frac{4}{x} = \frac{3x}{x} - \frac{4}{x} = \frac{3x-4}{x}$

37. $\frac{2}{ab^2} - \frac{1}{a^2b} = \frac{2(a)}{ab^2(a)} - \frac{1(b)}{a^2b(b)}$

$$= \frac{2a}{a^2b^2} - \frac{b}{a^2b^2} = \frac{2a-b}{a^2b^2}$$

39. $\dfrac{9a}{2a-3}+\dfrac{5}{3a-2}$

$=\dfrac{9a(3a-2)}{(2a-3)(3a-2)}+\dfrac{5(2a-3)}{(3a-2)(2a-3)}$

$=\dfrac{27a^2-18a}{(2a-3)(3a-2)}+\dfrac{10a-15}{(2a-3)(3a-2)}$

$=\dfrac{27a^2-8a-15}{(2a-3)(3a-2)}$

41. $\dfrac{1}{a-8}-\dfrac{2}{8-a}=\dfrac{1}{a-8}-\dfrac{2(-1)}{(8-a)(-1)}$

$=\dfrac{1}{a-8}-\dfrac{-2}{a-8}=\dfrac{3}{a-8}$

43. $\dfrac{3}{2x-4}+\dfrac{1}{x^2-4}=\dfrac{3}{2(x-2)}+\dfrac{1}{(x-2)(x+2)}$

$=\dfrac{3(x+2)}{2(x-2)(x+2)}+\dfrac{1(2)}{(x-2)(x+2)(2)}$

$=\dfrac{3x+6+2}{2(x+2)(x-2)}=\dfrac{3x+8}{2(x+2)(x-2)}$

45. $\dfrac{\frac{1}{2}-\frac{3}{4}}{\frac{2}{3}+\frac{1}{2}}=\dfrac{\left(\frac{1}{2}-\frac{3}{4}\right)(12)}{\left(\frac{2}{3}+\frac{1}{2}\right)(12)}=\dfrac{6-9}{8+6}=\dfrac{-3}{14}=-\dfrac{3}{14}$

47. $\dfrac{\frac{1}{a}+\frac{2}{3b}}{\frac{1}{2b}-\frac{3}{a}}=\dfrac{\left(\frac{1}{a}+\frac{2}{3b}\right)(6ab)}{\left(\frac{1}{2b}-\frac{3}{a}\right)(6ab)}=\dfrac{6b+4a}{3a-18b}$

49. $\dfrac{\left(\dfrac{1}{x-2}-\dfrac{3}{x+3}\right)(x-2)(x+3)}{\left(\dfrac{2}{x+3}+\dfrac{1}{x-2}\right)(x-2)(x+3)}$

$=\dfrac{x+3-3(x-2)}{2(x-2)+1(x+3)}=\dfrac{-2x+9}{3x-1}$

51. $\dfrac{\dfrac{x-1}{x-3}(x-3)(x+2)}{\left(\dfrac{1}{x^2-x-6}-\dfrac{4}{x+2}\right)(x-3)(x+2)}$

$=\dfrac{(x-1)(x+2)}{1-4(x-3)}=\dfrac{x^2+x-2}{-4x+13}$

53. $\dfrac{-2}{5}=\dfrac{3}{x}$

$-2x=15$

$x=-\dfrac{15}{2}$

The solution to the equation is $-\dfrac{15}{2}$.

55. $\dfrac{14}{a^2-1}+\dfrac{1}{a-1}=\dfrac{3}{a+1}$

$(a-1)(a+1)\dfrac{14}{a^2-1}+(a-1)(a+1)\dfrac{1}{a-1}$

$=(a-1)(a+1)\dfrac{3}{a+1}$

$14+a+1=3(a-1)$

$15+a=3a-3$

$18=2a$

$9=a$

The solution to the equation is 9.

57. $z-\dfrac{3z}{2-z}=\dfrac{6}{z-2}$

$(z-2)z-(z-2)\dfrac{3z}{2-z}=(z-2)\dfrac{6}{z-2}$

$z^2-2z-(-1)3z=6$

$z^2-2z+3z=6$

$z^2+z-6=0$

$(z+3)(z-2)=0$

$z+3=0$ or $z-2=0$

$z=-3$ or $z=2$

Since the value of $z-2$ is 0 when $z=2$, 2 is not a solution to the equation. The solution to the equation is -3.

59. $\dfrac{3}{x}=\dfrac{2}{7}$

$2x=21$

$x=\dfrac{21}{2}$

The solution to the equation is $\dfrac{21}{2}$.

61. $\dfrac{2}{w-3}=\dfrac{5}{w}$

$5w-15=2w$

$3w=15$

$w=5$

The solution to the proportion is 5.

63. Let $x=$ the number of private automobiles. Since the ratio of taxis to private automobiles was 15 to 2, we can write the following proportion.

$\dfrac{15}{2}=\dfrac{60}{x}$

$15x=120$

$x=8$

There were 8 private automobiles.

65. Let $x =$ the number of cups of rice and $x + 28 =$ the number of cups of water. Since the ratio of water to rice is 2 to 1, we can write the following proportion.

$$\frac{2}{1} = \frac{x + 28}{x}$$
$$2x = x + 28$$
$$x = 28$$
$$x + 28 = 56$$

He used 56 cups of water and 28 cups of rice.

67. $\frac{y - b}{m} = x$

$$y - b = mx$$
$$y = mx + b$$

69. $F = \frac{mv + 1}{m}$

$$Fm = mv + 1$$
$$Fm - mv = 1$$
$$m(F - v) = 1$$
$$m = \frac{1}{F - v}$$

71. $\frac{y + 1}{x - 3} = 4$

$$y + 1 = 4x - 12$$
$$y = 4x - 13$$

73. Let $x =$ the number of hours for Stacy or Tracy to assemble the puzzle working alone. Since Stacy works twice as fast as Fred, it takes Fred $2x$ hours to assemble the puzzle by himself. Since Stacy does $1/x$ of the puzzle per hour, Tracy does $1/x$ of the puzzle per hour and Fred does $1/(2x)$ of the puzzle per hour, we can write the following equation.

$$\frac{1}{x} + \frac{1}{x} + \frac{1}{2x} = \frac{1}{40}$$
$$40x \cdot \frac{1}{x} + 40x \cdot \frac{1}{x} + 40x \cdot \frac{1}{2x} = 40x \cdot \frac{1}{40}$$
$$40 + 40 + 20 = x$$
$$100 = x$$
$$200 = 2x$$

It would take Fred 200 hours to assemble the puzzle by himself.

75. Let $x =$ the number of cars owned by Ernie and $x + 10 =$ the number of cars owned by Bert. Ernie had $0.36x$ new cars and Bert had $0.25(x + 10)$ new cars. Since the total number of new cars was 33, we can write the following equation.

$$0.36x + 0.25(x + 10) = 33$$
$$0.36x + 0.25x + 2.5 = 33$$
$$0.61x = 30.5$$
$$x = \frac{30.5}{0.61} = 50$$
$$x + 10 = 60$$

Bert had 60 cars and Ernie had 50 cars before the merger.

77. Let $x =$ the amount used in air conditioning and $x + 30,000 =$ the amount used in cleaning agents.

$$\frac{x}{x + 30,000} = \frac{20\%}{24\%}$$
$$\frac{x}{x + 30,000} = \frac{5}{6}$$
$$6x = 5x + 150,000$$
$$x = 150,000$$

So 150,000 metric tons are used in vehicle air conditioning.

79. $\frac{5}{x} = \frac{5 \cdot 2}{x \cdot 2} = \frac{10}{2x}$

81. $\frac{2}{a - 5} = \frac{2(-1)}{(a - 5)(-1)} = \frac{-2}{5 - a}$

83. $3 = 3 \cdot \frac{x}{x} = \frac{3x}{x}$

85. $m \div \frac{1}{2} = m \cdot 2 = 2m$

87. $2a \div \frac{1}{6} = 2a \cdot 6 = 12a$

89. $\frac{a - 1}{a^2 - 1} = \frac{(a - 1)1}{(a - 1)(a + 1)} = \frac{1}{a + 1}$

91. $\frac{1}{a} - \frac{1}{5} = \frac{1 \cdot 5}{a \cdot 5} - \frac{1 \cdot a}{5 \cdot a} = \frac{5}{5a} - \frac{a}{5a} = \frac{5 - a}{5a}$

93. $\frac{a}{2} - 1 = \frac{a}{2} - \frac{2}{2} = \frac{a - 2}{2}$

95. $(a - b) \div (-1) = (a - b)(-1) = b - a$

97. $\frac{\frac{1}{5a}}{2} = \frac{1}{5a} \cdot \frac{1}{2} = \frac{1}{10a}$

99. $\frac{1}{x} + \frac{1}{2x} = \frac{1 \cdot 2}{x \cdot 2} + \frac{1}{2x} = \frac{2}{2x} + \frac{1}{2x} = \frac{3}{2x}$

101. $\frac{2}{3xy} + \frac{1}{6x} = \frac{2 \cdot 2}{3xy \cdot 2} + \frac{1 \cdot y}{6x \cdot y} = \frac{4}{6xy} + \frac{y}{6xy}$
$$= \frac{4 + y}{6xy}$$

103. $\frac{5}{a - 5} - \frac{3}{5 - a} = \frac{5}{a - 5} - \frac{3(-1)}{(5 - a)(-1)}$
$$= \frac{5}{a - 5} - \frac{-3}{a - 5} = \frac{8}{a - 5}$$

105.

$$\frac{2}{x-1} - \frac{2}{x} = 1$$

$$x(x-1)\frac{2}{x-1} - x(x-1)\frac{2}{x} = x(x-1)1$$

$$2x - 2(x-1) = x^2 - x$$
$$2 = x^2 - x$$
$$0 = x^2 - x - 2$$
$$0 = (x-2)(x+1)$$
$$x - 2 = 0 \quad \text{or} \quad x + 1 = 0$$
$$x = 2 \quad \text{or} \quad x = -1$$

The solutions to the equation are -1 and 2.

107. $\dfrac{-3}{x+2} \cdot \dfrac{5x+10}{9} = \dfrac{-3}{x+2} \cdot \dfrac{5(x+2)}{3 \cdot 3} = -\dfrac{5}{3}$

109.

$$\frac{1}{-3} = \frac{-2}{x}$$

$$1x = (-2)(-3)$$

$$x = 6$$

The solution to the equation is 6.

111. $\dfrac{ax + am + 3x + 3m}{a^2 - 9} \div \dfrac{2x + 2m}{a - 3}$

$$= \frac{(a+3)(x+m)}{(a-3)(a+3)} \cdot \frac{a-3}{2(x+m)} = \frac{1}{2}$$

113. $\dfrac{2}{x^2 - 25} + \dfrac{1}{x^2 - 4x - 5}$

$$= \frac{2}{(x-5)(x+5)} + \frac{1}{(x-5)(x+1)}$$

$$= \frac{2(x+1)}{(x-5)(x+5)(x+1)} + \frac{1(x+5)}{(x-5)(x+1)(x+5)}$$

$$= \frac{3x+7}{(x-5)(x+5)(x+1)}$$

115. $\dfrac{-3}{a^2 - 9} - \dfrac{2}{a^2 + 5a + 6}$

$$= \frac{-3}{(a-3)(a+3)} - \frac{2}{(a+2)(a+3)}$$

$$= \frac{-3(a+2)}{(a-3)(a+3)(a+2)} - \frac{2(a-3)}{(a+2)(a+3)(a-3)}$$

$$= \frac{-3a-6}{(a-3)(a+3)(a+2)} - \frac{2a-6}{(a-3)(a+3)(a+2)}$$

$$= \frac{-5a}{(a-3)(a+3)(a+2)}$$

117. $\dfrac{1}{a^2 - 1} + \dfrac{2}{1-a} = \dfrac{3}{a+1}$

$$(a+1)(a-1)\frac{1}{a^2-1} + (a+1)(a-1)\frac{2}{1-a}$$

$$= (a+1)(a-1)\frac{3}{a+1}$$

$$1 + (a+1)(-1)2 = (a-1)3$$
$$1 - 2a - 2 = 3a - 3$$
$$-1 - 2a = 3a - 3$$
$$2 = 5a$$
$$\frac{2}{5} = a$$

The solution to the equation is $\frac{2}{5}$.

CHAPTER 5 TEST

1. We cannot use any number for which $x^2 - 1 = 0$.

$$x^2 - 1 = 0$$
$$(x-1)(x+1) = 0$$
$$x - 1 = 0 \quad \text{or} \quad x + 1 = 0$$
$$x = 1 \quad \text{or} \quad x = -1$$

We cannot use -1 or 1 in place of x.

2. We cannot use any number for which $2 - 3x = 0$.

$$2 - 3x = 0$$
$$-3x = -2$$
$$x = \frac{2}{3}$$

We cannot use $\frac{2}{3}$ in place of x.

3. We cannot use 0 for x because if $x = 0$, then the denominator of $1/x$ is 0.

4. $\dfrac{2}{15} - \dfrac{4}{9} = \dfrac{2 \cdot 3}{15 \cdot 3} - \dfrac{4 \cdot 5}{9 \cdot 5} = \dfrac{6}{45} - \dfrac{20}{45} = -\dfrac{14}{45}$

5. $\dfrac{1}{y} + 3 = \dfrac{1}{y} + \dfrac{3y}{y} = \dfrac{1+3y}{y}$

6. $\dfrac{3}{a-2} - \dfrac{1}{2-a} = \dfrac{3}{a-2} - \dfrac{1(-1)}{(2-a)(-1)}$

$$= \frac{3}{a-2} - \frac{-1}{a-2} = \frac{4}{a-2}$$

7. $\dfrac{2}{x^2 - 4} - \dfrac{3}{x^2 + x - 2}$

$$= \frac{2}{(x-2)(x+2)} - \frac{3}{(x+2)(x-1)}$$

$$= \frac{2(x-1)}{(x-2)(x+2)(x-1)} - \frac{3(x-2)}{(x+2)(x-1)(x-2)}$$

$$= \frac{2x-2}{(x+2)(x-2)(x-1)} - \frac{3x-6}{(x+2)(x-2)(x-1)}$$

$$= \frac{-x+4}{(x+2)(x-2)(x-1)}$$

8. $\dfrac{m^2-1}{(m-1)^2} \cdot \dfrac{2m-2}{3m+3} = \dfrac{(m-1)(m+1)}{(m-1)^2} \cdot \dfrac{2(m-1)}{3(m+1)}$

$$= \dfrac{2}{3}$$

9. $\dfrac{a-b}{3} \div \dfrac{b^2-a^2}{6} = \dfrac{a-b}{3} \cdot \dfrac{2 \cdot 3}{(b-a)(b+a)}$

$$= \dfrac{-2}{a+b}$$

10. $\dfrac{5a^2b}{12a} \cdot \dfrac{2a^3b}{15ab^6} = \dfrac{5 \cdot 2a^5b^2}{2 \cdot 2 \cdot 3 \cdot 3 \cdot 5a^2b^6} = \dfrac{a^3}{18b^4}$

11. $\dfrac{\frac{2}{3}+\frac{4}{5}}{\frac{2}{5}-\frac{3}{2}} = \dfrac{\left(\frac{2}{3}+\frac{4}{5}\right)(30)}{\left(\frac{2}{5}-\frac{3}{2}\right)(30)} = \dfrac{20+24}{12-45} = \dfrac{44}{-33} = -\dfrac{4}{3}$

12. $\dfrac{\frac{2}{x}+\frac{1}{x-2}}{\frac{1}{x-2}-\frac{3}{x}} = \dfrac{\left(\frac{2}{x}+\frac{1}{x-2}\right)(x)(x-2)}{\left(\frac{1}{x-2}-\frac{3}{x}\right)(x)(x-2)}$

$$= \dfrac{2(x-2)+x}{x-3(x-2)} = \dfrac{3x-4}{-2x+6}$$

13. $\dfrac{3}{x} = \dfrac{7}{5}$

$$7x = 15$$

$$x = \dfrac{15}{7}$$

The solution to the equation is $\dfrac{15}{7}$.

14. $\dfrac{x}{x-1} - \dfrac{3}{x} = \dfrac{1}{2}$

$$2x(x-1)\dfrac{x}{x-1} - 2x(x-1)\dfrac{3}{x} = 2x(x-1)\dfrac{1}{2}$$

$$2x^2 - 6(x-1) = x(x-1)$$
$$2x^2 - 6x + 6 = x^2 - x$$
$$x^2 - 5x + 6 = 0$$
$$(x-2)(x-3) = 0$$
$$x-2 = 0 \quad \text{or} \quad x-3 = 0$$
$$x = 2 \quad \text{or} \quad x = 3$$

The solutions to the equation are 2 and 3.

15. $\dfrac{1}{x} + \dfrac{1}{6} = \dfrac{1}{4}$

$$12x \cdot \dfrac{1}{x} + 12x \cdot \dfrac{1}{6} = 12x \cdot \dfrac{1}{4}$$

$$12 + 2x = 3x$$
$$12 = x$$

The solution to the equation is 12.

16. $\dfrac{y-3}{x+2} = \dfrac{-1}{5}$

$$y - 3 = -\dfrac{1}{5}(x+2)$$

$$y - 3 = -\dfrac{1}{5}x - \dfrac{2}{5}$$

$$y = -\dfrac{1}{5}x - \dfrac{2}{5} + 3$$

$$y = -\dfrac{1}{5}x + \dfrac{13}{5}$$

17. $M = \dfrac{1}{3}b(c+d)$

$$3M = b(c+d)$$
$$3M = bc + bd$$
$$3M - bd = bc$$
$$\dfrac{3M-bd}{b} = c$$
$$c = \dfrac{3M-bd}{b}$$

18. Let x = the number of minutes it takes for them to do the job together. Reginald does 1/12 of the job per minute and Norman does 1/18 of the job per minute. Since together they get 1/x of the job done per minute, we can write the following equation.

$$\dfrac{1}{12} + \dfrac{1}{18} = \dfrac{1}{x}$$

$$36x \cdot \dfrac{1}{12} + 36x \cdot \dfrac{1}{18} = 36x \cdot \dfrac{1}{x}$$

$$3x + 2x = 36$$
$$5x = 36$$
$$x = \dfrac{36}{5} = 7.2$$

It takes them 7.2 minutes to complete the job when working together.

19. Let x = Brenda's rate and x + 5 = Randy's rate. Since Brenda rode 30 miles, her time was 30/x hours. Since Randy rode 60 miles, his time was 60/(x + 5) hours. Since Randy traveled one hour longer, we can write the following equation.

$$\dfrac{30}{x} + 1 = \dfrac{60}{x+5}$$

$$x(x+5)\dfrac{30}{x} + x(x+5)1 = x(x+5)\dfrac{60}{x+5}$$

$$30x + 150 + x^2 + 5x = 60x$$
$$x^2 - 25x + 150 = 0$$
$$(x-15)(x-10) = 0$$
$$x - 15 = 0 \quad \text{or} \quad x - 10 = 0$$
$$x = 15 \quad \text{or} \quad x = 10$$
$$x + 5 = 20 \quad \text{or} \quad x + 5 = 15$$

There are two possible answers to this problem. It could be that Brenda rode at 15 mph and Randy 20 mph, or it could be that Brenda rode at 10 mph and Randy 15 mph.

20. Let x = the dollar value of imports. Since the ratio of exports to imports was 2 to 3, we can write the following equation.

$$\frac{2}{3} = \frac{48}{x}$$
$$2x = 144$$
$$x = 72$$

The value of imports was 72 billion dollars.

Tying It All Together Chapters 1-5

1. $3x - 2 = 5$
$$3x = 7$$
$$x = \frac{7}{3}$$

The solution to the equation is $\frac{7}{3}$.

2. $\frac{3}{5}x = -2$
$$\frac{5}{3} \cdot \frac{3}{5}x = -2 \cdot \frac{5}{3}$$
$$x = -\frac{10}{3}$$

The solution to the equation is $-\frac{10}{3}$.

3. $2(x - 2) = 4x$
$$2x - 4 = 4x$$
$$-4 = 2x$$
$$-2 = x$$

The solution to the equation is -2.

4. $2(x - 2) = 2x$
$$2x - 4 = 2x$$
$$-4 = 0$$

There is no solution to the equation.

5. $2(x + 3) = 6x + 6$
$$2x + 6 = 6x + 6$$
$$2x = 6x$$
$$0 = 4x$$
$$0 = x$$

The solution to the equation is 0.

6. $2(3x + 4) + x^2 = 0$
$$6x + 8 + x^2 = 0$$
$$x^2 + 6x + 8 = 0$$
$$(x + 2)(x + 4) = 0$$
$$x + 2 = 0 \quad \text{or} \quad x + 4 = 0$$
$$x = -2 \quad \text{or} \quad x = -4$$

The solutions to the equation are -4 and -2.

7. $4x - 4x^3 = 0$
$$4x(1 - x^2) = 0$$
$$4x(1 - x)(1 + x) = 0$$
$$4x = 0 \quad \text{or} \quad 1 - x = 0 \quad \text{or} \quad 1 + x = 0$$
$$x = 0 \quad \text{or} \quad 1 = x \quad \text{or} \quad x = -1$$

The solutions to the equation are -1, 0, and 1.

8. $\frac{3}{x} = \frac{-2}{5}$
$$-2x = 15$$
$$x = -\frac{15}{2}$$

The solution to the equation is $-\frac{15}{2}$.

9. $\frac{3}{x} = \frac{x}{12}$
$$x^2 = 36$$
$$x^2 - 36 = 0$$
$$(x - 6)(x + 6) = 0$$
$$x - 6 = 0 \quad \text{or} \quad x + 6 = 0$$
$$x = 6 \quad \text{or} \quad x = -6$$

The solutions to the equation are -6 and 6.

10. $\frac{x}{2} = \frac{4}{x - 2}$
$$x(x - 2) = 8$$
$$x^2 - 2x = 8$$
$$x^2 - 2x - 8 = 0$$
$$(x - 4)(x + 2) = 0$$
$$x - 4 = 0 \quad \text{or} \quad x + 2 = 0$$
$$x = 4 \quad \text{or} \quad x = -2$$

The solutions to the equation are -2 and 4.

11. $18 \cdot \frac{w}{18} - 18 \cdot \frac{w - 1}{9} = 18 \cdot \frac{4 - w}{6}$
$$w - 2(w - 1) = 3(4 - w)$$
$$-w + 2 = 12 - 3w$$
$$2w = 10$$
$$w = 5$$

The solution to the equation is 5.

12. $8(x + 1)\left(\frac{x}{x + 1} + \frac{1}{2(x + 1)}\right) = 8(x + 1)\frac{7}{8}$
$$8x + 4 = 7x + 7$$
$$x = 3$$

The solutions to the equation is 3.

13. $2x + 3y = c$
$$3y = -2x + c$$
$$y = \frac{-2x + c}{3}$$
$$y = \frac{c - 2x}{3}$$

14. $\frac{y - 3}{x - 5} = \frac{1}{2}$
$$y - 3 = \frac{1}{2}(x - 5)$$
$$y - 3 = \frac{1}{2}x - \frac{5}{2}$$
$$y = \frac{1}{2}x + \frac{1}{2}$$

15.
$$2y = ay + c$$
$$2y - ay = c$$
$$y(2 - a) = c$$
$$y = \frac{c}{2-a}$$

16.
$$\frac{A}{y} = \frac{C}{B}$$
$$Cy = AB$$
$$y = \frac{AB}{C}$$

17.
$$\frac{A}{y} + \frac{1}{3} = \frac{B}{y}$$
$$3y \cdot \frac{A}{y} + 3y \cdot \frac{1}{3} = 3y \cdot \frac{B}{y}$$
$$3A + y = 3B$$
$$y = 3B - 3A$$

18.
$$\frac{A}{y} - \frac{1}{2} = \frac{1}{3}$$
$$6y \cdot \frac{A}{y} - 6y \cdot \frac{1}{2} = 6y \cdot \frac{1}{3}$$
$$6A - 3y = 2y$$
$$6A = 5y$$
$$\frac{6A}{5} = y$$
$$y = \frac{6A}{5}$$

19.
$$3y - 5ay = 8$$
$$y(3 - 5a) = 8$$
$$y = \frac{8}{3 - 5a}$$

20.
$$y^2 - By = 0$$
$$y(y - B) = 0$$
$$y = 0 \quad \text{or} \quad y - B = 0$$
$$y = 0 \quad \text{or} \qquad y = B$$

21.
$$A = \tfrac{1}{2}h(b + y)$$
$$2A = h(b + y)$$
$$2A = hb + hy$$
$$2A - hb = hy$$
$$\frac{2A - hb}{h} = y$$
$$y = \frac{2A - hb}{h}$$

22.
$$2(b + y) = b$$
$$2b + 2y = b$$
$$2y = -b$$
$$y = -\frac{b}{2}$$

23. $b^2 - 4ac = 2^2 - 4(1)(-15) = 4 - (-60) = 64$

24. $b^2 - 4ac = 8^2 - 4(1)(12) = 64 - 48 = 16$

25. $b^2 - 4ac = 5^2 - 4(2)(-3) = 25 - (-24) = 49$

26. $b^2 - 4ac = 7^2 - 4(6)(-3) = 49 - (-72) = 121$

27. $(3x - 5) - (5x - 3) = 3x - 5 - 5x + 3$
$$= -2x - 2$$

28. $(2a - 5)(a - 3) = 2a^2 - 5a - 6a + 15$
$$= 2a^2 - 11a + 15$$

29. $x^7 \div x^3 = x^{7-3} = x^4$

30. $\dfrac{x-3}{5} + \dfrac{x+4}{5} = \dfrac{x-3+x+4}{5} = \dfrac{2x+1}{5}$

31. $\dfrac{1}{2} \cdot \dfrac{1}{x} = \dfrac{1 \cdot 1}{2 \cdot x} = \dfrac{1}{2x}$

32. $\dfrac{1}{2} + \dfrac{1}{x} = \dfrac{1 \cdot x + 2 \cdot 1}{2 \cdot x} = \dfrac{x+2}{2x}$

33. $\dfrac{1}{2} \div \dfrac{1}{x} = \dfrac{1}{2} \cdot x = \dfrac{x}{2}$

34. $\dfrac{1}{2} - \dfrac{1}{x} = \dfrac{1 \cdot x - 2 \cdot 1}{2x} = \dfrac{x-2}{2x}$

35. $\dfrac{x-3}{5} - \dfrac{x+4}{5} = \dfrac{x-3-x-4}{5} = -\dfrac{7}{5}$

36. $\dfrac{3a}{2} \div 2 = \dfrac{3a}{2} \cdot \dfrac{1}{2} = \dfrac{3a}{4}$

37. $(x - 8)(x + 8) = x^2 - 8^2 = x^2 - 64$

38. $3x(x^2 - 7) = 3x^3 - 21x$

39. $2a^5 \cdot 5a^9 = 2 \cdot 5 \cdot a^{5+9} = 10a^{14}$

40. $x^2 \cdot x^8 = x^{2+8} = x^{10}$

41. $(k - 6)^2 = k^2 - 2 \cdot k \cdot 6 + 6^2 = k^2 - 12k + 36$

42. $(j + 5)^2 = j^2 + 2 \cdot j \cdot 5 + 5^2 = j^2 + 10j + 25$

43. $(g - 3) \div (3 - g) = \dfrac{g-3}{3-g} = \dfrac{-1(3-g)}{3-g} = -1$

44. $(6x^3 - 8x^2) \div (2x) = \dfrac{6x^3}{2x} - \dfrac{8x^2}{2x} = 3x^2 - 4x$

45. $P = \dfrac{1}{1+r} + \dfrac{1}{(1+r)^2}$

$$= \frac{1(1+r)}{(1+r)(1+r)} + \frac{1}{(1+r)^2} = \frac{r+2}{(1+r)^2}$$

If $r = 7\%$, then $P = \dfrac{0.07 + 2}{(1 + 0.07)^2} \approx 1.8080$

or $1.81.

93

6.1 WARM-UPS

1. False, because $2(4) - 3(2) \neq -8$.
2. False, because $(1, 5)$ satisfies $y = x + 4$, but $(5, 1)$ does not.
3. False, because the origin is not considered to be in any quadrant.
4. False, because the point $(4, 0)$ is on the x-axis.
5. True, because we have agreed that when a variable is multiplied by 0, it may be omitted.
6. True, because all of the points with an x-coordinate of -5 form a vertical line.
7. True, because the graph consists of all points with a y-coordinate of 6.
8. False, because if $(0, 2.5)$ is the y-intercept
9. False, because the point $(5, -3)$ is in quadrant IV.
10. True, because every point with a y-coordinate of 0 is on the x-axis.

6.1 EXERCISES

1. If $x = 0$ in $y = 3x + 9$, then $y = 3(0) + 9 = 9$.
If $y = 24$ in $y = 3x + 9$, then $24 = 3x + 9$, $15 = 3x$, or $x = 5$.
If $x = 2$ in $y = 3x + 9$, then $y = 3(2) + 9 = 15$.
The ordered pairs are $(0, 9)$, $(5, 24)$, and $(2, 15)$.

3. If $x = 0$ in $y = -3x - 7$, then $y = -3(0) - 7 = -7$. If $x = -4$ in $y = -3x - 7$, then $y = -3(-4) - 7 = 5$.
Replace y by -1 in $y = -3x - 7$:
$$-1 = -3x - 7$$
$$6 = -3x$$
$$-2 = x$$

The ordered pairs are $(0, -7)$, $(-4, 5)$ and $(-2, -1)$.

5. If $x = 0$ in $y = -12x + 5$, then $y = -12(0) + 5 = 5$.
If $x = 10$, in $y = -12x + 5$, then $y = -12(10) + 5 = -115$.
Replace y by 17 in $y = -12x + 5$:

$$17 = -12x + 5$$
$$12 = -12x$$
$$-1 = x$$

The ordered pairs are $(0, 5)$, $(10, -115)$, and $(-1, 17)$.

7. The ordered pairs are $(3, 0)$, $(0, -2)$, and $(12, 6)$.
9. The ordered pairs are $(5, -3)$, $(5, 5)$, and $(5, 0)$.
11. To plot $(1, 5)$, start at the origin and move 1 unit to the right and 5 units up.
13. To plot $(-2, 1)$ start at the origin and move 2 units to the left and then 1 unit up.
15. To plot $(3, -1/2)$, start at the origin and move 3 units to the right and 1/2 unit down.
17. To plot $(-2, -4)$, start at the origin and move 2 units to the left and 4 units down.
19. To plot $(0, 3)$ move 3 units up from the origin.
21. To plot $(-3, 0)$, move 3 units to the left of the origin.
23. To plot $(\pi, 1)$ move approximately 3.14 units to the right of the origin, and then 1 unit up.
25. To plot $(1.4, 4)$, start at the origin and move approximately 1.4 units to the right and then move 4 units up.

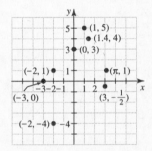

Graph for 11-25 odd.

27. Select five different values for x and use $y = x + 1$ to find the y-coordinates. The ordered pairs $(0, 1)$, $(1, 2)$, $(2, 3)$, $(-1, 0)$, and $(-2, -1)$ satisfy the equation. Plot these points and draw a line through them.

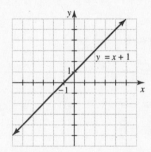

94

29. Select five different values for x and use $y = 2x + 1$ to find the y-coordinates. The ordered pairs $(0, 1)$, $(1, 3)$, $(2, 5)$, $(-1, -1)$, and $(-2, -3)$ satisfy the equation. Plot these points and draw a line through them.

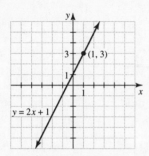

31. Select five different values for x and use $y = 3x - 2$ to find the y-coordinates. The ordered pairs $(0, -2)$, $(1, 1)$, $(2, 4)$, $(3, 7)$, and $(-1, -5)$, satisfy the equation. Plot these points and draw a line through them.

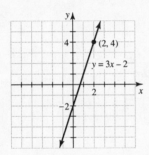

33. Select five different values for x and use $y = x$ to find the y-coordinates. The ordered pairs $(0, 0)$, $(1, 1)$, $(2, 2)$, $(-1, -1)$, and $(-2, -2)$ satisfy the equation. Plot these points and draw a line through them.

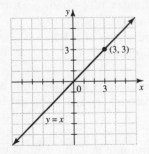

35. Select five different values for x and use $y = 1 - x$ to find the y-coordinates. The ordered pairs $(0, 1)$, $(1, 0)$, $(2, -1)$, $(-1, 2)$, and $(-2, 3)$ satisfy the equation. Plot these points and draw a line through them.

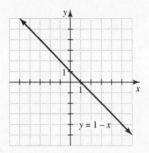

37. Select five different values for x and use $y = -2x + 3$ to find the y-coordinates. The ordered pairs $(0, 3)$, $(1, 1)$, $(2, -1)$, $(-1, 5)$, and $(-2, 7)$ satisfy the equation. Plot these points and draw a line through them.

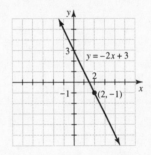

39. Select five different values for x and use $y = -3$ to find the y-coordinates. The ordered pairs $(0, -3)$, $(1, -3)$, $(2, -3)$, $(-1, -3)$, and $(-2, -3)$ satisfy the equation. Plot these points and draw a line through them.

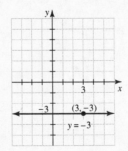

41. The equation $x = 2$ is equivalent to $0 \cdot y + x = 2$. The ordered pairs $(2, 1)$, $(2, 2)$, $(2, 3)$, $(2, 0)$, and $(2, -1)$ satisfy the equation. Plot these points and draw a line through them.

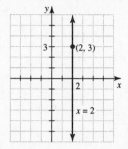

43. The equation $2x + y = 5$ is equivalent to $y = -2x + 5$. Select five different values for x and use $y = -2x + 5$ to find the y-coordinates. The ordered pairs $(0, 5)$, $(1, 3)$, $(2, 1)$, $(-1, 7)$, and $(3, -1)$ satisfy the equation. Plot these points and draw a line through them.

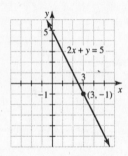

45. The equation $x + 2y = 4$ is equivalent to $y = -(1/2)x + 2$. The ordered pairs $(0, 2)$, $(2, 2)$, $(4, 0)$, $(-2, 3)$, and $(-4, 4)$ satisfy the equation. Plot these points and draw a line through them.

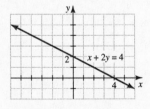

47. The equation $x - 3y = 6$ is equivalent to $y = \frac{1}{3}x - 2$. The ordered pairs $(0, -2)$, $(3, -1)$, $(6, 0)$, $(-3, -3)$, and $(-6, -4)$ satisfy the equation. Plot these points and draw a line through them.

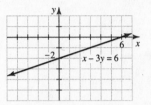

49. Select five different values for x and use $y = 0.36x + 0.4$ to find the y-coordinates. The ordered pairs $(0, 0.4)$, $(1, 0.76)$, $(2, 1.12)$, $(-1, 0.04)$, and $(-2, -0.32)$ satisfy the equation. Plot these points and draw a line through them.

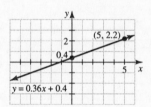

51. To get to $(-3, 45)$ we start at the origin and go to the left and then up. The point is in quadrant II.

53. Since the y-coordinate of $(-3, 0)$ is 0, the point is on the x-axis.

55. Since both coordinates of $(-2.36, -5)$ are negative, the point is in quadrant III.

57. Since both coordinates of $(3.4, 8.8)$ are positive, the point is in quadrant I.

59. Since the first coordinate is negative and the second coordinate positive, $(-1/2, 50)$ is in quadrant II.

61. Since the first coordinate is 0, $(0, -99)$ is on the y-axis.

63. Select five different values for x and use $y = x + 1200$ to find the y-coordinates. The ordered pairs $(0, 1200)$, $(1, 1201)$, $(2, 1202)$, $(-1, 1199)$, and $(-2, 1198)$ satisfy the equation. Plot these points and draw a line through them.

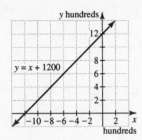

65. Select five different values for x and use $y = 50x - 2000$ to find the y-coordinates. The ordered pairs $(0, -2000)$, $(1, -1950)$, $(2, -1900)$, $(-1, -2050)$, and $(-2, -2100)$ satisfy the equation. Plot these points and draw a line through them.

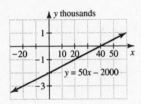

67. Select five different values for x and use $y = -400x + 2000$ to find the y-coordinates. The ordered pairs $(0, 2000)$, $(1, 1600)$, $(2, 1200)$, $(-1, 2400)$, and $(-2, 2800)$ satisfy the equation. Plot these points and draw a line through them.

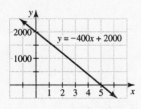

69. If $x = 0$ in $3x + 2y = 6$, then $0 + 2y = 6$, or $y = 3$. If $y = 0$ in $3x + 2y = 6$, then $3x + 0 = 6$, or $x = 2$. Draw a line through the intercepts $(0, 3)$ and $(2, 0)$.

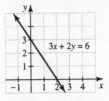

71. If $x = 0$ in $x - 4y = 4$, then $0 + 4y = 4$, or $y = 1$. If $y = 0$ in $x - 4y = 4$, then $x - 0 = 4$, or $x = 4$. Draw a line through the intercepts $(0, 1)$ and $(4, 0)$.

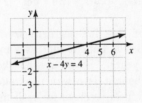

73. If $x = 0$ in $y = \frac{3}{4}x - 9$, then $y = -9$. If $y = 0$, then $0 = \frac{3}{4}x - 9$, or $x = 12$. Draw a line through the intercepts $(0, -9)$ and $(12, 0)$.

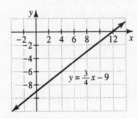

75. If $x = 0$ in $\frac{1}{2}x + \frac{1}{4}y = 1$, then $\frac{1}{2} \cdot 0 + \frac{1}{4}y = 1$, or $y = 4$. If $y = 0$, then $\frac{1}{2}x + 0 = 1$, or $x = 2$. Draw a line through $(0, 4)$ and $(2, 0)$.

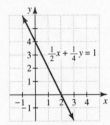

77. The variable x represents the number of radio ads and y represents the number of TV ads. The solutions to $300x + 400y = 24{,}000$ or $3x + 4y = 240$ that are whole numbers are $(0, 60)$, $(4, 57)$, $(8, 54)$, $(12, 51)$, $(16, 48)$, $(20, 45)$, $(24, 42)$, ... $(76, 3)$, $(80, 0)$. Counting these ordered pairs, we get 21 solutions.

79. Since the slope of the line segment from $(62, 70)$ to $(64, 80)$ is 5, the benefit increases by 5% per year for those years. So at age 63 the benefit is 75%. Full benefit is attained at age 67. More than full benefit is received for retiring at ages 68 and up.

6.2 WARM-UPS

1. True.

2. True, because that is the definition of slope.

3. False, because vertical lines do not have slope.

4. True, because to get to $(1, 1)$ from the origin you must rise 1 and run 1.

5. False, because slope can be negative.

6. False, because a line perpendicular to a line with slope 2 must have slope $-1/2$.

7. False, because for $(0, 3)$ and $(4, 0)$, $m = \frac{3 - 0}{0 - 4} = -\frac{3}{4}$.

8. False, because different parallel lines have the same slope.

9. True, because for $(1, 3)$ and $(-5, 3)$, $m = \frac{3 - 3}{1 - (-5)} = \frac{0}{6} = 0$.

10. True, because that is the definition of a parallelogram.

6.2 EXERCISES

1. The line goes through the points $(0, 2)$ and $(3, 0)$. Its slope is

$$m = \frac{2 - 0}{0 - 3} = -\frac{2}{3}.$$

3. The line goes through the points $(1, 1)$ and $(-1, -2)$. Its slope is

$$m = \frac{-2 - 1}{-1 - 1} = \frac{-3}{-2} = \frac{3}{2}.$$

5. The line goes through the points $(0, 0)$ and $(2, 4)$. Its slope is

$$m = \frac{4 - 0}{2 - 0} = 2.$$

7. The line goes through the points $(-2, 2)$ and $(2, 2)$. Its slope is

$$m = \frac{2 - 2}{2 - (-2)} = \frac{0}{4} = 0.$$

9. The line goes through the points $(-3, -3)$ and $(2, -1)$. Its slope is

$$m = \frac{-3 - (-1)}{-3 - 2} = \frac{-2}{-5} = \frac{2}{5}.$$

11. The line goes through the points $(-3, 1)$ and $(2, 2)$. Its slope is

$$m = \frac{2 - 1}{2 - (-3)} = \frac{1}{5}.$$

13. $m = \frac{2 - 6}{1 - 3} = \frac{-4}{-2} = 2$

15. $m = \frac{4 - (-1)}{2 - 5} = \frac{5}{-3} = -\frac{5}{3}$

17. $m = \frac{4 - 9}{-2 - 5} = \frac{-5}{-7} = \frac{5}{7}$

19. $m = \frac{-3 - 1}{-2 - (-5)} = \frac{-4}{3} = -\frac{4}{3}$

21. $m = \frac{4 - (-2)}{-3 - 3} = \frac{6}{-6} = -1$

23. $m = \frac{2 - \frac{1}{2}}{\frac{1}{2} - (-1)} = \frac{\frac{3}{2}}{\frac{3}{2}} = 1$

25. $m = \frac{3 - (-9)}{2 - 2} = \frac{12}{0}$

Since $x_1 = x_2$, the slope is undefined.

27. $m = \frac{-5 - (-5)}{-2 - 9} = \frac{0}{-11} = 0$

29. $m = \dfrac{0.9 - (-0.3)}{0.3 - (-0.1)} = \dfrac{1.2}{0.4} = 3$

31. Start at the point (1, 1) and rise 2 and go 3 units to the right to show a slope of 2/3. Draw a line through the two points.

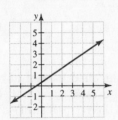

33. Start at the point (−2, 3) and go down 2 and then 1 unit to the right to show a slope of −2/1. Draw a line through the two points.

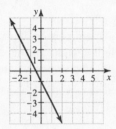

35. Start at the point (0, 0) and go down 2 units and then 5 units to the right to show a slope of −2/5. Draw a line through the two points.

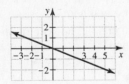

37. Start at the point (1, −2) and rise 1 and run 2 to get a second point on the line with slope 1/2. Draw the line. Start at (−1, 1) and rise 1 and run 2 to get a second point on the line. Draw the line.

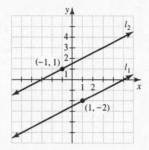

39. Start at (1, 2) and rise 1 and go 2 units to the right to indicate a slope of 1/2. Draw a line through the two points. Now start at (1, 2) and go down 2 units and then 1 unit to the right to indicate a slope of −2/1 = −2. Draw a line through the two points.

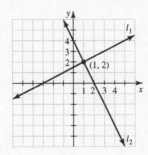

41. The slope of a line perpendicular to a line with slope 3/4 is the opposite of the reciprocal of 3/4. The reciprocal of 3/4 is 4/3 and the opposite of that is −4/3. Graphs may vary.

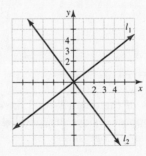

43. The line through $(-2, -3)$ and $(4, 0)$ has slope

$$m = \frac{-3 - 0}{-2 - 4} = \frac{-3}{-6} = \frac{1}{2}.$$

Any line parallel to it also has slope $\frac{1}{2}$.

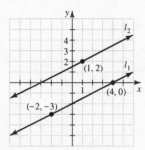

45. The line through $(-2, 4)$ and $(3, -1)$ has slope

$$m = \frac{4 - (-1)}{-2 - 3} = \frac{5}{-5} = -1.$$

The slope of any line perpendicular to this line is the opposite of the reciprocal of -1. The reciprocal of -1 is -1. The opposite of -1 is 1.

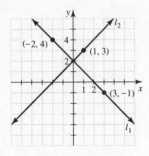

47. Plot the points and find the slope of each side of the quadrilateral.

$$m_1 = \frac{2 - 3}{-3 - 2} = \frac{-1}{-5} = \frac{1}{5}$$

$$m_2 = \frac{3 - 0}{2 - 3} = \frac{3}{-1} = -3$$

$$m_3 = \frac{-1 - 0}{-2 - 3} = \frac{-1}{-5} = \frac{1}{5}$$

$$m_4 = \frac{2 - (-1)}{-3 - (-2)} = \frac{3}{-1} = -3$$

Since the opposite sides of the quadrilateral have the same slopes, the opposite sides are parallel. The quadrilateral is a parallelogram.

49. Plot the points and find the slope of each side of the quadrilateral.

$$m_1 = \frac{0 - 1}{-2 - 0} = \frac{-1}{-2} = \frac{1}{2}$$

$$m_2 = \frac{-1 - 1}{1 - 0} = \frac{-2}{1} = -2$$

$$m_3 = \frac{-2 - (-1)}{-1 - 1} = \frac{-1}{-2} = \frac{1}{2}$$

$$m_4 = \frac{0 - (-2)}{-2 - (-1)} = \frac{2}{-1} = -2$$

Since the opposite sides of the quadrilateral have the same slopes, the opposite sides are parallel. Since 1/2 is the opposite of the reciprocal of -2, the adjacent sides are perpendicular. The quadrilateral is a rectangle.

51. Plot the points and find the slope of each side of the triangle.

$$m_1 = \frac{-3 - 4}{-4 - 5} = \frac{-7}{-9} = \frac{7}{9}$$

$$m_2 = \frac{-4 - 4}{-3 - 5} = \frac{-8}{-8} = 1$$

$$m_3 = \frac{-3 - (-4)}{-4 - (-3)} = \frac{1}{-1} = -1$$

The sides with slopes 1 and -1 are perpendicular to each other. So the triangle is a right triangle.

53. Plot the points and find the slopes of the two line segments.

$$m_1 = \frac{-8 - 2}{-3 - 1} = \frac{-10}{-4} = \frac{5}{2}$$

$$m_2 = \frac{2 - 9}{1 - 4} = \frac{-7}{-3} = \frac{7}{3}$$

Since the slopes are not the same, these three points are not on the same straight line.

55. $m = \dfrac{1,000,000 - 700,000}{1995 - 1990} = \dfrac{300,000}{5}$

$= 60,000$

The slope of 60,000 means that the cost is increasing on the average by $60,000 per year. The cost in 1993 was $880,000. The cost in the year 2000 will be $1,300,000.

6.3 WARM-UPS

1. True, because the slope determines a second point on the line and two points determine a line.

2. False, because $(1, 2)$ does not satisfy $y = 3x + 2$.

3. True, because $x = -2$ runs parallel to the y-axis.

4. True, because any equation of the form $x = k$ for some real number k has a graph that is a vertical line.

5. True, because the slope of $y = x - 3$ is 1 and the slope of $y = 5 - x$ is -1.

6. False, because nonvertical parallel lines must have the same slope.

7. False, because $2y = 3x - 8$ is equivalent to $y = \frac{3}{2}x - 4$ and has slope $\frac{3}{2}$.

8. True, because even vertical lines that do not have slope can be written in standard form.

9. True, because $x = 2$ is vertical and $y = 5$ is horizontal.

10. False, because the y-intercept for $y = x$ is $(0, 0)$.

6.3 EXERCISES

1. The line goes through the points $(0, 1)$ and $(2, 4)$. The slope of the line is $m = \frac{4-1}{2-0} = \frac{3}{2}$ and the y-intercept is $(0, 1)$. The equation of the line is $y = \frac{3}{2}x + 1$.

3. The line goes through the points $(0, 2)$ and $(2, -2)$. The slope of the line is $m = \frac{-2-2}{2-0} = -\frac{4}{2} = -2$ and the y-intercept is $(0, 2)$. The equation of the line is $y = -2x + 2$.

5. The line goes through the points $(0, -2)$ and $(3, 1)$. The slope of the line is $m = \frac{-2-1}{0-3} = \frac{-3}{-3} = 1$ and the y-intercept is $(0, -2)$. The equation of the line is $y = x - 2$.

7. The line goes through the points $(0, 0)$ and $(2, -2)$. The slope of the line is $m = \frac{-2-0}{2-0} = \frac{-2}{2} = -1$ and the y-intercept is $(0, 0)$. The equation of the line is $y = -x$.

9. The line goes through the points $(0, -1)$ and $(-3, -1)$. The slope of the line is $m = \frac{-1-(-1)}{2-(-3)} = \frac{0}{5} = 0$ and the y-intercept is $(0, -1)$. The equation of the line is $y = -1$.

11. The line goes through the points $(-2, 0)$ and $(-2, 2)$. The line is vertical and so it has no slope. There is no slope-intercept form for this line. The equation of the line is $x = -2$.

13. For the line $y = 3x - 9$, the slope is 3 and the y-intercept is $(0, -9)$.

15. For the line $y = 4$, the slope is 0 and the y-intercept is $(0, 4)$.

17. For the line $y = -3x$, the slope is -3 and the y-intercept is $(0, 0)$.

19. The equation $x + y = 5$ is equivalent to $y = -x + 5$. The slope is -1 and the y-intercept is $(0, 5)$.

21. The equation $x - 2y = 4$, is equivalent to $y = \frac{1}{2}x - 2$. The slope is $\frac{1}{2}$ and the y-intercept is $(0, -2)$.

23. The equation $2x - 5y = 10$ is equivalent to $y = \frac{2}{5}x - 2$. The slope is $\frac{2}{5}$ and the y-intercept is $(0, -2)$.

25. The equation $2x - y + 3 = 0$ is equivalent to $y = 2x + 3$. The slope is 2 and the y-intercept is $(0, 3)$.

27. For the line $x = -3$, there is no y-intercept because the line is vertical. The slope is undefined for a vertical line.

29. $y = -x + 2$
          $x + y = 2$

31. $y = \frac{1}{2}x + 3$
          $2y = x + 6$
          $-x + 2y = 6$
          $x - 2y = -6$

33. $y = \frac{3}{2}x - \frac{1}{3}$
          $6y = 6(\frac{3}{2}x - \frac{1}{3})$
          $6y = 9x - 2$
          $-9x + 6y = -2$
          $9x - 6y = 2$

35. $y = -\frac{3}{5}x + \frac{7}{10}$
          $10y = 10(-\frac{3}{5}x + \frac{7}{10})$
          $10y = -6x + 7$
          $6x + 10y = 7$

37. $\frac{3}{5}x + 6 = 0$

$3x + 30 = 0$

$x = -10$

39. $\frac{3}{4}y = \frac{5}{2}$

$3y = 10$

41. $\frac{x}{2} = \frac{3y}{5}$

$5x = 6y$

$5x - 6y = 0$

43. $y = 0.02x + 0.5$

$100y = 2x + 50$

$-2x + 100y = 50$

$x - 50y = -25$

45. The line $y = 2x - 1$ has a y-intercept of $(0, -1)$ and a slope of $2 = 2/1$. Start at $(0, -1)$, rise 2 and then go 1 unit to the right to locate a second point on the line. Draw a line through the two points as shown.

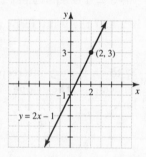

47. The line $y = -3x + 5$ has a y-intercept of $(0, 5)$ and a slope of $-3 = -3/1$. Start at $(0, 5)$, go down 3 units, and then 1 unit to the right to locate a second point on the line. Draw a line through the two points as shown.

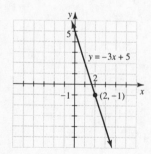

49. The line $y = \frac{3}{4}x - 2$ has a y-intercept of $(0, -2)$ and a slope of 3/4. Start at $(0, -2)$, go up 3 units, and then 4 units to the right to locate a second point on the line. Draw a line as shown.

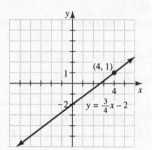

51. The equation $2y + x = 0$ is equivalent to $y = -\frac{1}{2}x$. The y-intercept is $(0, 0)$ and the slope is $-1/2$. Start at $(0, 0)$, go down 1 unit, and then go 2 units to the right to locate a second point on the line. Draw the line as shown.

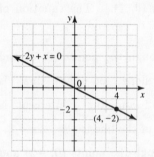

53. The equation $3x - 2y = 10$ is equivalent to $y = (3/2)x - 5$. The y-intercept is $(0, -5)$ and the slope is 3/2. Start at $(0, -5)$, go up 3 units, and then 2 units to the right to locate a second point on the line. Draw a line through the two points as shown.

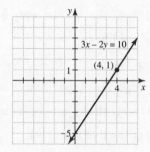

55. The equation $y - 2 = 0$ is equivalent to $y = 2$. The y-intercept is (0, 2) and the slope is 0. Draw a horizontal line through (0, 2).

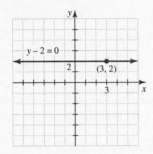

57. The line $y = 3x - 5$ has slope 3. Any line perpendicular to it has slope $-1/3$. The line through (0, 6) with slope $-1/3$ has equation $y = -\frac{1}{3}x + 6$.

59. The equation $2x + y = 5$ is equivalent to $y = -2x + 5$. Any line parallel to it has slope -2. The line through (0, 3) with slope -2 has equation $y = -2x + 3$.

61. Any line parallel to the x-axis has 0 slope. A line through (2, 3) with 0 slope has a y-intercept of (0, 3). The equation of the line through (0, 3) with 0 slope is $y = 3$.

63. The line through (0, 4) and (5, 0) has slope $m = \frac{4-0}{0-5} = -\frac{4}{5}$. The line through (0, 4) with slope $-4/5$ has equation $y = -\frac{4}{5}x + 4$.

65. If $n = 5000$, then

$C = 200(5000) + 150,000 = \$1,150,000$.

If $n = 5001$, then

$C = 200(5001) + 150,000 = \$1,150,200$.

The extra mower increased the cost by \$200.

6.4 WARM-UPS

1. False, because the point-slope formula is $y - y_1 = m(x - x_1)$.
2. False, because we can find the equation of a line through any two points.
3. True, because the line is vertical and its slope is undefined.
4. True, because the line through the origin with slope 1 is $y = 1 \cdot x + 0$, or $y = x$.

5. False, because $5x + y = 4$ is equivalent to $y = -5x + 4$, which has slope -5.
6. True, because the line $y = 4x - 3$ has slope 4 and any line perpendicular to it has slope $-1/4$.
7. True, because $x + y = 1$ is equivalent to $y = -x + 1$, which has slope -1.
8. True, because $2(-2) - (-3) = -1$ is correct.
9. True, because $2x + y = 4$ is equivalent to $y = -2x + 4$ and both lines have slope -2.
10. True, because the slope of $y = x$ is 1 and the slope of $y = -x$ is -1.

6.4 EXERCISES

1. $y - 1 = 5(x + 2)$

$\quad y - 1 = 5x + 10$

$\qquad y = 5x + 11$

3. $\qquad 3x - 4y = 80$

$\qquad -4y = -3x + 80$

$\qquad\quad y = \frac{3}{4}x - 20$

5. $\qquad y - \frac{1}{2} = \frac{2}{3}\left(x - \frac{1}{4}\right)$

$\qquad y - \frac{1}{2} = \frac{2}{3}x - \frac{1}{6}$

$\qquad\quad y = \frac{2}{3}x - \frac{1}{6} + \frac{1}{2}$

$\qquad\quad y = \frac{2}{3}x + \frac{1}{3}$

7. Use the point (2, 3) and slope 1/3 in point-slope form $y - y_1 = m(x - x_1)$.

$\qquad y - 3 = \frac{1}{3}(x - 2)$

$\qquad y - 3 = \frac{1}{3}x - \frac{2}{3}$

$\qquad\quad y = \frac{1}{3}x - \frac{2}{3} + 3$

$\qquad\quad y = \frac{1}{3}x + \frac{7}{3}$

9. Use the point $(-2, 5)$ and slope $-1/2$ in point-slope form $y - y_1 = m(x - x_1)$.

$\qquad y - 5 = -\frac{1}{2}(x - (-2))$

$\qquad y - 5 = -\frac{1}{2}x - 1$

$\qquad\quad y = -\frac{1}{2}x + 4$

11. Use the point $(-1, -7)$ and slope -6 in point-slope form $y - y_1 = m(x - x_1)$.
$$y - (-7) = -6(x - (-1))$$
$$y + 7 = -6x - 6$$
$$y = -6x - 13$$

13. $\quad y - 3 = 2(x - 5)$
$$y - 3 = 2x - 10$$
$$-2x + y = -7$$
$$2x - y = 7$$

15. $\quad\quad y = \frac{1}{2}x - 3$
$$2y = x - 6$$
$$-x + 2y = -6$$
$$x - 2y = 6$$

17. $\quad\quad y - 2 = \frac{2}{3}(x - 4)$
$$3y - 6 = 2x - 8$$
$$-2x + 3y = -2$$
$$2x - 3y = 2$$

19. Find the slope of the line through the points $(1, 2)$ and $(5, 8)$.

$$m = \frac{2 - 8}{1 - 5} = \frac{-6}{-4} = \frac{3}{2}$$

Use the point $(1, 2)$ and slope $3/2$ in the point-slope form $y - y_1 = m(x - x_1)$.

$$y - 2 = \frac{3}{2}(x - 1)$$
$$2y - 4 = 3(x - 1)$$
$$2y - 4 = 3x - 3$$
$$-3x + 2y = 1$$
$$3x - 2y = -1$$

21. Find the slope of the line through the points $(-2, -1)$ and $(3, -4)$.

$$m = \frac{-1 - (-4)}{-2 - 3} = \frac{3}{-5} = -\frac{3}{5}$$

Use the point $(3, -4)$ and slope $-3/5$ in the point-slope form $y - y_1 = m(x - x_1)$.

$$y - (-4) = -\frac{3}{5}(x - 3)$$
$$y + 4 = -\frac{3}{5}x + \frac{9}{5}$$
$$5y + 20 = -3x + 9$$
$$3x + 5y = -11$$

23. Find the slope of the line through the points $(-2, 0)$ and $(0, 2)$.

$$m = \frac{0 - 2}{-2 - 0} = \frac{-2}{-2} = 1$$

Use the point $(0, 2)$ and slope 1 in the slope-intercept form $y = mx + b$.
$$y = 1x + 2$$
$$-x + y = 2$$
$$x - y = -2$$

25. The slope of the dashed line is 1. The slope of the solid line is -1 and it goes through $(5, -1)$.
$$y - (-1) = -1(x - 5)$$
$$y + 1 = -x + 5$$
$$y = -x + 4$$

27. The slope of the dashed line is $-3/5$. The solid line has slope $5/3$ and goes through $(3, 4)$.

$$y - 4 = \frac{5}{3}(x - 3)$$
$$y - 4 = \frac{5}{3}x - 5$$
$$y = \frac{5}{3}x - 1$$

29. Any line perpendicular to $y = 3x - 1$ has slope $-1/3$. Use the point $(3, 4)$ and the slope $-1/3$ in the point-slope form $y - y_1 = m(x - x_1)$.

$$y - 4 = -\frac{1}{3}(x - 3)$$
$$y - 4 = -\frac{1}{3}x + 1$$
$$y = -\frac{1}{3}x + 5$$

31. Any line parallel to $y = x - 9$ has slope 1. Use the point $(7, 10)$ and slope 1 in the point-slope form.

$$y - 10 = 1(x - 7)$$
$$y = x + 3$$

33. The line $3x - 2y = 10$ is equivalent to $y = \frac{3}{2}x - 5$, and so its slope is $3/2$. Any line perpendicular to $3x - 2y = 10$ has slope $-2/3$. Use the point $(1, 1)$ and slope $-2/3$ in the point-slope form.

$$y - 1 = -\frac{2}{3}(x - 1)$$
$$y - 1 = -\frac{2}{3}x + \frac{2}{3}$$
$$y = -\frac{2}{3}x + \frac{5}{3}$$

35. The equation $2x + y = 8$ is equivalent to $y = -2x + 8$, which has slope -2. Any line parallel to $2x + y = 8$ also has slope -2. Use the point $(-1, -3)$ and slope -2 in the point-slope form.

$$y - (-3) = -2(x - (-1))$$
$$y + 3 = -2x - 2$$
$$y = -2x - 5$$

37. The equation $3x + y = 5$ is equivalent to $y = -3x + 5$. It has slope of -3. Any line perpendicular to it has slope $1/3$. Use the point $(-1, 2)$ and slope $1/3$ in the point-slope form.

$$y - 2 = \tfrac{1}{3}(x - (-1))$$
$$y - 2 = \tfrac{1}{3}x + \tfrac{1}{3}$$
$$y = \tfrac{1}{3}x + \tfrac{7}{3}$$

39. The equation $-2x + y = 6$ is equivalent to $y = 2x + 6$. It has slope 2. Any line parallel to $-2x + y = 6$ also has slope 2. Use the point $(2, 3)$ and slope 2 in the point-slope form.

$$y - 3 = 2(x - 2)$$
$$y - 3 = 2x - 4$$
$$y = 2x - 1$$

41. $m = \dfrac{7 - 4.5}{92 - 88} = \dfrac{2.5}{4} = 0.625$

Use slope 0.625 and point $(88, 4.5)$:

$$y - 4.5 = 0.625(x - 88)$$
$$y - 4.5 = 0.625x - 55$$
$$y = 0.625x - 50.5$$

For the year 2000, x would be 100:

$$y = 0.625(100) - 50.5 = 12$$

So in 2000 there will be 12 billion transactions.

45. a) The intercepts are $(0, -300)$ and $(15, 0)$.

b) The intercepts are $(0, 500)$ and $(50/3, 0)$.

c) The intercepts are $(0, -2000)$ and $(3000, 0)$.

47. Use the viewing window $-1 \leq x \leq 1$ and $-1 \leq y \leq 1$.

6.5 WARM-UPS

1. True, because any equation of the form $y = mx + b$ is a linear function.

2. True, because the formula is $C = 2\pi r$.

3. False, because the formula for the area of a circle is $A = \pi r^2$.

4. True, because $D = 8R$ where R is her speed and D is her distance.

5. True, because the formula is $C = \tfrac{5}{9}(F - 32)$.

6. False, because the slope of the line through $(1980, 3000)$ and $(1990, 2000)$ is

$$m = \frac{3000 - 2000}{1980 - 1990} = \frac{1000}{-10} = -100.$$

7. True, because the cost C is determined by the equation $C = 90t$ where t is the number of hours worked.

8. False, because the formula for the area of a square is $A = s^2$ where s is the length of a side.

9. True, because the formula for the perimeter of a square is $P = 4s$ where s is the length of a side.

10. True, because the perimeter of a rectangle with a length of 5 meters is $P = 2W + 2(5)$ where W is the width.

6.5 EXERCISES

1. Since there are 3 feet in 1 yard, the length of an object in feet is 3 times its length in yards, $F = 3Y$, where F is length in feet and Y is length in yards.

3. Since $D = RT$, when the car is traveling 65 mph, we have $D = 65T$.

5. The formula for the circumference of a circle is $C = \pi D$.

7. Rodney's pay P is \$7.80 times the number of hours H that he works, $P = 7.8H$.

9. The graph of $P = 40n + 300$ has a P-intercept of $(0, 300)$ and a slope of 40. The graph also goes through the point $(200, 8300)$.

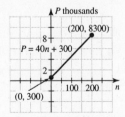

11. The graph of R = 30t + 1000 goes through (100, 4000) and (900, 28000).

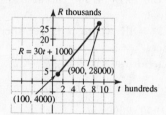

13. The graph of C = 2πr for 1 ≤ r ≤ 10 goes through the points (1, 2π) = (1, 6.28) and (10, 20π) = (10, 62.8).

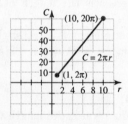

15. The graph of h = −7.5d + 350 for 0 ≤ d ≤ 40 goes through the points (0, 350) and (40, 50).

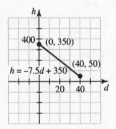

17. The profit per share in 1987 was P = 0.35(7) + 4.60 = $7.05. The graph goes through the points (7, 7.05) and (0, 4.6).

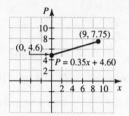

19. We want the equation of the line through the points (2, 70) and (4, 110). The slope is

$$m = \frac{110 - 70}{4 - 2} = \frac{40}{2} = 20.$$

Use the point (2, 70) and slope 20 in the point-slope form, where C is the charge and n is the number of hours worked.

$$C - 70 = 20(n - 2)$$
$$C - 70 = 20n - 40$$
$$C = 20n + 30$$

If n = 7 at Fred's house, then C = 20(7) + 30 = $170.

21. We want the equation of the line through the points (7.75, 13) and (5.75, 7) where the first coordinate is the length of the foot L and the second coordinate is the shoe size S. First find the slope.

$$m = \frac{13 - 7}{7.75 - 5.75} = \frac{6}{2} = 3$$

Now use the point (5.75, 7) and the slope 3 in the point-slope form of the line.

$$S - 7 = 3(L - 5.75)$$
$$S - 7 = 3L - 17.25$$
$$S = 3L - 10.25$$
$$S = 3L - \frac{41}{4}$$

If L = 6.25, then S = 3(6.25) − 10.25 = 8.5.
A child with a 6.25 inch foot wears a size 8.5.

106

23. We want the equation of the line through the points (1, 42) and (2, 74) where the first coordinate is time t in seconds, and the second coordinate is velocity v in feet per second. The slope of the line is

$$m = \frac{74 - 42}{2 - 1} = 32.$$

Use a slope of 32 and the point (1, 42) in the point-slope form.

$$v - 42 = 32(t - 1)$$
$$v - 42 = 32t - 32$$
$$v = 32t + 10$$

If $t = 3.5$ seconds, then the velocity is $v = 32(3.5) + 10 = 122$ feet per second.

25. We want the equation of the line through the points (90, 0.75) and (30, 1.25) where the first coordinate is temperature t and the second coordinate is width w. The slope is

$$m = \frac{1.25 - 0.75}{30 - 90} = \frac{0.5}{-60} = -\frac{1}{120}.$$

Use a slope of $-1/120$ and the point (90, 0.75) in the point-slope form.

$$w - 0.75 = -\frac{1}{120}(t - 90)$$

$$w - 0.75 = -\frac{1}{120}t + \frac{3}{4}$$

$$w = -\frac{1}{120}t + \frac{3}{2}$$

If $t = 80$, then the width is

$$w = -\frac{1}{120}(80) + \frac{3}{2} = -\frac{2}{3} + \frac{3}{2} = \frac{5}{6}.$$

When the temperature is 80° F, the width is $\frac{5}{6}$ inch.

27. We want the equation of the line through the two points (3, 1.8) and (5, 3) where the first coordinate is weight w in pounds and the second coordinate is amount A in inches that the spring stretches. The slope is

$$m = \frac{3 - 1.8}{5 - 3} = \frac{1.2}{2} = 0.6.$$

Use the slope 0.6 and the point (5, 3) in the point-slope form.

$$A - 3 = 0.6(w - 5)$$
$$A - 3 = 0.6w - 3$$
$$A = 0.6w$$

If $w = 6$, the $A = 0.6(6) = 3.6$. With a weight of 6 pounds the spring will stretch 3.6 inches.

29. Find the equation of the line through (2, 0.16) and (5, 0.40).

$$m = \frac{0.40 - 0.16}{5 - 2} = \frac{0.24}{3} = 0.08$$

$$a - 0.16 = 0.08(c - 2)$$
$$a - 0.16 = 0.08c - 0.16$$
$$a = 0.08c$$

If $c = 3$, then $a = 0.08(3) = 0.24$. If the absorption is 0.50, then the concentration is 6.25.

31. Use the graphs to answer the questions.
a) At $A = 25$ the female has a higher basal energy requirement.
b) At $A = 72$ the male has a higher basal energy requirement.
c) Basal energy requirements are equal at about age 65.
d) If $B = 0$, then $1623 - 6.9A = 0$ or $A = 235$.

6.6 WARM-UPS
1. False, because there must be no ordered pairs with the same first coordinate and different second coordinates.
2. True, because the area is determined from the side by the formula $A = s^2$.
3. False, because the ordered pairs $(-1, 3)$ and $(-1, -3)$ have the same first coordinate and different second coordinates.
4. True, because no two ordered pairs have the same first coordinate and different second coordinates.
5. True, because any real number can be use for x in the expression x^3.
6. False, because the domain is the set of all real numbers.
7. False, because the range is the set of nonnegative real numbers.
8. True, because the equation $x = 2y$ is equivalent to $y = (1/2)x$ and this equation guarantees that no ordered pairs will have the same x-coordinate and different y-coordinates.

9. False, because the ordered pairs (4, 2) and (4, −2) both satisfy $x = y^2$.

10. True, because $f(-2) = (-2)^2 - 5 = 4 - 5 = -1$.

6.6 EXERCISES

1. If each topping costs $0.50 and there are t toppings, then the total cost of the toppings is 0.50t dollars. The total cost of the pizza is $C = 0.50t + 5$, where C is in dollars.

3. The amount of sales tax is 0.09S. The total cost is $T = S + 0.09S$, or $T = 1.09S$.

5. The circumference of a circle is determined from its radius according to the formula $C = 2\pi r$.

7. The perimeter of a square is four times as large as one of the equal sides, $P = 4s$.

9. Since $A = \frac{1}{2}bh$ for a triangle and $b = 10$, $A = \frac{1}{2}(10)h = 5h$. The formula $A = 5h$ expresses the area as a function of the height.

11. Yes, this table defines y as a function of x.

13. Yes, this table defines V as a function of t.

15. No, this table does not define P as a function of a, because of the ordered pairs (2, 2) and (2, −2).

17. Yes, this table defines q as a function of b.

19. This set of ordered pairs is a function because no two ordered pairs have the same first coordinate and different second coordinates.

21. This set of ordered pairs is a function because no two ordered pairs have the same first coordinate and different second coordinates.

23. This set of ordered pairs is not a function because the ordered pairs (0, −1) and (0, 1) have the same first coordinate and different second coordinates.

25. This set of ordered pairs is a function because no two ordered pairs have the same first coordinate and different second coordinates.

27. This set of ordered pairs is a function because the equation $y = x - 3$ is used to determine the second coordinate from the first. So two ordered pairs with the same x-coordinate could not have different y-coordinates.

29. This set of ordered pairs is not a function because the ordered pairs (1, 1) and (1, −1) both satisfy $x = |y|$ and therefore they belong to the set.

31. The equation $x = y + 1$ is equivalent to $y = x - 1$. So the y-coordinate is determined from the x-coordinate by the formula $y = x + 1$. Two ordered pairs with the same x-coordinate could not have different y-coordinates. So the set is a function.

33. This set is not a function because the ordered pairs (0, 1) and (0, −1) both satisfy $x = y^2 - 1$ and so they are in the set.

35. The equation $x = 4y$ is equivalent to $y = \frac{1}{4}x$. So two ordered pairs with the same x-coordinate could not have different y-coordinates. The equation does define a function.

37. The y-coordinate is determined by the equation $y = 2/x$. So two ordered pairs with the same x-coordinate could not have different y-coordinates. The equation defines a function.

39. The y-coordinate is determined by the equation $y = x^3 - 1$. So two ordered pairs with the same x-coordinate could not have different y-coordinates. Every number x determines a unique y. So the equation defines a function.

41. The ordered pairs (0, 5) and (0, −5) both satisfy $x^2 + y^2 = 25$. So the equation does not define a function.

43. This graph is the graph of a function because no vertical line crosses it more than once.

45. This graph is not the graph of a function because you can draw a vertical line that crosses it more than once.

47. This graph is not the graph of a function because you can draw a vertical line that crosses it more than once.

49. The domain is the set of first coordinates {1, 2, 3}, and the range is the set of second coordinates {3, 5, 7}.

51. In the equation $y = |x + 3|$ any real number could be used in place of x. So the domain is the set of all real numbers R. Since the absolute value of every real number is nonnegative, the values of y are nonnegative. The range is the set of nonnegative real numbers.

53. Since any real number could be used in place of x in $y = x$, the domain is the set of all real numbers. Since any real number could occur as a y-coordinate, the range is the set of all real numbers.

55. In the equation $y = x^2$ any real number could be used in place of x. So the domain is the set of all real numbers. Since the square of any real number is nonnegative, the values of y are nonnegative. The range is the set of nonnegative real numbers.

57. The condition that $s > 0$ specifies the domain. The domain is $\{s \,|\, s > 0\}$. Since $A = s^2$, A will be a positive number. The range is $\{A \,|\, A > 0\}$.

59. $f(0) = 2(0) - 1 = 0 - 1 = -1$

61. $f(\frac{1}{2}) = 2(\frac{1}{2}) - 1 = 0$

63. $g(4) = 4^2 - 3 = 16 - 3 = 13$

65. $g(0.5) = 0.5^2 - 3 = 0.25 - 3 = -2.75$

67. $h(3) = |3 - 1| = |2| = 2$

69. $h(0) = |0 - 1| = |-1| = 1$

71. $f(5.68) = (5.68)^3 - (5.68)^2 = 150.988$

73. $g(3.5) = (3.5)^2 - 4.2(3.5) + 2.76 = 0.31$

75. $W(70) = 0.000534(70)^3 \approx 183$ pounds
$W(74) = 0.000534(74)^3 \approx 216$ pounds

77. Since the credit card company receives 4% of each charge, the retailer receives 96% of each charge. So $r = 0.96C$ or $C = \frac{r}{0.96}$.

6.7 WARM-UPS

1. True, because of the definition of varies directly.

2. False, because if a varies inversely as b, then $a = k/b$ for some constant k.

3. True, because if y varies directly as x, then $y = kx$. If $y = 8$ when $x = 2$, then $8 = k \cdot 2$. So $k = 4$.

4. False, because if y varies inversely as x, then $y = k/x$. If $y = 8$ when $x = 2$, $8 = k/2$, or $k = 16$.

5. False, because if C varies jointly as h and t, then $C = kht$ for some constant k.

6. True, because if k is the sales tax rate, then the amount of tax is kP where P is the price of the car.

7. True, because if z varies inversely as w, then $z = k/w$. If $z = 10$ when $w = 2$, then $10 = k/2$, or $k = 20$. So $z = 20/w$.

8. True, because $T = D/R$ and if D is fixed, then T varies inversely with the rate.

9. False, because if m varies directly as w, then $m = kw$ for some constant k.

10. False, because if y varies jointly as x and z, then $y = kxz$.

6.7 EXERCISES

1. T varies directly as h means T is a constant multiple of h, $T = kh$.

3. If y varies inversely as r, then y is a constant divided by r, $y = \frac{k}{r}$.

5. If R is jointly proportional to t and s, then R is a constant multiple of the product of t and s, $R = kts$.

7. If i is directly proportional to b, then i is a constant multiple of b, $i = kb$.

9. If A is jointly proportional to y and m, then A is a constant multiple of the product of y and m, $A = kym$.

11. If y varies directly as x, then $y = kx$. If $y = 5$ when $x = 3$, then $5 = k \cdot 3$, or $k = \frac{5}{3}$. So $y = \frac{5}{3}x$.

13. If A varies inversely as B, then $A = \frac{k}{B}$. If $A = 3$ when $B = 2$, then $3 = \frac{k}{2}$. So $k = 6$ and $A = \frac{6}{B}$.

15. If m varies inversely as p, then $m = \frac{k}{p}$. If $m = 22$ when $p = 9$, then $22 = \frac{k}{9}$. So $k = 198$ and $m = \frac{198}{p}$.

17. If A varies jointly as t and u, then $A = ktu$. If $A = 24$ when $t = 6$ and $u = 2$, then $24 = k \cdot 6 \cdot 2$. So $k = 2$ and $A = 2tu$.

19. If T varies directly as u, then $T = ku$. If $T = 9$ when $u = 2$, then $9 = k \cdot 2$. So $k = \frac{9}{2}$ and $T = \frac{9}{2}u$.

21. If Y varies directly as x, then $Y = kx$. If $Y = 100$ when $x = 20$, then $100 = k \cdot 20$ and $k = 5$. So the formula is $Y = 5x$. If $x = 5$, then $Y = 5(5) = 25$.

23. If a varies inversely as b, then $a = \frac{k}{b}$. If $a = 3$ when $b = 4$, then $3 = \frac{k}{4}$ and $k = 12$. So the formula is $a = \frac{12}{b}$. If $b = 12$, then $a = \frac{12}{12} = 1$.

25. If P varies jointly as s and t, then $P = kst$. If $P = 56$ when $s = 2$ and $t = 4$, then $56 = k \cdot 2 \cdot 4$ and $k = 7$. So the formula is $P = 7st$. If $s = 5$ and $t = 3$, then $P = 7 \cdot 5 \cdot 3 = 105$.

27. If the weight w varies directly as the length L, then $w = kL$. If a 12-foot boat weighs 86 pounds, then $86 = k \cdot 12$ and $k = \frac{86}{12} = \frac{43}{6}$. So the formula is $w = \frac{43}{6}L$. If the length is 14 feet, then $w = \frac{43}{6} \cdot 14 = \frac{301}{3} = 100.3$ pounds.

29. If the time t varies inversely as the number of elephants n, then $t = \frac{k}{n}$. If $t = 75$ when $n = 4$, then $75 = \frac{k}{4}$ and $k = 300$. So the formula is $t = \frac{300}{n}$. If $n = 6$, then $t = \frac{300}{6} = 50$ minutes.

31. If the cost C is jointly proportional to the length L and the diameter D, then $C = kLD$. If $C = \$5.80$ when $L = 10$ feet and $D = 1$ inch, then $5.80 = k \cdot 10 \cdot 1$ and $k = 0.58$. So the formula is $C = 0.58LD$. If $L = 15$ and $D = 2$, then $C = 0.58 \cdot 15 \cdot 2 = \$17.40$.

33. Since the approach speed A is directly proportional to the landing speed L, we have $A = kL$. Use $A = 90$ and $L = 75$ to find k:

$$k75 = 90$$
$$k = \frac{90}{75} = 1.2$$

So $A = 1.2L$. If $A = 96$, then $96 = 1.2L$ or $L = \frac{96}{1.2} = 80$ mph.

CHAPTER 6 REVIEW

1. To locate the point $(-2, 5)$ we start at the origin and move 2 units to the left and then 5 units up. This point is in quadrant II.

3. Since the y-coordinate is zero in $(3, 0)$, the point is on the x-axis.

5. Since the x-coordinate is zero in $(0, -6)$, the point is on the y-axis.

7. To locate the point $(1.414, -3)$ we start at the origin and move approximately 1.414 units to the right and then 3 units down. This point is in quadrant IV.

9. If $x = 0$, then $y = 3(0) - 5 = -5$.
If $x = -3$, then $y = 3(-3) - 5 = -14$.
If $x = 4$, then $y = 3(4) - 5 = 7$.
The points are $(0, -5)$, $(-3, -14)$, and $(4, 7)$.

11. Solve $2x - 3y = 8$ for y.

$$y = \frac{2}{3}x - \frac{8}{3}$$

If $x = 0$, then $y = \frac{2}{3}(0) - \frac{8}{3} = -\frac{8}{3}$.

If $x = 3$, then $y = \frac{2}{3}(3) - \frac{8}{3} = -\frac{2}{3}$.

If $x = -6$, then $y = \frac{2}{3}(-6) - \frac{8}{3} = -\frac{20}{3}$.

The points are $(0, -\frac{8}{3})$, $(3, -\frac{2}{3})$, and $(-6, -\frac{20}{3})$.

13. Three ordered pairs that satisfy $y = -3x + 4$ are $(0, 4)$, $(1, 1)$, and $(2, -2)$. Draw a line through these points as shown.

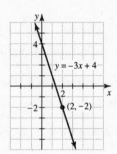

15. Three ordered pairs that satisfy $x + y = 7$ are $(1, 6)$, $(2, 5)$, and $(3, 4)$. Draw a line through these points as shown.

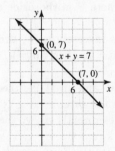

17. $m = \frac{1-0}{1-0} = 1$

19. $m = \frac{-3-0}{-2-0} = \frac{-3}{-2} = \frac{3}{2}$

21. $m = \frac{-2-1}{-4-3} = \frac{-3}{-7} = \frac{3}{7}$

23. Plot the points and find the slope of each line segment.

$$m_1 = \frac{1-(-3)}{-4-(-3)} = \frac{4}{-1} = -4$$

$$m_2 = \frac{-3-(-1)}{-3-5} = \frac{-2}{-8} = \frac{1}{4}$$

$$m_3 = \frac{1-(-1)}{-4-5} = \frac{2}{-9} = -\frac{2}{9}$$

Since m_1 is the opposite of the reciprocal of m_2, the two line segments are perpendicular and the triangle is a right triangle.

25. Plot the points and draw the quadrilateral with the diagonals. The slope of the diagonal joining $(7, 6)$ and $(2, 1)$ is $m = \frac{6-1}{7-2} = \frac{5}{5} = 1$. The slope of the diagonal joining $(3, 5)$ and $(6, 2)$ is $m = \frac{5-2}{3-6} = \frac{3}{-3} = -1$. Since 1 is the opposite of the reciprocal of -1, the diagonals are perpendicular.

27. The equation $y = 3x - 18$ is in slope-intercept form. So the slope is 3 and the y-intercept is $(0, -18)$.

29. Rewrite $2x - y = 3$ in slope-intercept form.

$$-y = -2x + 3$$
$$y = 2x - 3$$

The slope is 2 and the y-intercept is $(0, -3)$.

31. Rewrite $4x - 2y - 8 = 0$ in slope-intercept form.

$$-2y = -4x + 8$$
$$y = 2x - 4$$

The slope is 2 and the y-intercept is $(0, -4)$.

33. To sketch $y = \frac{2}{3}x - 5$, start at the y-intercept $(0, -5)$ and use the slope $\frac{2}{3}$ to locate a second point on the line. From $(0, -5)$ rise 2 and run 3. Draw a line through the two points.

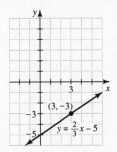

35. Rewrite $-2x + y = -6$ in slope-intercept form.

$$y = -2x + 6$$

Start at the y-intercept $(0, 6)$ and use the slope $-2 = -2/1$ to locate a second point on the line. From $(0, 6)$ rise -2 and run 1. Draw a line through the two points.

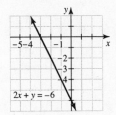

37. The line $y = -4$ has slope 0 and y-intercept $(0, -4)$. Draw a horizontal line through $(0, -4)$.

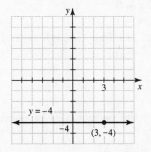

111

39. Use slope-intercept form with slope 1/3 and y-intercept (0, 4).

$$y = \frac{1}{3}x + 4$$
$$3y = x + 12$$
$$-x + 3y = 12$$
$$x - 3y = -12$$

41. Any line perpendicular to $y = 2x - 1$ has slope $-\frac{1}{2}$. The line through (0, 0) with slope $-\frac{1}{2}$ has the following equation.

$$y = -\frac{1}{2}x$$
$$2y = -x$$
$$x + 2y = 0$$

43. Any line parallel to the x-axis has slope 0. The line through (3, 5) with slope 0 has y-intercept (0, 5).

$$y = 0 \cdot x + 5$$
$$y = 5$$

45.
$$y - 3 = \frac{2}{3}(x + 6)$$
$$y - 3 = \frac{2}{3}x + 4$$
$$y = \frac{2}{3}x + 7$$

47.
$$3x - 7y - 14 = 0$$
$$-7y = -3x + 14$$
$$y = \frac{3}{7}x - 2$$

49.
$$y - 5 = -\frac{3}{4}(x + 1)$$
$$y - 5 = -\frac{3}{4}x - \frac{3}{4}$$
$$y = -\frac{3}{4}x + \frac{17}{4}$$

51. Use the point-slope form with (−4, 7) and slope −2.
$$y - 7 = -2(x - (-4))$$
$$y - 7 = -2x - 8$$
$$y = -2x - 1$$

53. Find the slope of the line through (−2, 1) and (3, 7).

$$m = \frac{7 - 1}{3 - (-2)} = \frac{6}{5}$$

Use the slope 6/5 and the point (3, 7) in point-slope form.
$$y - 7 = \frac{6}{5}(x - 3)$$

$$y - 7 = \frac{6}{5}x - \frac{18}{5}$$
$$y = \frac{6}{5}x + \frac{17}{5}$$

55. Any line parallel to $y = 3x - 1$ has slope 3. Use point-slope form with slope 3 and the point (3, −5).
$$y - (-5) = 3(x - 3)$$
$$y + 5 = 3x - 9$$
$$y = 3x - 14$$

57. The ordered pairs (10, 370), (20, 340), and (90, 130) satisfy the equation $P = -3t + 400$.

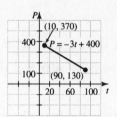

59. The ordered pairs (0.1, 35), (0.2, 40), and (0.9, 75) satisfy the equation $v = 50n + 30$.

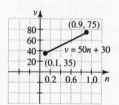

61. We want the equation of a line through the points (2, 113) and (5, 209). The slope is $m = \frac{209 - 113}{5 - 2} = \frac{96}{3} = 32$. Use slope 32 and the point (2, 113) in the point-slope form.
$$C - 113 = 32(n - 2)$$
$$C - 113 = 32n - 64$$
$$C = 32n + 49$$

For 4 days the charge would be $C = 32(4) + 49 = \$177$.

112

63. Find the equation of the line through (0.9, 0.1) and (0.8, 0.2).

$$m = \frac{0.2 - 0.1}{0.8 - 0.9} = \frac{0.1}{-0.1} = -1$$
$$q - 0.1 = -1(p - 0.9)$$
$$q - 0.1 = -p + 0.9$$
$$q = -p + 1$$
$$q = 1 - p$$

65. The set of ordered pairs is a function because no two ordered pairs have the same first coordinate and different second coordinates.

67. The set of ordered pairs is not a function because (3, 4) and (3, 5) have the same first coordinate and different second coordinates.

69. The equation $y = 45x$ determines the y-coordinate from the x-coordinate. So no two ordered pairs could have the same x-coordinate and different y-coordinates. The set of ordered pairs is a function.

71. The equation $y = x^2 + 10$ determines a unique y for every value of x. So this equation defines a function.

73. The ordered pairs (0, 1) and (0, −1) both satisfy the equation $x^2 + y^2 = 1$. So the equation does not define a function.

75. In the formula $f(x) = 2x - 3$, any real number could be used in place of x and any real number could occur as a result of multiplying by 2 and subtracting 3. So the domain is the set of all real numbers and the range is the set of all real numbers.

77. The domain is the set of first coordinates, {1, 2, 3} and the range is the set of second coordinates, {0, 2}.

79. In the equation $y = x^2$, we can use any real number for x. So the domain is the set of all real numbers. Since the square of a real number is nonnegative, the range is the set of nonnegative real numbers.

81. If y varies directly as w, then $y = kw$. If $y = 48$ when $w = 4$, then $48 = k \cdot 4$, or $k = 12$. The formula is $y = 12w$. If $w = 11$, then $y = 12 \cdot 11 = 132$.

83. If y varies inversely as v, then $y = k/v$. If $y = 8$ when $v = 6$, then $8 = k/6$, or $k = 48$. So the formula is $y = 48/v$. If $v = 24$, then $y = 48/24 = 2$.

85. If y varies jointly as u and v, then $y = kuv$. If $y = 72$ when $u = 3$ and $v = 4$, then $72 = k \cdot 3 \cdot 4$, or $k = 6$. So the equation is $y = 6uv$. If $u = 5$ and $v = 2$, then $y = 6 \cdot 5 \cdot 2 = 60$.

87. If the cost C (in dollars) varies directly with the length n of the ride (in minutes), then $C = kn$. If a 12-minute ride costs \$9.00, then $9 = k \cdot 12$, or $k = 0.75$. So the equation is $C = 0.75n$. If $n = 20$ minutes, then $C = 0.75(20) = \$15$.

CHAPTER 6 TEST

1. Since the x-coordinate is negative and the y-coordinate is positive, the point (−2, 7) lies in quadrant II.

2. Since the y-coordinate is 0, $(-\pi, 0)$ lies on the x-axis.

3. Since the x-coordinate is positive and the y-coordinate is negative, (3, −6) lies in quadrant IV.

4. Since the x-coordinate is 0, (0, 1785) lies on the y-axis.

5. $m = \frac{4-3}{4-3} = \frac{1}{1} = 1$

6. $m = \frac{-3-(-8)}{-2-4} = \frac{5}{-6} = -\frac{5}{6}$

7. Use the slope −1/2 and y-intercept (0, 3) in the slope-intercept form: $y = -\frac{1}{2}x + 3$.

8. Use slope 3/7 and the point (−1, −2) in point-slope form.

$$y - (-2) = \frac{3}{7}(x - (-1))$$
$$y + 2 = \frac{3}{7}x + \frac{3}{7}$$
$$y = \frac{3}{7}x - \frac{11}{7}$$

9. The line $y = -3x + 12$ has slope −3. Any line perpendicular to it has slope 1/3. Use slope 1/3 and the point (2, −3) in the point slope form.

$$y - (-3) = \frac{1}{3}(x - 2)$$
$$y + 3 = \frac{1}{3}x - \frac{2}{3}$$
$$3y + 9 = x - 2$$
$$-x + 3y = -11$$
$$x - 3y = 11$$

10. The equation $5x + 3y = 9$ is equivalent to $y = -\frac{5}{3}x + 3$. Its slope is $-\frac{5}{3}$ and any line parallel to it has slope $-\frac{5}{3}$. Use slope $-\frac{5}{3}$ and the point $(3, 4)$ in the point-slope form.

$$y - 4 = -\frac{5}{3}(x - 3)$$
$$y - 4 = -\frac{5}{3}x + 5$$
$$3y - 12 = -5x + 15$$
$$5x + 3y = 27$$

11. The graph of $y = \frac{1}{2}x - 3$ has y-intercept of $(0, -3)$ and a slope of 1/2. Start at $(0, -3)$ and rise 1 and run 2 to locate a second point. Draw a line through the two points.

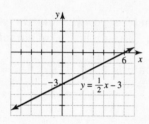

12. If $x = 0$ in $2x - 3y = 6$ then $y = -2$, and if $y = 0$ in $2x - 3y = 6$, then $x = 3$. So the x-intercept is $(3, 0)$ and the y-intercept is $(0, -2)$. Draw a line through the two intercepts.

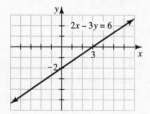

13. The graph of $y = 4$ is a horizontal line with y-intercept of $(0, 4)$.

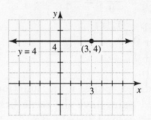

14. The graph of $x = -2$ is a vertical line with x-intercept $(-2, 0)$.

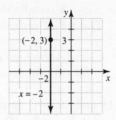

15. Yes, the set is a function.

16. No, the set is not a function, because the ordered pairs $(1, 1)$ and $(1, -1)$ both belong to the set.

17. In the formula $f(x) = |x| + 1$, we can use any real number in place of x. So the domain is the set of all real numbers. Since $|x|$ is greater than or equal to 0, $|x| + 1$ has a value that is greater than or equal to 1. So the range is $\{y \mid y \geq 1\}$.

18. In the equation $y = x^2$, x can be any real number. So the domain is R. Since y is equal to the square of x, y must also be nonnegative. So the range is $\{y \mid y \geq 0\}$.

19. If $f(x) = 2x + 5$, then $f(-2) = 2(-2) + 5 = 1$.

20. If $g(x) = x^2 - 4$, then $g(3) = 3^2 - 4 = 5$

21. Plot the three points and find the slope of each line segment.

$$m_1 = \frac{2 - (-3)}{3 - (-2)} = \frac{5}{5} = 1 \qquad m_2 = \frac{2 - 0}{3 - 4} = \frac{2}{-1} = -2$$
$$m_3 = \frac{0 - (-3)}{4 - (-2)} = \frac{3}{6} = \frac{1}{2}$$

Since m_2 is the opposite of the reciprocal of m_3, the two line segments are perpendicular and the triangle is a right triangle.

22. To find the total shipping and handling fee S, we multiply the number of CDs n by $0.75 and add on the $2.50 fee: $S = 0.75n + 2.50$.

23. We want the equation of the line containing the two points (10, 50) and (16, 68). First find the slope.

$$m = \frac{68 - 50}{16 - 10} = \frac{18}{6} = 3$$

Use slope 3 and the point (10, 50) in the point-slope form where P is the price in cents and v is the volume of the cup in ounces.

$$P - 50 = 3(v - 10)$$
$$P - 50 = 3v - 30$$
$$P = 3v + 20$$

If v = 20 ounces, then $P = 3(20) + 20 = 80$ cents.

24. If the price P (in dollars) varies directly with the weight w (in pounds), then $P = kw$. If a 30-pound watermelon sells for $4.20, then $4.20 = k \cdot 30$, or $k = 0.14$. So the formula is $P = 0.14w$. If w = 20 pounds, then $P = 0.14(20) = \$2.80$.

25. Since the study time T is inversely proportional to the score s, $T = k/s$. If $T = 3$ when $s = 60$, then $3 = k/60$, or $k = 180$. The formula is $T = 180/s$. If $s = 90$, then his study time is $T = 180/90 = 2$ hours.

Tying It All Together   Chapters 1-6

1. $3^2 - 2^3 = 9 - 8 = 1$

2. $3^2 \cdot 2^3 = 9 \cdot 8 = 72$

3. $10^4 \cdot 10^9 = 10^{13}$

4. $(2^{12}) \div (2^{10}) = 2^{12 - 10} = 2^2 = 4$

5. $(34 \cdot 258)^0 = 1$

6. $(8^0 - 3^2)^3 = (1 - 9)^3 = (-8)^3 = -512$

7. $\left(\frac{1}{2}\right)^3 + \left(\frac{2}{3}\right)^2 = \frac{1}{8} + \frac{4}{9} = \frac{9}{72} + \frac{32}{72} = \frac{41}{72}$

8. $\left(-\frac{3}{2}\right)^3 - \left(-\frac{3}{4}\right)^2 = -\frac{27}{8} - \frac{9}{16} = -\frac{54}{16} - \frac{9}{16}$

$$= -\frac{63}{16}$$

9. $\left(\frac{3}{5}\right)^3 \cdot \left(\frac{5}{6}\right)^2 = \frac{27}{125} \cdot \frac{25}{36} = \frac{9 \cdot 3}{25 \cdot 5} \cdot \frac{25}{9 \cdot 4} = \frac{3}{20}$

10. $\left(-\frac{3}{5}\right)^3 \div \left(-\frac{6}{5}\right)^4 = -\frac{27}{125} \cdot \frac{625}{1296}$

$$= -\frac{27}{125} \cdot \frac{125 \cdot 5}{27 \cdot 48} = -\frac{5}{48}$$

11. $\dfrac{\left(\frac{1}{4} - \frac{1}{8}\right)8}{\left(\frac{3}{4} + \frac{1}{2}\right)8} = \dfrac{2 - 1}{6 + 4} = \dfrac{1}{10}$

12. $\dfrac{\left(\frac{1}{3} - \frac{1}{5}\right)60}{\left(\frac{3}{10} + \frac{1}{20}\right)60} = \dfrac{20 - 12}{18 + 3} = \dfrac{8}{21}$

13. $-3(2x - 7) = -3 \cdot 2x - (-3)7 = -6x + 21$

14. $x - 3(2x - 7) = x - 6x + 21 = -5x + 21$

15. $(x - 3)(2x - 7) = 2x^2 - 6x - 7x + 21$
$$= 2x^2 - 13x + 21$$

16. $(2x - 1)^2 = (2x)^2 - 2 \cdot 2x \cdot 1 + 1^2$
$$= 4x^2 - 4x + 1$$

17. $(z + 5)^2 = z^2 + 2 \cdot z \cdot 5 + 5^2 = z^2 + 10z + 25$

18. $(w - 7)(w + 7) = w^2 - 49$

19. $3x^2y^3 \cdot 12xy^4 = 36x^3y^7$

20. $(2x^2y)^3 \cdot 5x^2y^6 = 8x^6y^3 \cdot 5x^2y^6 = 40x^8y^9$

21. The graph of $y = \frac{1}{3}x$ has a y-intercept of (0, 0) and a slope of 1/3. From the origin, rise 1 and run 3 to locate the point (3, 1). Draw a line through (0, 0) and (3, 1).

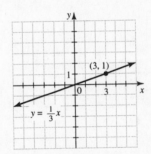

22. The graph of $y = 3x$ has a y-intercept of (0, 0) and a slope of 3. From the origin rise 3 and run 1 to locate the point (1, 3). Draw a line through (0, 0) and (1, 3).

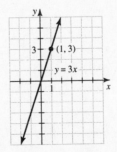

23. The graph of $y = -3x$ has slope of $-3 = -3/1$ and a y-intercept of $(0, 0)$. From the origin go down 3 units and 1 unit to the right to locate the point $(1, -3)$. Draw a line through $(0, 0)$ and $(1, -3)$.

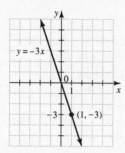

24. The graph of $y = -\frac{1}{3}x$ has a y-intercept of $(0, 0)$ and a slope of $-1/3$. From the origin go down 1 unit and then 3 units to the right to locate the point $(3, -1)$. Draw a line through $(0, 0)$ and $(3, -1)$.

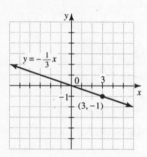

25. The graph of $y = 3x + 1$ has a y-intercept of $(0, 1)$ and a slope of $3 = 3/1$. Start at $(0, 1)$ and go up 3 units and then 1 unit to the right to locate the point $(1, 4)$. Draw a line through $(0, 1)$ and $(1, 4)$.

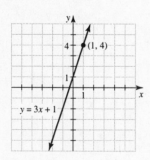

26. The graph of $y = 3x - 2$ has a y-intercept of $(0, -2)$ and a slope of $3 = 3/1$. Start at $(0, -2)$ and go up 3 units and then 1 unit to the right to locate the point $(1, 1)$. Draw a line through $(0, -2)$ and $(1, 1)$.

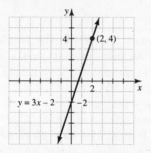

27. The graph of $y = 3$ is a line with slope 0 and y-intercept $(0, 3)$. Draw a horizontal line through $(0, 3)$.

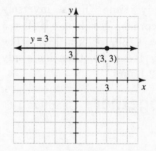

28. The graph of $x = 3$ is a vertical line with x-intercept $(3, 0)$. Draw a vertical line through $(3, 0)$.

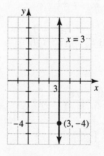

29.
$$3\pi y + 2 = t$$
$$3\pi y = t - 2$$
$$y = \frac{t - 2}{3\pi}$$

116

30.
$$x = \frac{y - b}{m}$$
$$mx = y - b$$
$$mx + b = y$$
$$y = mx + b$$

31. $3x - 3y - 12 = 0$
$$-3y = -3x + 12$$
$$y = x - 4$$

32. $2y - 3 = 9$
$$2y = 12$$
$$y = 6$$

33. $y^2 - 3y - 40 = 0$
$$(y - 8)(y + 5) = 0$$
$$y - 8 = 0 \quad \text{or} \quad y + 5 = 0$$
$$y = 8 \quad \text{or} \quad y = -5$$

34.
$$\frac{y}{2} - \frac{y}{4} = \frac{1}{5}$$
$$20\left(\frac{y}{2} - \frac{y}{4}\right) = 20\left(\frac{1}{5}\right)$$
$$10y - 5y = 4$$
$$5y = 4$$
$$y = \frac{4}{5}$$

35.
$$5 = 4x - 7$$
$$-4x = -7 - 5$$
$$-4x = -12$$
$$x = 3$$
The solution to the equation is 3.

36.
$$5 = 4x^2 - 11$$
$$16 = 4x^2$$
$$0 = 4x^2 - 16$$
$$x^2 - 4 = 0$$
$$(x - 2)(x + 2) = 0$$
$$x - 2 = 0 \quad \text{or} \quad x + 2 = 0$$
$$x = 2 \quad \text{or} \quad x = -2$$
The solutions to the equation are −2 and 2.

37. $(3x - 4)(x + 9) = 0$
$$3x - 4 = 0 \quad \text{or} \quad x + 9 = 0$$
$$x = \frac{4}{3} \quad \text{or} \quad x = -9$$
The solutions to the equation are −9 and $\frac{4}{3}$.

38.
$$\frac{2}{3} - \frac{x}{6} = \frac{1}{2} + \frac{x}{4}$$
$$12 \cdot \frac{2}{3} - 12 \cdot \frac{x}{6} = 12 \cdot \frac{1}{2} + 12 \cdot \frac{x}{4}$$
$$8 - 2x = 6 + 3x$$
$$8 = 6 + 5x$$
$$2 = 5x$$
$$\frac{2}{5} = x$$
The solution to the equation is $\frac{2}{5}$.

39. $2x^2 - 7x = 0$
$$x(2x - 7) = 0$$
$$x = 0 \quad \text{or} \quad 2x - 7 = 0$$
$$x = 0 \quad \text{or} \quad x = \frac{7}{2}$$
The solutions to the equation are 0 and $\frac{7}{2}$.

40.
$$\frac{3}{x} = \frac{x - 1}{2}$$
$$x(x - 1) = 6$$
$$x^2 - x - 6 = 0$$
$$(x - 3)(x + 2) = 0$$
$$x - 3 = 0 \quad \text{or} \quad x + 2 = 0$$
$$x = 3 \quad \text{or} \quad x = -2$$
The solutions to the equation are −2 and 3.

41. a) $m = \frac{3 - 1}{50 - 35} = \frac{2}{15}$

b) $m = \frac{6 - 3}{65 - 50} = \frac{3}{15} = \frac{1}{5}$

c) A 38 year-old man should save twice his income during the years 35 to 50, or $80,000. To do this in 15 years he would have to save $80,000 ÷ 15 or $5333 dollars per year.
$$\frac{5333}{40000} \approx 0.133$$
So he should save about 13% of his income. Since this calculation does not include interest it is very approximate and other answers might also be reasonable.

d) The 58-year old women needs to get from $180,000 to $360,000 in 15 years, or she needs to save about $12,000 per year. So at age 58 she should have saved

$3(60,000) + 8(12,000) = \$276,000.$

These figures do not include interest and are approximate.

7.1 WARM-UPS

1. True, because $-3 = 2(1) - 5$ is correct.
2. False, because $-3 = -2(1) - 5$ is incorrect.
3. False, because it does not satisfy both equations.
4. False, because the lines have different slopes and will intersect in exactly one point.
5. True, because the two lines have the same slope (so they are parallel) but different y-intercepts.
6. True, because the system has no solution.
7. True, because the first equation is equivalent to $y = 9 - x$.
8. True, because all ordered pairs that satisfy $y = 9 - x$ also satisfy $x + y = 9$.
9. True.
10. False, because the exact solution cannot usually be found be graphing the coordinates are not integers.

7.1 EXERCISES

1. The point $(3, -2)$ satisfies both equations because $2(3) + (-2) = 4$ and $3 - (-2) = 5$ are both correct. The point $(6, 1)$ does not satisfy $2x + y = 4$, and $(2, 4)$ does not satisfy $x - y = 5$. So $(3, -2)$ is the only solution to the system.

3. Check $(0, -2)$: $\quad 6(0) - 2(-2) = 4 \quad$ Correct.
$\qquad\qquad\qquad -2 = 3(0) - 2 \quad$ Correct.
Check $(2, 4)$: $\quad 6(2) - 2(4) = 4 \quad$ Correct.
$\qquad\qquad\quad 4 = 3(2) - 2 \quad$ Correct.
Check $(3, 7)$: $\quad 6(3) - 2(7) = 4 \quad$ Correct.
$\qquad\qquad\quad 7 = 3(3) - 2 \quad$ Correct.
All three points $(0, -2)$, $(2, 4)$, and $(3, 7)$ are solutions to the system.

5. Check $(3, 3)$: $\quad 2(3) - 3 = 3 \quad$ Correct.
$\qquad\qquad\quad 2(3) - 3 = 2 \quad$ Incorrect.
Check $(5, 7)$: $\quad 2(5) - 7 = 3 \quad$ Correct.
$\qquad\qquad\quad 2(5) - 7 = 2 \quad$ Incorrect.
Check $(7, 11)$: $\quad 2(7) - 11 = 3 \quad$ Correct.
$\qquad\qquad\quad 2(7) - 11 = 2 \quad$ Incorrect.
None of the three points satisfies both equations of the system.

7. The graphs appear to intersect at $(-2, 3)$.
$$3 = 3(-2) + 9$$
$$2(-2) + 3(3) = 5$$
Since $(-2, 3)$ satisfies both equations, it is the solution to the system.

9. The graph of $y = 2x$ is a straight line with y-intercept $(0, 0)$ and slope 2. The graph of $y = -x + 6$ is a straight line with y-intercept $(0, 6)$ and slope -1. It appears that the lines intersect at $(2, 4)$. By checking $(2, 4)$ in both equations we can be sure that the solution to the system is $(2, 4)$.

11. The graph of $y = 3x - 1$ has a y-intercept of $(0, -1)$ and a slope of 3. The equation $2y - 3x = 1$ is equivalent to $y = \frac{3}{2}x + \frac{1}{2}$. The y-intercept is $(0, 1/2)$ and the slope is $3/2$. It appears that the lines intersect at $(1, 2)$. By checking $(1, 2)$ in both equations we can be certain that $(1, 2)$ is the solution to the system.

13. The line $y = x - 5$ has a y-intercept of $(0, -5)$ and a slope of 1. The equation $x + y = -5$ is equivalent to $y = -x - 5$, which has a y-intercept of $(0, -5)$ and a slope of -1. The lines have the same y-intercept $(0, -5)$. The solution to the system is $(0, -5)$.

15. The equation $2y + x = 4$ is equivalent to $y = -\frac{1}{2}x + 2$, which has y-intercept $(0, 2)$ and slope $-1/2$. The equation $2x - y = -7$ is equivalent to $y = 2x + 7$, which has y-intercept $(0, 7)$ and slope 2. It appears that $(-2, 3)$ is the intersection of the two lines. Check $(-2, 3)$ in both of the original equations to be sure that it is the solution to the system.

17. The graph of $y = x$ has a y-intercept of $(0, 0)$ and a slope of 1. The equation $x + y = 0$ is equivalent to $y = -x$, which has a y-intercept of $(0, 0)$ and a slope of -1. The lines intersect at $(0, 0)$, the y-intercept of each line. The solution to the system is $(0, 0)$.

19. The line $y = 2x - 1$ has y-intercept $(0, -1)$ and slope 2. The equation $x - 2y = -4$ is equivalent to $y = \frac{1}{2}x + 2$, which has y-intercept $(0, 2)$ and slope $1/2$. It appears that the lines intersect at $(2, 3)$. Check that $(2, 3)$ satisfies both of the original equations to be sure that $(2, 3)$ is the solution to the system.

21. The equation $x - y = 3$ is equivalent to $y = x - 3$, which has y-intercept $(0, -3)$ and slope 1. The equation $3x = 3y + 12$ is equivalent to $y = x - 4$, which has y-intercept $(0, -4)$ and slope 1. Since the two lines are parallel, there is no solution to the system. The system is inconsistent.

23. The equation $x - y = 3$ is equivalent to $y = x - 3$, which has y-intercept $(0, -3)$ and slope 1. The equation $3x = y + 5$ is equivalent to $y = 3x - 5$, which has y-intercept $(0, -5)$ and slope 3. The lines are not parallel and appear to intersect at $(1, -2)$. Check $(1, -2)$ in both equations to be sure that $(1, -2)$ is the solution to the system. The system is independent.

25. The equation $2x + y = 3$ is equivalent to $y = -2x + 3$, which has a y-intercept of $(0, 3)$ and slope -2. The equation $6x - 9 = -3y$ is equivalent to $y = -2x + 3$, which has a y-intercept of $(0, 3)$ and slope -2. The two equations have the same graph. The set of points that satisfy both equations is the same as the set of points that satisfy one of the equations, $\{(x, y) \mid 2x + y = 3\}$. The system is dependent.

27. The equation $x - y = 0$ is equivalent to $y = x$. The equation $5x = 5y$ is equivalent to $y = x$. The two equations have the same graph. The set of points that satisfy both equations is the same as the set of points that satisfy one of the equations, $\{(x, y) \mid y = x\}$. The system is dependent.

29. The equation $x - y = -1$ is equivalent to $y = x + 1$, which has y-intercept $(0, 1)$ and slope 1. The line $y = \frac{1}{2}x - 1$ has y-intercept $(0, -1)$ and slope $1/2$. The lines appear to intersect at $(-4, -3)$. Check $(-4, -3)$ in both of the original equations to be sure that $(-4, -3)$ is the solution to the system of equations. Since the system has only one solution, the system is independent.

31. The equation $y - 4x = 4$ is equivalent to $y = 4x + 4$, which has y-intercept $(0, 4)$ and slope 4. The equation $y + 4x = -4$ is equivalent to $y = -4x - 4$, which has y-intercept $(0, -4)$ and slope -4. The lines are not parallel and they appear to intersect at $(-1, 0)$. Check $(-1, 0)$ in both of the original equations to be sure that $(-1, 0)$ is the solution to the system. Since the system has only one solution, the system is independent.

33. It appears that the graphs intersect at $(0, 0)$ and $(2, 4)$. Since both of these ordered pairs satisfy both equations, they are both solutions to the system.

35. From the graph it appears that movie spending and home video spending were equal in 1986. From the graph it appears that about \$4 billion was spent in 1986 on each source. It appears that in 1993 home video spending was twice that of movie spending.

37. The graphs intersect when $x = 15,000$ copies. So the total cost is the same for either machine at 15,000 copies. If she buys another at 10,000 copies, then the Panasonic copier is cheaper.

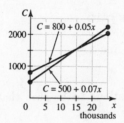

7.2 WARM-UPS

1. True.   **2.** True, because the solution to $2x + 3(x - 7) = 4$ is 5.

3. True, because if $x = 5$, then $y = 5 - 7 = -2$.

4. True, because $\frac{1}{2} + 2 \cdot \frac{1}{4} = 1$ and $2 \cdot \frac{1}{2} - 4 \cdot \frac{1}{4} = 0$ are both correct.

5. True, because $(1/2, 1/4)$ would be difficult to read from a graph.

6. True, because we can always solve one equation for one of the variables and substitute into the other.

7. False, because it has only one solution.

8. True, because a false statement indicates an inconsistent system.

9. True, because an identity indicates a dependent system.

10. True, because we can always solve one of the equations for one variable and substitute into the other equation.

7.2 EXERCISES

1. Replace y by $x + 3$ in the equation $2x - 3y = -11$.

$$2x - 3(x + 3) = -11$$
$$2x - 3x - 9 = -11$$
$$-x = -2$$
$$x = 2$$

Use $x = 2$ in the equation $y = x + 3$ to find y.

$$y = 2 + 3$$
$$y = 5$$

Check $(2, 5)$ in both of the original equations to be sure that the ordered pair $(2, 5)$ is the solution to the system.

3. Since $x = 2y - 4$, we can replace x by $2y - 4$ in the equation $2x + y = 7$.

$$2(2y - 4) + y = 7$$
$$4y - 8 + y = 7$$
$$5y = 15$$
$$y = 3$$

Use $y = 3$ in $x = 2y - 4$ to find x.

$$x = 2(3) - 4$$
$$x = 2$$

Check $(2, 3)$ in both equations to be sure that $(2, 3)$ is the solution to the system.

5. The equation $2x + y = 5$ is equivalent to $y = 5 - 2x$. Replace y by $5 - 2x$ in the equation $5x + 2y = 8$.

$$5x + 2(5 - 2x) = 8$$
$$5x + 10 - 4x = 8$$
$$x = -2$$

Use $x = -2$ in $y = 5 - 2x$ to find y.

$$y = 5 - 2(-2)$$
$$y = 9$$

Check that $(-2, 9)$ satisfies both equations. The ordered pair $(-2, 9)$ is the solution to the system.

7. The equation $x + y = 0$ is equivalent to $y = -x$. Replace y by $-x$ in the equation $3x + 2y = -5$.

$$3x + 2(-x) = -5$$
$$3x - 2x = -5$$
$$x = -5$$

Use $x = -5$ in $y = -x$ to find y.

$$y = -(-5) = 5$$

Check that $(-5, 5)$ satisfies both of the original equations. The solution is $(-5, 5)$.

9. The equation $x + y = 1$ is equivalent to $x = 1 - y$. Replace x by $1 - y$ in the equation $4x - 8y = -4$.

$$4(1 - y) - 8y = -4$$
$$4 - 4y - 8y = -4$$
$$-12y = -8$$
$$y = \frac{2}{3}$$

Use $y = 2/3$ in the equation $x = 1 - y$ to find x.

$$x = 1 - \frac{2}{3} = \frac{3}{3} - \frac{2}{3} = \frac{1}{3}$$

Check that the pair $\left(\frac{1}{3}, \frac{2}{3}\right)$ satisfies both of the original equations to be sure that it is the solution to the system.

11. Solve $2x + 3y = 2$ for y.

$$3y = -2x + 2$$
$$y = -\frac{2}{3}x + \frac{2}{3}$$

Replace y by $-\frac{2}{3}x + \frac{2}{3}$ in the equation $4x - 9y = -1$.

$$4x - 9\left(-\frac{2}{3}x + \frac{2}{3}\right) = -1$$
$$4x + 6x - 6 = -1$$
$$10x = 5$$
$$x = \frac{1}{2}$$

Use $x = 1/2$ in the equation $y = -\frac{2}{3}x + \frac{2}{3}$.

$$y = -\frac{2}{3} \cdot \frac{1}{2} + \frac{2}{3} = -\frac{1}{3} + \frac{2}{3} = \frac{1}{3}$$

Check that $\left(\frac{1}{2}, \frac{1}{3}\right)$ is the solution to the system.

13. The equation $x - 2y = -2$ is equivalent to $x = 2y - 2$. Replace x by $2y - 2$ in the equation $x + 2y = 8$.

$$2y - 2 + 2y = 8$$
$$4y = 10$$
$$y = \frac{5}{2}$$

Use $y = \frac{5}{2}$ in $x = 2y - 2$ to find x.

$$x = 2\left(\frac{5}{2}\right) - 2 = 5 - 2 = 3$$

Check that $\left(3, \frac{5}{2}\right)$ satisfies both equations of the system. Since there is only one solution, the system is independent.

15. Use $x = 4 - 2y$ to eliminate x from the equation $4y + 2x = -8$.
$$4y + 2(4 - 2y) = -8$$
$$4y + 8 - 4y = -8$$
$$8 = -8$$
Since the last equation is false, there is no solution to the system of equations. The system is inconsistent.

17. Use $y = 2x + 3$ to eliminate y from the equation $y - 3 = 2(x - 1)$.
$$2x - 3 - 3 = 2(x - 1)$$
$$2x - 6 = 2x - 2$$
$$-6 = -2$$
Since the last equation is false, there is no solution to the system. The system is inconsistent.

19. Solve $3x - 2y = 7$ for y.
$$-2y = -3x + 7$$
$$y = \frac{3}{2}x - \frac{7}{2}$$
Replace y by $\frac{3}{2}x - \frac{7}{2}$ to eliminate y in $3x + 2y = 7$.
$$3x + 2\left(\frac{3}{2}x - \frac{7}{2}\right) = 7$$
$$3x + 3x - 7 = 7$$
$$6x = 14$$
$$x = \frac{7}{3}$$
Use $x = 7/3$ in $y = \frac{3}{2}x - \frac{7}{2}$ to find y.
$$y = \frac{3}{2} \cdot \frac{7}{3} - \frac{7}{2} = \frac{7}{2} - \frac{7}{2} = 0$$
Check that $\left(\frac{7}{3}, 0\right)$ satisfies both equations of the system. Since the system has only one solution, the system is independent.

21. The equation $x + 5y = 4$ is equivalent to $x = 4 - 5y$. Replace x by $4 - 5y$ in the equation $x + 5y = 4y$.
$$4 - 5y + 5y = 4y$$
$$4 = 4y$$
$$1 = y$$
Use $y = 1$ in $x = 4 - 5y$ to find x.
$$x = 4 - 5(1) = -1$$

Check that $(-1, 1)$ satisfies both equations of the system. Since there is only one solution to the system, the system is independent.

23. Let x = the amount invested at 10% and y = the amount invested at 5%. She receives 0.10x interest on the first investment and 0.05y interest on the second investment. We can write one equation about her total investment and the other about her total interest.
$$x + y = 20,000$$
$$0.10x + 0.05y = 1600$$
The equation $x + y = 20,000$ is equivalent to $x = 20,000 - y$. Substitute this expression into the second equation to eliminate x.
$$0.10(20,000 - y) + 0.05y = 1600$$
$$2,000 - 0.10y + 0.05y = 1600$$
$$-0.05y = -400$$
$$y = 8,000$$
$$x = 20,000 - 8,000$$
$$x = 12,000$$
She invested $12,000 at 10% and $8,000 at 5%.

25. Let x = Spielberg's earnings and y = Culkin's earnings both in millions of dollars.
$$x + y = 58$$
$$x - y = 26$$
The equation $x - y = 26$ is equivalent to $x = y + 26$. Substitute this expression into $x + y = 58$ to eliminate x.
$$y + 26 + y = 58$$
$$2y = 32$$
$$y = 16$$
$$x = 16 + 26 = 42$$
So Spielberg earned $42 million and Culkin earned $16 million.

27. Let x = the amount of money he gets for mowing the lawn and y = the amount of money he gets for shoveling the snow. He made $50 for mowing the lawn 3 times and shoveling the sidewalk 2 times. So $3x + 2y = 50$. Last year he made $45 for mowing the lawn 2 times and shoveling the sidewalk 3 times. So $2x + 3y = 45$. Solve the first equation for y.

$$3x + 2y = 50$$
$$2y = -3x + 50$$
$$y = -\frac{3}{2}x + 25$$

Substitute this expression for y into the other equation.

$$2x + 3\left(-\frac{3}{2}x + 25\right) = 45$$
$$2x - \frac{9}{2}x + 75 = 45$$

$$-\frac{5}{2}x = -30$$

$$x = -\frac{2}{5}(-30) = 12$$

Use $x = 12$ in the equation $y = -\frac{3}{2}x + 25$.

$$y = -\frac{3}{2}(12) + 25 = 7$$

He makes $12 for mowing the lawn and $7 for shoveling the sidewalk.

29. Let $x =$ the weight on the left rear and $y =$ the weight on the left front. The right front weight is 264 pounds.

$$x + y = 0.50(1150)$$
$$x + 264 = 0.48(1150)$$
$$x + 264 = 552$$
$$x = 288$$
$$288 + y = 0.50(1150)$$
$$y = 287$$

So the left rear should weight 288 pounds and the left front should weight 287 pounds. The right rear should weigh $1150 - 288 - 287 - 264$ or 311 pounds. Since $0.52(1150) = 598$ pounds, and $311 + 288 = 599$, this weight distribution violates the rules.

31. Substitute $200x + 600$ for y in the equation $y = -150x + 900$:

$$200x + 60 = -150x + 900$$
$$350x = 840$$
$$x = 2.40$$

The price at which supply equals demand is $2.40 per pound.

33. a) Life expectancy of a male born in 1975 is

$$y = 0.159(1975) - 244.3 = 69.7 \text{ years.}$$

Life expectancy of a female born in 1975 is

$$y = 0.203(1975) - 325.7 = 75.2 \text{ years.}$$

c) No, because the slope of the line for females is larger than the slope for males.

d) Life expectancies are equal for x that satisfies

$$0.203x - 325.7 = 0.159x - 244.3$$
$$0.044x = 81.4$$
$$x = 1850$$

7.3 WARM-UPS

1. False, we multiply the first by -4 and the second by 3 and then add.
2. True, because we can multiply each equation by an appropriate number to cause the desired variable to be eliminated.
3. True, because $3(1) + 2(2) = 7$ and $4(1) - 5(2) = -6$ are both correct.
4. True, but first the equations should be written in the same form.
5. True, because $2 = -3(0) + 2$ and $-1 = -3(1) + 2$ are correct, and $2(2) + 6(0) - 4 = 0$ and $2(-1) + 6(1) - 4 = 0$ are correct. **6.** False, because $y = x - 5$ and $y = x - 6$ are inconsistent. The graphs are two different parallel lines. **7.** False, because because system (c) has no solution. **8.** False, because system (b) is dependent. **9.** False, system (a) is independent. **10.** True, because the system is inconsistent.

7.3 EXERCISES

1.
$$2x + y = 5$$
$$\underline{3x - y = 10}$$
$$5x \qquad = 15$$
$$x = 3$$

Use $x = 3$ in $2x + y = 5$ to find y.
$$2(3) + y = 5$$
$$y = -1$$
The solution to the system is $(3, -1)$.

3.
$$x + 2y = 7$$
$$\underline{-x + 3y = 18}$$
$$5y = 25$$
$$y = 5$$

Use $y = 5$ in $x + 2y = 7$ to find x.
$$x + 2(5) = 7$$
$$x = -3$$
The solution to the system is $(-3, 5)$.

5.
$$x + 2y = 2$$
$$-4x + 3y = 25$$
Multiply each side of the first equation by 4 and add the result to the second equation.
$$4x + 8y = 8$$
$$\underline{-4x + 3y = 25}$$
$$11y = 33$$
$$y = 3$$

Use $y = 3$ in $x + 2y = 2$ to find x.
$$x + 2(3) = 2$$
$$x = -4$$
The solution to the system is $(-4, 3)$.

7.
$$x + 3y = 4$$
$$2x - y = -1$$
Multiply each side of the second equation by 3 and add the result to the first equation.
$$x + 3y = 4$$
$$\underline{6x - 3y = -3}$$
$$7x \quad\quad = 1$$
$$x = \frac{1}{7}$$
Use $x = 1/7$ in $x + 3y = 4$ to find y.
$$\frac{1}{7} + 3y = 4$$
$$1 + 21y = 28$$
$$21y = 27$$
$$y = \frac{9}{7}$$
The solution to the system is $\left(\frac{1}{7}, \frac{9}{7}\right)$.

9.
$$y = 4x - 1$$
$$y = 3x + 7$$
Multiply the second equation by -1 and add the result to the first equation.
$$y = 4x \ - \ 1$$
$$\underline{-y = -3x - 7}$$
$$0 = \ x \ - \ 8$$
$$8 = x$$
Use $x = 8$ in $y = 4x - 1$ to find y.
$$y = 4(8) - 1 - 31$$
The solution to the system is $(8, 31)$.

11.
$$4x = 3y + 1$$
$$2x = y - 1$$
Multiply the second equation by -2 and add the result to the first equation.
$$4x = 3y + 1$$
$$\underline{-4x = -2y + 2}$$
$$0 = y + 3$$
$$-3 = y$$
Use $y = -3$ in $2x = y - 1$ to find x.
$$2x = -3 - 1$$
$$2x = -4$$
$$x = -2$$
The solution to the system is $(-2, -3)$.

13.
$$2x - 5y = -22$$
$$-6x + 3y = 18$$
Multiply the first equation by 3 and add the result to the second equation.
$$6x - 15y = -66$$
$$\underline{-6x + 3y = 18}$$
$$-12y = -48$$
$$y = 4$$
Use $y = 4$ in $2x - 5y = -22$ to find x.

$$2x - 5(4) = -22$$
$$2x - 20 = -22$$
$$2x = -2$$
$$x = -1$$
The solution to the system is $(-1, 4)$.

15.
$$2x + 3y = 4$$
$$-3x + 5y = 13$$
Multiply the first equation by 3 and the second equation by 2.
$$6x + 9y = 12$$
$$\underline{-6x + 10y = 26}$$
$$19y = 38$$
$$y = 2$$
Use $y = 2$ in $2x + 3y = 4$ to find x.
$$2x + 3(2) = 4$$
$$2x = -2$$
$$x = -1$$
The solution to the system is $(-1, 2)$.

17.
$$2x - 5y = 11$$
$$3x - 2y = 11$$
Multiply the first equation by -3 and the second by 2.
$$-6x + 15y = -33$$
$$\underline{6x - 4y = 22}$$
$$11y = -11$$
$$y = -1$$
Use $y = -1$ in $2x - 5y = 11$.
$$2x - 5(-1) = 11$$
$$2x + 5 = 11$$
$$2x = 6$$
$$x = 3$$
The solution to the system is $(3, -1)$.

19.
$$5x + 4y = 13$$
$$2x + 3y = 8$$
Multiply the first equation by -3 and the second by 4.
$$-15x - 12y = -39$$
$$\underline{8x + 12y = 32}$$
$$-7x \quad\quad = -7$$
$$x = 1$$
Use $x = 1$ in $5x + 4y = 13$ to find y.
$$5(1) + 4y = 13$$
$$4y = 8$$
$$y = 2$$
The solution to the system is $(1, 2)$.

21.
$$x + y = 5$$
$$x + y = 6$$

Multiply the first equation by -1 and add the result to the second equation.

$$-x - y = -5$$
$$\underline{x + y = 6}$$
$$0 = 1$$

Since the result of the addition is a false statement, there is no solution to the system. The system is inconsistent.

23.
$$x + y = 5$$
$$2x + 2y = 10$$

Multiply the first equation by -2 and add the result to the second equation.
$$-2x - 2y = -10$$
$$\underline{2x + 2y = 10}$$
$$0 = 0$$

Since the result of the addition is an identity, the system is dependent. In fact, the second equation is a multiple of the first equation. So the set of solutions to the system is the same as the set of solutions to one of the equations, $\{(x, y) \mid x + y = 5\}$.

25.
$$2x = y + 3$$
$$2y = 4x - 6$$

Rewrite the first equation in the same order as the second.
$$-y = -2x + 3$$
$$2y = 4x - 6$$

Multiply the first equation by 2.
$$-2y = -4x + 6$$
$$\underline{2y = 4x - 6}$$
$$0 = 0$$

Since the result of the addition is an identity, the system is dependent. The set of solutions to the system is $\{(x, y) \mid 2x = y + 3 \}$.

27.
$$x + 3y = 3$$
$$5x = 15 - 15y$$

Rewrite the second equation as $5x + 15y = 15$. Multiplying the first equation by -5 gives us $-5x - 15y = -15$. Add these equations.
$$5x + 15y = 15$$
$$\underline{-5x - 15y = -15}$$
$$0 = 0$$

Since the result of the addition is an identity, the system is dependent. The set of solutions to the system is $\{(x, y) \mid x + 3y = 3\}$.

29.
$$6x - 2y = -2$$
$$\tfrac{1}{3}y = x + \tfrac{4}{3}$$

Multiply the second equation by 3 to get $y = 3x + 4$. Then substitute this into the first equation to eliminate y.

$$6x - 2(3x + 4) = -2$$
$$6x - 6x - 8 = -2$$
$$-8 = -2$$

Since the substitution resulted in a false statement, there is no solution to the system. The system is inconsistent.

31. Multiply the first equation by 6 and the second equation by 4.

$$6\left(\tfrac{1}{2}x - \tfrac{2}{3}y\right) = 6(-6)$$
$$4\left(-\tfrac{3}{4}x - \tfrac{1}{2}y\right) = 4(-18)$$
$$3x - 4y = -36$$
$$\underline{-3x - 2y = -72}$$
$$-6y = -108$$
$$y = 18$$

Use $y = 18$ in $3x - 4y = -36$ to find x.
$$3x - 4(18) = -36$$
$$3x - 72 = -36$$
$$3x = 36$$
$$x = 12$$

The solution to the system is (12, 18). The system is independent.

33. Multiply the first equation by 100 and the second by -4.

$$100(0.04x + 0.09y) = 100(7)$$
$$-4(x + y) = -4(100)$$

$$4x + 9y = 700$$
$$\underline{-4x - 4y = -400}$$
$$5y = 300$$
$$y = 60$$

Use $y = 60$ in $x + y = 100$ to get $x = 40$. The solution to the system is (40, 60) and the system is independent.

35. Multiply the first equation by -30 and the second by 10.

$$-30(0.1x - 0.2y) = -30(-0.01)$$
$$10(0.3x + 0.5y) = 10(0.08)$$

$$-3x + 6y = 0.3$$
$$\underline{3x + 5y = 0.8}$$
$$11y = 1.1$$
$$y = 0.1$$

Use $y = 0.1$ in $3x + 5y = 0.8$ to find x.
$$3x + 5(0.1) = 0.8$$
$$3x + 0.5 = 0.8$$
$$3x = 0.3$$
$$x = 0.1$$

The solution to the system is (0.1, 0.1) and the system is independent.

37.
$$2.33x - 4.58y = 16.319$$
$$4.98x + 3.44y = -2.162$$

Multiply the first equation by 3.44 and the second by 4.58:

$$8.0152x - 15.7552y = 56.13736$$
$$\underline{22.8084x + 15.7552y = -9.90196}$$
$$30.8236x \qquad\qquad = 46.2354$$
$$x = 1.5$$
$$2.33(1.5) - 4.58y = 16.319$$
$$-4.58y = 12.824$$
$$y = -2.8$$

The solution is $(1.5, -2.8)$.

39. Let $x =$ the number of cars and $y =$ the number of trucks. Since the dealer has 250 vehicles, $x + y = 250$. Since the total tax is $850, $3x + 4y = 850$. Multiply the first equation by -3 and add the result to the second equation.

$$-3x - 3y = -750$$
$$\underline{3x + 4y = 850}$$
$$y = 100$$

Since $x + y = 250$ and $y = 100$, we have $x = 150$. The dealer had 150 cars and 100 trucks.

41. Let $x =$ the number of adults and $y =$ the number of children. Since the total number of people was 30, $x + y = 30$. Since the total bill was \$99, we have the equation $5.50x + 2.75y = 99$. Write the first equation as $y = 30 - x$ and use substitution.

$$5.50x + 2.75(30 - x) = 99$$
$$5.50x + 82.5 - 2.75x = 99$$
$$2.75x + 82.5 = 99$$
$$2.75x = 16.5$$
$$x = 6$$

Since $y = 30 - x$ and $x = 6$, we have $y = 24$. There were 6 adults and 24 children on the trip, and a bus driver who didn't enter the zoo.

43. Let $x =$ the number of men in the survey and $y =$ the number of women in the survey. Since there were 300 people in the survey, we have $x + y = 300$. Since one-half of the men and one-third of the women were smokers, we have $\frac{1}{2}x + \frac{1}{3}y = 130$. Write the first equation as $y = 300 - x$ and use substitution.

$$\frac{1}{2}x + \frac{1}{3}(300 - x) = 130$$
$$\frac{1}{2}x + 100 - \frac{1}{3}x = 130$$
$$\frac{1}{2}x - \frac{1}{3}x = 30$$
$$\frac{1}{6}x = 30$$
$$x = 180$$

Since $y = 300 - x$ and $x = 180$, $y = 120$. There were 180 men and 120 women in the survey.

7.4 WARM-UPS

1. True, because $4 > 3(-1) + 1$ is correct.
2. True, because $3(2) - 2(-3) \geq 12$ is correct.
3. True, because the inequality $y > x + 9$ indicates the region above the line $y = x + 9$.
4. False, because the inequality $x < y + 2$ is equivalent to $y > x - 2$ and that indicates the region above the line.
5. False, because the graph of $x = 3$ is a vertical line.
6. False, because the graph of $y \leq 5$ is the region on or below the line $y = 5$.
7. True, because $x < 3$ indicates all points that have an x-coordinate smaller than 3.
8. False, because for the inequality $\geq$ we use a solid boundary line.
9. True, because $(0, 0)$ satisfies $y \geq x$.
10. False, because any point above $y = 2x + 1$ satisfies $y > 2x + 1$ and $(0, 0)$ does not satisfy $y > 2x + 1$.

7.4 EXERCISES

1. Check each point $(2, 3)$, $(-3, -9)$, and $(8, 3)$ in the inequality $x - y > 5$.

$\quad 2 - 3 > 5 \quad$ Incorrect.
$\quad -3 - (-9) > 5 \quad$ Correct.
$\quad 8 - 3 > 5 \quad$ Incorrect.

Only $(-3, -9)$ satisfies the inequality.

3. Check each point $(3, 0)$, $(1, 3)$, and $(-2, 5)$ in the inequality $y \geq -2x + 5$.

$\quad 0 \geq -2(3) + 5 \quad$ Correct.
$\quad 3 \geq -2(1) + 5 \quad$ Correct.
$\quad 5 \geq -2(-2) + 5 \quad$ Incorrect.

The points $(3, 0)$ and $(1, 3)$ satisfy the inequality.

5. Check each point $(2, 3)$, $(7, -1)$, and $(0, 5)$ in the inequality $x > -3y + 4$.

$\quad 2 > -3(3) + 4 \quad$ Correct.
$\quad 7 > -3(-1) + 4 \quad$ Incorrect.
$\quad 0 > -3(5) + 4 \quad$ Correct.

The points $(2, 3)$ and $(0, 5)$ satisfy the inequality.

7. To graph $y < x + 4$ first graph $y = x + 4$. The graph of $y = x + 4$ is a line with y-intercept (0, 4) and slope 1. Draw it dashed because of the inequality symbol. Shade the region below the line to indicate points that satisfy $y < x + 4$.

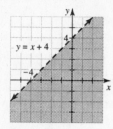

9. To graph $y > -x + 3$ first graph the equation $y = -x + 3$. The graph of $y = -x + 3$ is a line with y-intercept (0, 3) and slope -1. Draw the line dashed because of the inequality symbol. Shade the region above the line to indicate points that satisfy $y > -x + 3$.

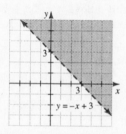

11. To graph $y > \frac{2}{3}x - 3$ first graph the equation $y = \frac{2}{3}x - 3$. The graph of $y = \frac{2}{3}x - 3$ is a line with y-intercept (0, -3) and slope 2/3. Draw the line dashed because of the inequality symbol. Shade the region above the line to indicate points that satisfy $y > \frac{2}{3}x - 3$.

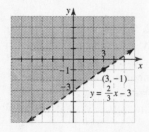

13. To graph $y \leq -\frac{2}{5}x + 2$ first graph $y = -\frac{2}{5}x + 2$. The graph of $y = -\frac{2}{5}x + 2$ is a line with y-intercept (0, 2) and slope $-2/5$. Draw a solid line because of the inequality symbol. Shade the region below the line to indicate the points that satisfy $y \leq -\frac{2}{5}x + 2$.

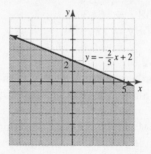

15. To graph $y - x \geq 0$ first graph $y - x = 0$, or $y = x$. The graph of $y = x$ is a line through the origin with slope 1. Draw the line solid because of the inequality symbol. The inequality $y - x \geq 0$ is equivalent to $y \geq x$. So shade the region above the line to indicate the points that satisfy the inequality.

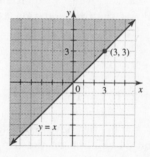

17. Write $x > y - 5$ as $y < x + 5$. Next, graph $y = x + 5$ with y-intercept (0, 5) and slope 1. Draw the line dashed because of the inequality symbol. Shade the region below the line to indicate the points that satisfy $y < x + 5$.

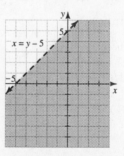

19. Solve the inequality $x - 2y + 4 \leq 0$ for y.

$$-2y \leq -x - 4$$

$$y \geq \tfrac{1}{2}x + 2$$

First graph $y = \tfrac{1}{2}x + 2$ with y-intercept $(0, 2)$ and slope 1/2. Draw the line solid because of the inequality symbol. Shade the region above the line to indicate the points that satisfy $y \geq \tfrac{1}{2}x + 2$.

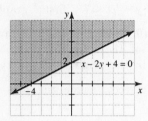

21. To graph $y \geq 2$ first graph $y = 2$. The graph of $y = 2$ is a horizontal line with y-intercept $(0, 2)$. Draw the line solid because of the inequality symbol. Shade the region above the line to indicate the points that satisfy $y \geq 2$.

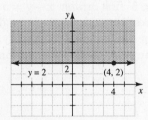

23. To graph $x > 9$ first graph $x = 9$, the vertical line with x-intercept at $(9, 0)$. Draw the line dashed because of the inequality symbol. Shade the region to the right to indicate the points that satisfy the inequality $x > 9$.

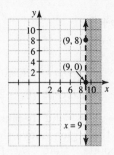

25. The inequality $x + y \leq 60$ is equivalent to $y \leq -x + 60$. First graph the equation $y = -x + 60$. The graph of $y = -x + 6$ is a line with y-intercept $(0, 60)$ and slope -1. Draw the line solid because the inequality symbol includes equality. Shade the region below the line to indicate the points that satisfy $y \leq -x + 60$.

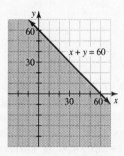

27. To graph $x \leq 100y$, or $y \geq \tfrac{1}{100}x$, first graph the equation $y = \tfrac{1}{100}x$. The graph of $y = \tfrac{1}{100}x$ goes through $(0, 0)$ and has slope $\tfrac{1}{100}$. Draw the line solid because the inequality symbol includes equality. Shade the region above the line to indicate the points that satisfy $y \geq \tfrac{1}{100}x$.

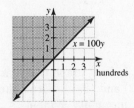

29. First solve the inequality for y.

$$3x - 4y \le 8$$
$$-4y \le -3x + 8$$
$$y \ge \frac{3}{4}x - 2$$

First graph the equation $y = \frac{3}{4}x - 2$. The graph of $y = \frac{3}{4}x - 2$ is a line with y-intercept $(0, -2)$ and slope 3/4. Draw the line solid because $\ge$ includes the line. Shade the region above the line to indicate the points that satisfy $y \ge \frac{3}{4}x - 2$.

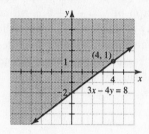

31. To graph $2x - 3y < 6$ first graph $2x - 3y = 6$. The line $2x - 3y = 6$ goes through $(0, -2)$ and $(3, 0)$. Draw the line dashed because of the inequality symbol. Use $(0, 0)$ as a test point. Since $2(0) - 3(0) < 6$ is correct, shade the side of the line that contains the point $(0, 0)$.

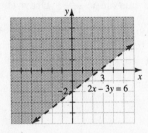

33. To graph $x - 4y \le 8$ first graph the line $x - 4y = 8$ through $(0, -2)$ and $(8, 0)$. Draw the line solid because of the inequality symbol $\le$. Test $(0, 0)$. Since $0 - 4 \cdot 0 \le 8$ is correct, shade the region containing $(0, 0)$.

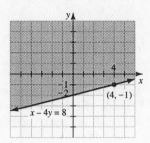

35. To graph $y - \frac{7}{2}x \le 7$ first graph $2y - 7x = 14$ through $(0, 7)$ and $(-2, 0)$. Draw the line solid because of the inequality symbol $\le$. Test $(0, 0)$ in the original inequality. Since $0 - \frac{7}{2} \cdot 0 \le 7$ is correct, shade the region containing $(0, 0)$.

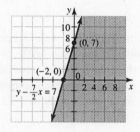

37. To graph $x - y < 5$ first graph $x - y = 5$. The line $x - y = 5$ goes through $(0, -5)$ and $(5, 0)$. Test $(0, 0)$ in the original inequality. Since $0 - 0 < 5$ is correct, shade the region containing $(0, 0)$.

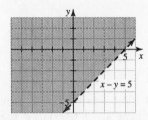

39. To graph $3x - 4y < -12$ first graph $3x - 4y = -12$ through $(0, 3)$ and $(-4, 0)$. Draw the line dashed because of the inequality symbol $<$. Test the point $(0, 0)$ in the original inequality. Since $3 \cdot 0 - 4 \cdot 0 < -12$ is incorrect, shade the region that does not contain $(0, 0)$.

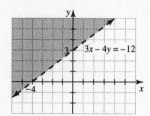

41. To graph $x < 5y - 100$ first graph $x = 5y - 100$ through $(0, 20)$ and $(-100, 0)$. Draw the line dashed because of the inequality symbol $<$. Test $(0, 0)$ in the original inequality. Since $0 < 5 \cdot 0 - 100$ is incorrect, we shade the region that does not contain $(0, 0)$.

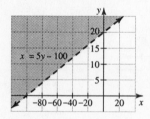

43. Let $x =$ the number of round tables and $y =$ the number of rectangular tables. Since there is at most 3850 cubic feet of storage space available, $25x + 35y \leq 3850$, or $5x + 7y \leq 770$. Graph $5x + 7y = 770$ through $(0, 110)$ and $(154, 0)$. Shade the region below this line in the first quadrant, because x and y cannot be negative.

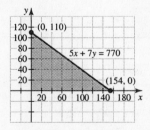

45. To graph $a > 0.07c + 0.02$, graph the dashed line through $(0, 0.02)$ and $(5, 0.37)$. Shade above this line.

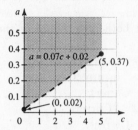

7.5 WARM-UPS

1. False, $(2, -3)$ does not satisfy $y > -3x + 5$.
2. True, because $0 > -3(5) + 5$ and $0 < 2(5) - 3$.
3. True, because $0 > 2(0) - 3$ and $0 < 2(0) + 3$.
4. True, because the region is below the line $y = 2x + 3$ and above the line $y = 2x - 3$.
5. False, because $(0, 0)$ is on the line $x - y = 0$.
6. False, because $2 + 2 > 4$ is incorrect.
7. True, because $4 + 5 > 4$ and $4 - 5 < 0$ are both correct.
8. False, $x + y > 4$ is equivalent to $y > -x + 4$
9. True, because in general the graph of $y < mx + b$ is the region below the line $y = mx + b$.
10. True, because a point would have to be above the line $y = 2x + 3$ and also below the line $y = 2x - 3$ to be a solution to the system.

7.5 EXERCISES

1. Check each point $(4, 3)$, $(8, 2)$, and $(-3, 0)$ in both inequalities of the system.
Check $(4, 3)$: $\quad 4 - 3 < 5 \quad$ Correct.
$\qquad\qquad 2(4) + 3 > 3 \quad$ Correct.
Check $(8, 2)$: $\quad 8 - 2 < 5 \quad$ Incorrect.
$\qquad\qquad 2(8) + 2 > 3 \quad$ Correct.
Check $(-3, 0)$: $\quad -3 - 0 < 5 \quad$ Correct.
$\qquad\qquad\quad 2(-3) + 0 > 3 \quad$ Incorrect.

Only $(4, 3)$ satisfies both inequalities. So $(4, 3)$ is a solution to the system.

3. Check each point in both inequalities of the system.
Check $(-3, 2)$: $\quad 2 > -2(-3) + 1 \quad$ Incorrect.
$\qquad\qquad\quad 2 < 3(-3) + 5 \quad$ Incorrect.

Check $(-1, 5)$: $5 > -2(-1) + 1$ Correct.
$5 < 3(-1) + 5$ Incorrect.
Check $(3, 6)$: $6 > -2(3) + 1$ Correct.
$6 < 3(3) + 5$ Correct.
Only $(3, 6)$ satisfies both inequalities. So $(3, 6)$ is a solution to the system.

5. Check each point in both inequalities of the system.
Check $(-5, 4)$: $-5 > 3$ Incorrect.
$4 < -2$ Incorrect.
Check $(9, -5)$: $9 > 3$ Correct.
$-5 < -2$ Correct.
Check $(6, 0)$: $6 > 3$ Correct.
$0 < -2$ Incorrect.
Only $(9, -5)$ is a solution to the system.

7. First graph the lines $y = -x - 1$ and $y = x + 1$. Use dashed lines. Select a test point in each of the four regions, say $(0, 0)$, $(-4, 0)$, $(0, 3)$, and $(0, -3)$. Only $(0, 3)$ satisfies both $y > -x - 1$ and $y > x + 1$. Shade the region containing $(0, 3)$.

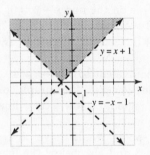

9. First graph the lines $y = 2x - 3$ and $y = -x + 2$. Use dashed lines. Select a test point in each of the regions, say $(0, 0)$, $(0, 5)$, $(0, -5)$, and $(5, 0)$. Only $(5, 0)$ satisfies both $y < 2x - 3$ and $y > -x + 2$. Shade the region containing $(5, 0)$.

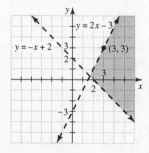

11. First graph the lines $x + y = 5$ and $x - y = 3$. Use dashed lines. Select a test point in each region, say $(0, 0)$, $(4, 0)$, $(6, 0)$, and $(0, 6)$. Only $(0, 6)$ satisfies both $x + y > 5$ and $x - y < 3$. Shade the region containing $(0, 6)$.

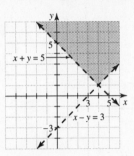

13. First graph the lines $2x - 3y = 6$ and $x - y = 3$. Use dashed lines. Select a test point in each of the regions, say $(0, 0)$, $(-2, -4)$, $(0, -4)$, and $(6, 2.5)$. Only $(6, 2.5)$ satisfies both $2x - 3y < 6$ and $x - y > 3$. Shade the region containing $(6, 2.5)$.

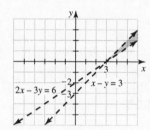

15. First graph the lines $x = 5$ and $y = 5$. Use dashed lines. Select a test point in each region, say $(0, 0)$, $(6, 0)$, $(0, 6)$, and $(6, 6)$. Only $(6, 6)$ satisfies both $x > 5$ and $y > 5$. Shade the region containing $(6, 6)$.

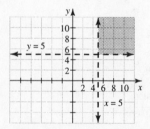

17. Graph the lines $y = -1$ and $x = -3$. Use dashed lines. Select a test point in each region, say $(0, 0)$, $(-4, 0)$, $(-4, -2)$, and $(0, -3)$. Only $(0, -3)$ satisfies both $y < -1$ and $x > -3$. Shade the region containing $(0, -3)$.

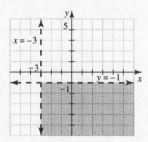

19. First graph the lines $y = 2x - 4$ and $y = 2x + 1$. Use dashed lines. Select a test point in each region, say $(-4, 0)$, $(0, 0)$, and $(4, 0)$. Only $(0, 0)$ satisfies both $y > 2x - 4$ and $y < 2x + 1$. Shade the region containing $(0, 0)$.

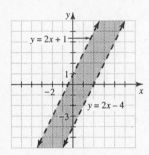

21. First graph the lines $y = x$ and $x = 3$. Use dashed lines. Select a test point in each region, say $(-4, 0)$, $(1, 0)$, $(4, 0)$, and $(4, 5)$. Only $(4, 5)$ satisfies both $y > x$ and $x > 3$. Shade the region containing $(4, 5)$.

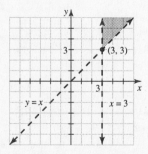

23. First graph the lines $y = -x$ and $x = -1$. Use dashed lines. Select a test point in each region, say $(-4, 0)$, $(0, -1)$, $(2, 0)$, and $(-4, 5)$. Only $(-4, 5)$ satisfies both $y > -x$ and $x < -1$. Shade the region containing $(-4, 5)$.

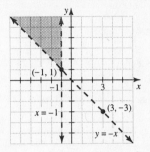

25. First graph the lines $x = 1$ and $y - 2x = 3$. Use dashed lines. Select a test point in each region, say $(0, 0)$, $(0, 4)$, $(2, 0)$, and $(2, 7)$. Only $(2, 0)$ satisfies both $x > 1$ and $y - 2x < 3$. Shade the region containing $(2, 0)$.

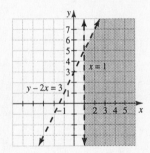

27. First graph the lines $2x - 5y = 5$ and $x + 2y = 4$. Use dashed lines. Select a test point in each region, say $(0, 0)$, $(0, -4)$, $(5, 0)$, and $(0, 4)$. Only $(0, 4)$ satisfies both $2x - 5y < 5$ and $x + 2y > 4$. Shade the region containing $(0, 4)$.

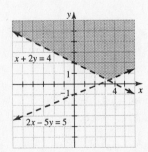

29. First graph the lines $x + y = 3$ and $x + y = 1$. Use dashed lines. Select a test point in each region, say $(0, -2)$, $(0, 2)$ and $(0, 5)$. Only $(0, 5)$ satisfies both $x + y > 3$ and $x + y > 1$. Shade the region containing $(0, 5)$.

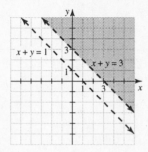

31. First graph the lines $y = 3x + 2$ and $y = 3x + 3$. Use dashed lines. Select a test point in each region, say $(0, 0)$, $(0, 5)$, and $(0, 2.5)$. Only $(0, 2.5)$ satisfies both $y > 3x + 2$ and $y < 3x + 3$. Shade the region containing $(0, 2.5)$.

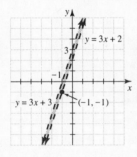

33. First graph the equations $x + y = 5$ and $x - y = -1$. Use dashed lines. Select a test point in each region, say $(-4, 0)$, $(0, 6)$, $(6, 0)$, and $(0, 0)$. Only $(0, 0)$ satisfies both $x + y < 5$ and $x - y > -1$. Shade the region containing $(0, 0)$.

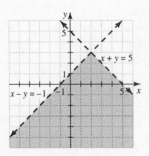

35. First graph the equations $2x - 3y = 6$ and $3x + 4y = 12$. Use dashed lines. Select a test point in each region, say $(6, 0)$, $(0, -6)$, $(0, 6)$, and $(0, 0)$. Only $(0, 0)$ satisfies both $2x - 3y < 6$ and $3x + 4y < 12$. Shade the region containing $(0, 0)$.

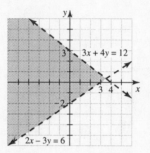

37. First graph the lines $3x - 5y = 15$ and $3x + 2y = 12$. Use dashed lines. Select a test point in each region, say $(6, 0)$, $(5, 1)$, $(0, -5)$, and $(0, 0)$. Only $(0, 0)$ satisfies both $3x - 5y < 15$ and $3x + 2y < 12$. Shade the region containing $(0, 0)$.

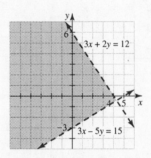

39. First graph the line $h = 0.65(220 - a)$ through the points $(20, 130)$ and $(70, 97.5)$. Next graph the line $h = 0.75(220 - a)$ through $(20, 150)$ and $(70, 112.5)$. Now shade the region between the lines.

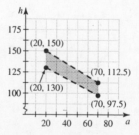

41. Let x = the number of small yard barns and y = the number of large yard barns.

$$250x + 400y \leq 4000$$
$$20x + 30y \leq 300$$

$$5x + 8y \leq 80$$
$$2x + 3y \leq 30$$

Graph the equations $5x + 8y = 80$ and $2x + 3y = 30$. Since $(0, 0)$ satisfies both inequalities we shade the region containing the origin. Only shade in the first quadrant because the number of barns cannot be negative.

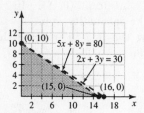

CHAPTER 7 REVIEW

1. The line $y = 2x + 1$ has a y-intercept of $(0, 1)$ and slope of 2. The line $x + y = 4$ goes through $(0, 4)$ and $(4, 0)$. The lines appear to intersect at $(1, 3)$. Check $(1, 3)$ in both of the equations to be certain that $(1, 3)$ is the solution of the system.

3. The line $y = 2x + 3$ has a y-intercept of $(0, 3)$ and a slope of 2. The line $y = -2x - 1$ has a y-intercept of $(0, -1)$ and a slope of -2. The lines appear to intersect at $(-1, 1)$. Check that $(-1, 1)$ satisfies both equations to be certain that $(-1, 1)$ is the solution to the system.

5. Use $y = 3x$ to eliminate y in the equation $2x + 3y = 22$.

$$2x + 3(3x) = 22$$
$$11x = 22$$
$$x = 2$$

Use $x = 2$ in $y = 3x$ to get $y = 3(2) = 6$.
The solution to the system is $(2, 6)$.

7. Use $x = y - 5$ to eliminate x in the equation $2x - 3y = -7$.

$$2(y - 5) - 3y = -7$$
$$2y - 10 - 3y = -7$$
$$-y = 3$$
$$y = -3$$

Use $y = -3$ in $x = y - 5$ to get $x = -3 - 5 = -8$.
The solution to the system is $(-8, -3)$.

9.
$$x - y = 4$$
$$\underline{2x + y = 5}$$
$$3x = 9$$
$$x = 3$$

Use $x = 3$ in $2x + y = 5$ to find y.
$$2(3) + y = 5$$
$$y = -1$$

The solution to the system is $(3, -1)$. Since there is only one solution to the system, the system is independent.

11.
$$2x - 4y = 8$$
$$x - 2y = 4$$

Multiply the second equation by -2 and add the result to the first equation.

$$2x - 4y = 8$$
$$\underline{-2x + 4y = -8}$$
$$0 = 0$$

Since the addition resulted in an identity, the system is dependent and the set of all solutions to the system is $\{(x, y) \mid x - 2y = 4\}$.

13.
$$y = 3x - 5$$
$$2y = -x - 3$$

Multiply the first equation by -2 and add the result to the second equation.

$$-2y = -6x + 10$$
$$\underline{2y = -x - 3}$$
$$0 = -7x + 7$$
$$7x = 7$$
$$x = 1$$

If $x = 1$ and $y = 3x - 5$, then $y = 3(1) - 5 = -2$. The solution to the system is $(1, -2)$ and the system is independent.

15.
$$2x + 7y = 0$$
$$7x + 2y = 0$$

Multiply the first equation by -7 and the second by 2.

$$-14x - 49y = 0$$
$$\underline{14x + 4y = 0}$$
$$-45y = 0$$
$$y = 0$$

Use $y = 0$ in $2x + 7y = 0$ to find x.
$$2x + 7(0) = 0$$
$$2x = 0$$
$$x = 0$$

The solution to the system is $(0, 0)$ and the system is independent.

17. $x - y = 6$
          $2x - 12 = 2y$
Rewrite the second equation as $2x - 2y = 12$.
          $x - y = 6$
          $2x - 2y = 12$
Multiply the first equation by -2 and add the result to the second.

$$\begin{array}{r} -2x + 2y = -12 \\ 2x - 2y = 12 \\ \hline 0 = 0 \end{array}$$

Since the addition resulted in an identity, the system is dependent and the solution set is $\{(x, y) \mid x - y = 6\}$.

19. $y = 4x$
          $y = 4x + 3$
Multiply the first equation by -1 and add the result to the second equation.

$$\begin{array}{r} -y = -4x \\ y = 4x + 3 \\ \hline 0 = 3 \end{array}$$

Since the result of the addition is false, there is no solution to the system and the system is inconsistent.

21. First graph the line $y = \frac{1}{3}x - 5$, which has y-intercept $(0, -5)$ and slope $1/3$. Use a dashed line because the inequality symbol is $>$. Shade the region above the line to indicate the points that satisfy $y > \frac{1}{3}x - 5$.

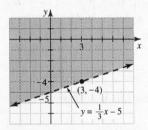

23. First graph the line $y = -2x + 7$, which has y-intercept $(0, 7)$ and slope -2. Use a solid line because of the inequality symbol $\leq$. Shade the region below the line to indicate the points that satisfy $y \leq -2x + 7$.

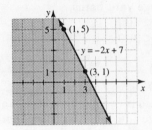

25. First graph the line $y = 8$. The line $y = 8$ is a horizontal line with y-intercept $(0, 8)$. Use a solid line because of the inequality symbol $\leq$. Shade the region below the line to indicate the points that satisfy $y \leq 8$.

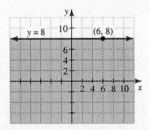

27. First graph the line $2x + 3y = -12$. The line goes through the points $(0, -4)$ and $(-6, 0)$. Draw a solid line because of the inequality symbol $\leq$. Test $(0, 0)$ in the inequality. Since $2(0) + 3(0) \leq -12$ is incorrect, we shade the region that does not contain $(0, 0)$ to indicate the points that satisfy $2x + 3y \leq -12$.

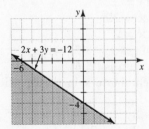

29. First graph the lines $x = 5$ and $y = 4$. Use dashed lines. Select a test point in each region, say $(0, 5)$, $(6, 6)$, $(6, 0)$, and $(0, 0)$. Only $(0, 0)$ satisfies both $x < 5$ and $y < 4$. Shade the region containing $(0, 0)$.

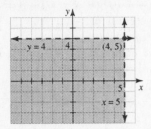

31. First graph the equations $x + y = 2$ and $y = 2x - 3$. Use dashed lines. Select a test point in each region, say $(5, 0)$, $(2, 5)$, $(2, -5)$, and $(0, 0)$. Only $(0, 0)$ satisfies both $x + y < 2$ and $y > 2x - 3$. Shade the region containing $(0, 0)$.

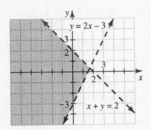

33. First graph $y = 5x - 7$ and $y = 5x + 1$. Use dashed lines. Select a test point in each region, say $(-5, 0)$, $(5, 0)$, and $(0, 0)$. Only $(0, 0)$ satisfies both $y > 5x - 7$ and $y < 5x + 1$. Shade the region containing $(0, 0)$.

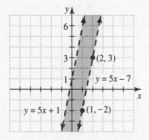

35. First graph the lines $y = 3x + 5$ and $y = 3x$. Use dashed lines. Select a test point in each region, say $(-5, 0)$, $(-1, 0)$, and $(4, 0)$. Only $(4, 0)$ satisfies both $y < 3x + 5$ and $y < 3x$. Shade the region containing $(4, 0)$.

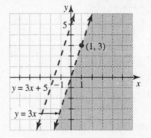

37. Let $x =$ the cost of one apple and $y =$ the cost of one orange. We can write one equation about the total cost of 2 apples and 3 oranges, and the other equation about the total cost of 3 apples and 2 oranges.

$$2x + 3y = 1.95$$
$$3x + 2y = 2.05$$

Multiply the first equation by -3 and the second equation by 2.

$$-3(2x + 3y) = -3(1.95)$$
$$2(3x + 2y) = 2(2.05)$$

$$\begin{array}{r} -6x - 9y = -5.85 \\ 6x + 4y = 4.10 \\ \hline -5y = -1.75 \\ y = 0.35 \end{array}$$

Use $y = 0.35$ in $2x + 3y = 1.95$ to find x.

$$\begin{array}{r} 2x + 3(0.35) = 1.95 \\ 2x + 1.05 = 1.95 \\ 2x = 0.90 \\ x = 0.45 \end{array}$$

The cost of one apple is $0.45 and the cost of one orange is $0.35.

39. Let $x =$ the actual number of $5 bills and $y =$ the actual number of $10 bills. Since he actually had $380, we can write $5x + 10y = 380$. When he thought he had x $10 bills and y $5 bills, he thought he had $430. So

135

$10x + 5y = 430$. Multiply the first equation by -2 and add the result to the second equation.

$$-2(5x + 10y) = -2(380)$$
$$10x + 5y = 430$$

$$-10x - 20y = -760$$
$$\underline{10x + 5y = 430}$$
$$-15y = -330$$
$$y = 22$$

Use $y = 22$ in $10x + 5y = 430$ to find x.
$$10x + 5(22) = 430$$
$$10x + 110 = 430$$
$$10x = 320$$
$$x = 32$$

He actually had 32 fives and 22 tens.

41. Let $x =$ the number of servings of green beans and $y =$ the number of servings of chicken soup. Since each serving of green beans contains 1 gram of protein and each serving of chicken soup contains 3 grams of protein, we have the equation $x + 3y = 13$. Since each serving of green beans contains 4 grams of carbohydrates and each serving of chicken soup contains 9 grams of carbohydrates, we have the equation $4x + 9y = 43$. Write the first equation as $x = 13 - 3y$ and use substitution.

$$4(13 - 3y) + 9y = 43$$
$$52 - 12y + 9y = 43$$
$$-3y = -9$$
$$y = 3$$

Use $y = 3$ in $x = 13 - 3y$ to get $x = 13 - 3(3) = 4$. To get the recommended amount of protein and carbohydrates you should have 4 servings of green beans and 3 servings of chicken soup.

CHAPTER 7 TEST

1. The graph of $x + y = 2$ contains the points $(0, 2)$ and $(2, 0)$. The graph of $y = 2x + 5$ has a y-intercept of $(0, 5)$ and a slope of 2. From the graph it appears that the lines intersect at $(-1, 3)$. Check $(-1, 3)$ in both equations to be sure that it is the solution to the system.

2. Substitute $y = 2x - 3$ into $2x + 3y = 7$.
$$2x + 3(2x - 3) = 7$$
$$2x + 6x - 9 = 7$$
$$8x = 16$$
$$x = 2$$
Use $x = 2$ in $y = 2x - 3$ to get $y = 2(2) - 3 = 1$. The solution to the system is $(2, 1)$.

3. Write $x - y = 4$ as $x = y + 4$ and substitute into the equation $3x - 2y = 11$.
$$3(y + 4) - 2y = 11$$
$$3y + 12 - 2y = 11$$
$$y + 12 = 11$$
$$y = -1$$
Use $y = -1$ in $x = y + 4$ to get $x = -1 + 4 = 3$. The solution to the system is $(3, -1)$.

4. Multiply the first equation by -2 and add the result to the second equation.
$$-2(2x + 5y) = -2(19)$$
$$4x - 3y = -1$$

$$-4x - 10y = -38$$
$$\underline{4x - 3y = -1}$$
$$-13y = -39$$
$$y = 3$$
Use $y = 3$ in $4x - 3y = -1$ to find x.
$$4x - 3(3) = -1$$
$$4x - 9 = -1$$
$$4x = 8$$
$$x = 2$$
The solution to the system is $(2, 3)$.

5. Multiply the first equation by -2 and the second by 3.
$$-2(3x - 2y) = -2(10)$$
$$3(2x + 5y) = 3(13)$$

$$-6x + 4y = -20$$
$$\underline{6x + 15y = 39}$$
$$19y = 19$$
$$y = 1$$
Use $y = 1$ in $3x - 2y = 10$ to find x.
$$3x - 2(1) = 10$$
$$3x = 12$$
$$x = 4$$
The solution to the system is $(4, 1)$.

6. The graphs of $y = 4x - 9$ and $y = 4x + 8$ are parallel lines with different y-intercepts. So there is no solution to the system. The system is inconsistent.

7. Substitute $y = x - 4$ into $3x - 3y = 12$.
$$3x - 3(x - 4) = 12$$
$$3x - 3x + 12 = 12$$
$$12 = 12$$
Since the substitution results in an identity, the system is dependent.

8. Substitute $y = 2x$ into $y = 5x$.

$$2x = 5x$$
$$0 = 3x$$
$$0 = x$$

If $x = 0$ and $y = 2x$, then $y = 2(0) = 0$. The solution to the system is $(0, 0)$. So the system is independent.

9. First graph the line $y = 3x - 5$, which has a y-intercept of $(0, -5)$ and a slope of 3. Use a dashed line. Shade the region above the line to indicate the points that satisfy $y > 3x - 5$.

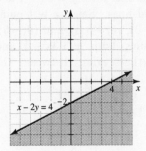

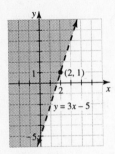

10. First graph the line $x - y = 3$. The line goes through the points $(3, 0)$ and $(0, -3)$. Test $(0, 0)$ in the inequality. Since $0 - 0 < 3$ is correct, we shade the region containing $(0, 0)$ to indicate the points that satisfy $x - y < 3$.

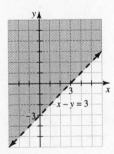

11. First graph $x - 2y = 4$ through the points $(4, 0)$ and $(0, -2)$. Test $(0, 0)$ in the inequality. Since $0 - 2(0) \geq 4$ is incorrect, we shade the region that does not contain $(0, 0)$ to indicate the points that satisfy $x - 2y \geq 4$.

12. First graph the lines $x = 6$ and $y = -1$. Use dashed lines. Select a test point in each region, say $(0, -3)$, $(7, 0)$, $(7, -2)$, and $(0, 0)$. Only $(0, 0)$ satisfies both $x < 6$ and $y > -1$. Shade the region containing $(0, 0)$.

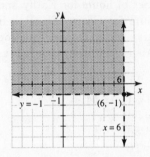

13. First graph $2x + 3y = 6$ and $3x - y = 3$. Draw dashed lines. Select a test point in each region, say $(0, 0)$, $(2, 0)$, $(4, 0)$, and $(0, 4)$. Only $(0, 4)$ satisfies both $2x + 3y > 6$ and $3x - y < 3$. Shade the region containing $(0, 4)$ to indicate the solution to the system of inequalities.

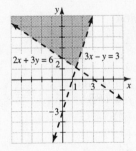

14. First graph the parallel lines $y = 3x - 4$ and $y = 3x - 3$. Use dashed lines. Select a test point in each region, say $(0, 0)$, $(0, -3.5)$, and $(0, -5)$. Only $(0, -3.5)$ satisfies both $y > 3x - 4$ and $3x - y > 3$. Shade the region between the parallel lines.

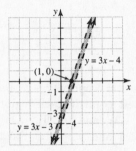

15. Let $x =$ the number of hours for Kathy and $y =$ the number of hours for Chris. Since their total time is 54 hours, we have the equation $x + y = 54$. Since Chris studied only one-half as many hours as Kathy, we have $y = \frac{1}{2}x$ or $x = 2y$.

Substitute $x = 2y$ into $x + y = 54$.
$$2y + y = 54$$
$$3y = 54$$
$$y = 18$$
If $y = 18$ and $x = 2y$, then $x = 2(18) = 36$. Kathy studied 36 hours and Chris studied 18 hours.

16. Let $x =$ the price of a single and $y =$ the price of a double. For Monday we have the equation $5x + 3y = 188$. For Tuesday we have the equation $3x + 4y = 170$. Multiply the first equation by -3 and the second by 5.
$$-3(5x + 3y) = -3(188)$$
$$5(3x + 4y) = 5(170)$$

$$\begin{array}{r} -15x - 9y = -564 \\ 15x + 20y = 850 \\ \hline 11y = 286 \\ y = 26 \end{array}$$
Use $y = 26$ in $5x + 3y = 188$ to find x.
$$5x + 3(26) = 188$$
$$5x + 78 = 188$$
$$5x = 110$$
$$x = 22$$
Since singles rent for \$22 and doubles rent for \$26, on Wednesday night the motel made \$48 by renting one of each.

Tying It All Together Chapters 1-7

1.
$$2(x - 5) + 3x = 25$$
$$2x - 10 + 3x = 25$$
$$5x = 35$$
$$x = 7$$
The solution to the equation is 7.

2.
$$3x - 5 = 0$$
$$3x = 5$$
$$x = \frac{5}{3}$$
The solution to the equation is $\frac{5}{3}$.

3.
$$\frac{x}{3} - \frac{2}{5} = \frac{x}{2} - \frac{12}{5}$$
$$30\left(\frac{x}{3} - \frac{2}{5}\right) = 30\left(\frac{x}{2} - \frac{12}{5}\right)$$
$$10x - 12 = 15x - 72$$
$$60 = 5x$$
$$12 = x$$
The solution to the equation is 12.

4.
$$x^2 + 2x = 24$$
$$x^2 + 2x - 24 = 0$$
$$(x + 6)(x - 4) = 0$$
$$x + 6 = 0 \quad \text{or} \quad x - 4 = 0$$
$$x = -6 \quad \text{or} \quad x = 4$$

The solutions to the equation are -6 and 4.

5.
$$\frac{x + 5}{x - 1} = \frac{2}{3}$$
$$3x + 15 = 2x - 2$$
$$x = -17$$

The solution to the equation is -17.

6.
$$2x^2 - 7x + 3 = 0$$
$$(2x - 1)(x - 3) = 0$$
$$2x - 1 = 0 \quad \text{or} \quad x - 3 = 0$$
$$x = \frac{1}{2} \quad \text{or} \quad x = 3$$

The solutions to the equation are $\frac{1}{2}$ and 3.

7.
$$3(2 - x) < -6$$
$$6 - 3x < -6$$
$$-3x < -12$$
$$x > 4$$

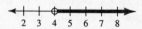

8. $-3 \le 2x - 4 \le 6$

$1 \le 2x \le 10$

$\frac{1}{2} \le x \le 5$

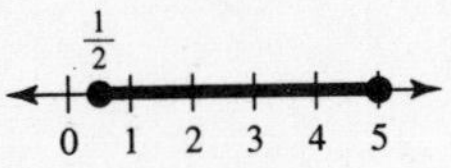

9. $x > 1$

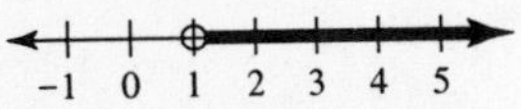

10. The graph of $y = 3x - 7$ is a line with y-intercept at $(0, -7)$ and slope $3 = 3/1$. To draw the graph start at $(0, -7)$, rise 3, and run 1.

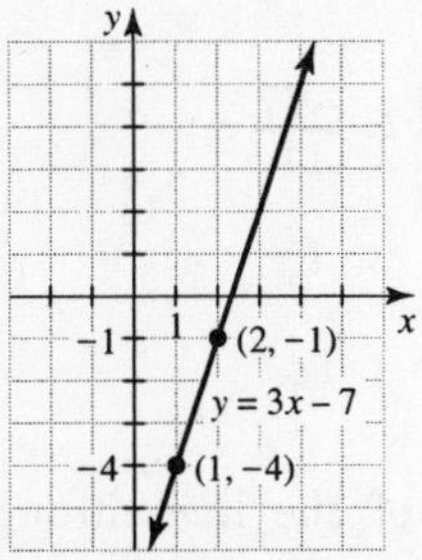

11. The graph of $y = -x + 5$ is a line with y-intercept $(0, 5)$ and slope $-1 = -1/1$. To draw the line, start at $(0, 5)$, go down 1 unit, and then 1 unit to the right.

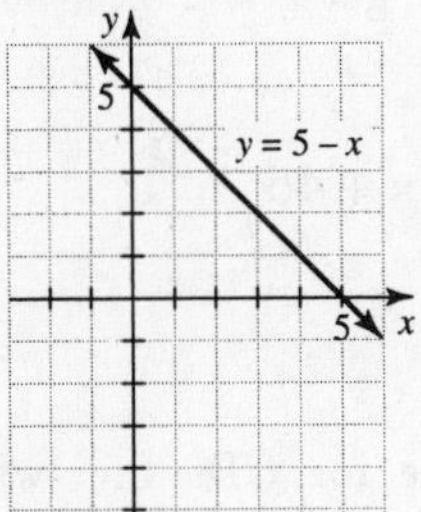

12. The graph of $y = x - 1$ is a line with y-intercept $(0, -1)$ and slope $1 = 1/1$. To draw the line, start at $(0, -1)$, go up 1 unit, and 1 unit to the right.

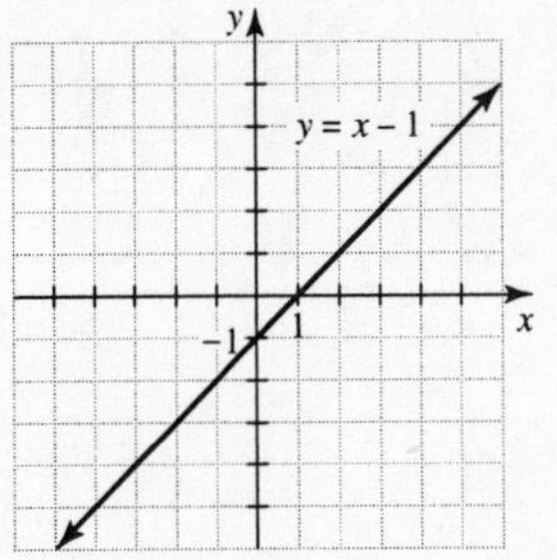

13. The graph of $y = x + 1$ is a line with y-intercept $(0, 1)$ and slope $1 = 1/1$. To draw the line, start at $(0, 1)$, rise 1 unit, and go 1 unit to the right.

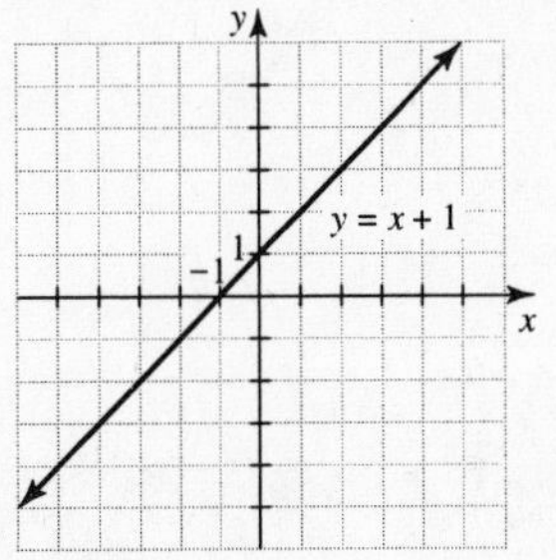

14. The graph of $y = -2x + 4$ is a line with y-intercept $(0, 4)$ and slope $-2 = -2/1$. To draw the line, start at $(0, 4)$, go down 2 units, and go 1 unit to the right.

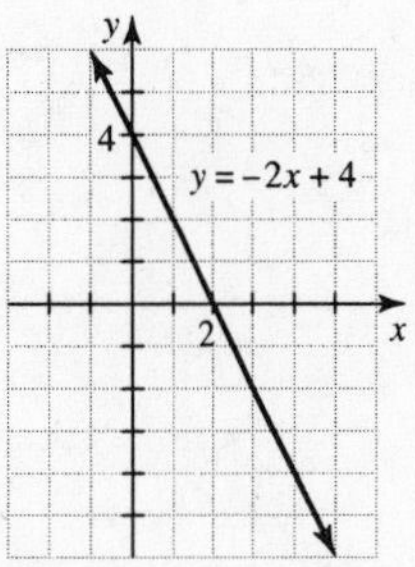

15. The graph of $y = -4x - 1$ is a line with y-intercept $(0, -1)$ and slope $-4 = -4/1$. To draw the line, start at $(0, -1)$, go down 4 units, and go 1 unit to the right.

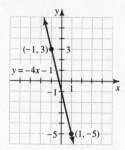

16. First graph the line $y = 3x - 7$. Use a solid line through $(0, -7)$ with slope 3. Shade the region above the line to indicate the points that satisfy $y \geq 3x - 7$.

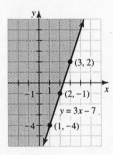

17. First graph the line $x - 2y = 6$. Draw a dashed line through $(0, -3)$ and $(6, 0)$. Test $(0, 0)$ in the inequality. Since $0 - 2(0) < 6$ is correct, shade the region containing $(0, 0)$ to indicate the points that satisfy $x - 2y < 6$.

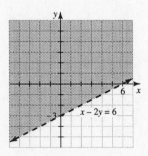

18. First graph the line $x = 1$. Draw a vertical dashed line through $(1, 0)$. Shade the region to the right to indicate the points that satisfy $x > 1$.

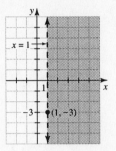

19. The y-intercept is $(0, 36)$. The slope is

$$m = \frac{84 - 36}{8 - 0} = 6.$$

The equation is $y = 6x + 36$.

20. The slope is $m = \frac{11 - 88}{12 - 1} = \frac{-77}{11} = -7$.

$$y - 88 = -7(x - 1)$$
$$y - 88 = -7x + 7$$
$$y = -7x + 95$$

21. a) Find the equation of the line through $(0, 45)$ and $(4, 25)$.

$$m = \frac{25 - 45}{4 - 0} = -5$$

Using slope-intercept form we get $p = -5x + 45$.

b) Find the equation of the line through $(0, 30)$ and $(4, 40)$.

$$m = \frac{40 - 30}{4 - 0} = 2.5$$

Using slope-intercept form we get $p = 2.5x + 30$.

c) By substitution we get

$$2.5x + 30 = -5x + 45$$
$$7.5x = 15$$
$$x = 2$$

So in 1996 the average price for CDs Inc. will equal the industry average.

8.1 WARM-UPS

1. True, because $10^{-2} = \frac{1}{10^2} = \frac{1}{100}$.

2. False, because $\left(-\frac{1}{5}\right)^{-1} = -5$.

3. False, $3^{-2} \cdot 2^{-1} = \frac{1}{9} \cdot \frac{1}{2} = \frac{1}{18} \neq 6^{-3}$.

4. True, $\frac{3^{-2}}{3^{-1}} = 3^{-2-(-1)} = 3^{-1} = \frac{1}{3}$.

5. False, $2.37 \times 10^{-1} = 2.37 \times \frac{1}{10} = 0.237$.

6. True, because multiplying by 10^{-5} moves the decimal point 5 places to the left.

7. True, because $25 \times 10^7 = 2.5 \times 10^1 \times 10^7$
$= 2.5 \times 10^8$.

8. True, because 0.442×10^{-3}
$= 4.42 \times 10^{-1} \times 10^{-3} = 4.42 \times 10^{-4}$.

9. True, because $(3 \times 10^{-9})^2 = 3^2 \times (10^{-9})^2$
$= 9 \times 10^{-18}$.

10. False, because $(2 \times 10^{-5})(4 \times 10^4)$
$= 8 \times 10^{-5+4} = 8 \times 10^{-1}$.

8.1 EXERCISES

1. $3^{-1} = \frac{1}{3}$

3. $(-2)^{-4} = \frac{1}{(-2)^4} = \frac{1}{16}$

5. $-4^{-2} = -\frac{1}{4^2} = -\frac{1}{16}$

7. $\frac{5^{-2}}{10^{-2}} = \frac{10^2}{5^2} = \frac{100}{25} = 4$

9. $\left(\frac{5}{2}\right)^{-3} = \left(\frac{2}{5}\right)^3 = \frac{8}{125}$

11. $6^{-1} + 6^{-1} = \frac{1}{6} + \frac{1}{6} = \frac{1}{3}$

13. $\frac{10}{5^{-3}} = 10 \cdot 5^3 = 10 \cdot 125 = 1250$

15. $\frac{1}{4^{-3}} + \frac{3^2}{2^{-1}} = 4^3 + 9 \cdot 2 = 64 + 18 = 82$

17. $x^{-1}x^2 = x^1 = x$

19. $-2x^2 \cdot 8x^{-6} = -16x^{-4} = -\frac{16}{x^4}$

21. $-3a^{-2}(-2a^{-3}) = 6a^{-5} = \frac{6}{a^5}$

23. $\frac{u^{-5}}{u^3} = u^{-8} = \frac{1}{u^8}$

25. $\frac{8t^{-3}}{-2t^{-5}} = -4t^{-3-(-5)} = -4t^2$

27. $\frac{-6x^5}{-3x^{-6}} = 2x^{5-(-6)} = 2x^{11}$

29. $(x^2)^{-5} = x^{-10} = \frac{1}{x^{10}}$

31. $(a^{-3})^{-3} = a^{(-3)(-3)} = a^9$

33. $(2x^{-3})^{-4} = 2^{-4}x^{12} = \frac{x^{12}}{2^4} = \frac{x^{12}}{16}$

35. $(4x^2y^{-3})^{-2} = 4^{-2}x^{-4}y^6 = \frac{y^6}{16x^4}$

37. $\left(\frac{2x^{-1}}{y^{-3}}\right)^{-2} = \frac{2^{-2}(x^{-1})^{-2}}{(y^{-3})^{-2}} = \frac{x^2}{4y^6}$

39. $\left(\frac{2a^{-3}}{ac^{-2}}\right)^{-4} = \frac{2^{-4}a^{12}}{a^{-4}c^8} = \frac{a^{16}}{2^4c^8} = \frac{a^{16}}{16c^8}$

41. $2^{-1} \cdot 3^{-1} = \frac{1}{2} \cdot \frac{1}{3} = \frac{1}{6}$

43. $(2 \cdot 3^{-1})^{-1} = 2^{-1} \cdot 3^1 = \frac{3}{2}$

45. $(x^{-2})^{-3} + 3x^7(-5x^{-1}) = x^6 - 15x^6 = -14x^6$

47. $\frac{a^3b^{-2}}{a^{-1}} + \left(\frac{b^6a^{-2}}{b^5}\right)^{-2} = \frac{a^4}{b^2} + \frac{b^{-12}a^4}{b^{-10}}$
$= \frac{a^4}{b^2} + \frac{a^4}{b^2} = \frac{2a^4}{b^2}$

49. The exponent 9 indicates that the decimal point moves 9 places to the right.
$9.86 \times 10^9 = 9,860,000,000$

51. The exponent -3 indicates that the decimal point moves 3 places to the left.

$1.37 \times 10^{-3} = 0.00137$

53. The exponent -6 indicates that the decimal point moves 6 places to the left.

$1 \times 10^{-6} = 0.000001$

55. The exponent 5 indicates that the decimal point moves 5 places to the right.
$6 \times 10^5 = 600,000$

57. Move the decimal point 3 places. Since 9000 is larger than 10 the exponent is positive.
$9000 = 9 \times 10^3$

59. Move the decimal point 4 places. Since 0.00078 is smaller than 1, the exponent is negative.
$0.00078 = 7.8 \times 10^{-4}$

61. Move the decimal point 6 places. Since 0.0000085 is smaller than 1, the exponent is negative.
$0.0000085 = 8.5 \times 10^{-6}$

63. $525 \times 10^9 = 5.25 \times 10^2 \times 10^9 = 5.25 \times 10^{11}$

65. $(3 \times 10^5)(2 \times 10^{-15}) = 6 \times 10^{5 + (-15)}$
$$= 6 \times 10^{-10}$$

67. $\dfrac{4 \times 10^{-8}}{2 \times 10^{30}} = 2 \times 10^{-8 - 30} = 2 \times 10^{-38}$

69. $\dfrac{3 \times 10^{20}}{6 \times 10^{-8}} = 0.5 \times 10^{28} = 5 \times 10^{-1} \times 10^{28}$
$$= 5 \times 10^{27}$$

71. $(3 \times 10^{12})^2 = 3^2 \times (10^{12})^2 = 9 \times 10^{24}$

73. $(5 \times 10^4)^3 = 125 \times 10^{12} = 1.25 \times 10^2 \times 10^{12}$
$$= 1.25 \times 10^{14}$$

75. $(4 \times 10^{32})^{-1} = 4^{-1} \times 10^{-32} = \frac{1}{4} \times 10^{-32}$
$$= 0.25 \times 10^{-32} = 2.5 \times 10^{-1} \times 10^{-32}$$
$$= 2.5 \times 10^{-33}$$

77. $(4300)(2,000,000) = 4.3 \times 10^3 \cdot 2 \times 10^6$
$$= 8.6 \times 10^9$$

79. $(4,200,000)(0.00005) = 4.2 \times 10^6 \cdot 5 \times 10^{-5}$
$$= 21 \times 10^1 = 2.1 \times 10^1 \times 10^1 = 2.1 \times 10^2$$

81. $(300)^3(0.000001)^5 = (3 \times 10^2)^3(1 \times 10^{-6})^5$
$$= 27 \times 10^6 \cdot 1 \times 10^{-30} = 27 \times 10^{-24}$$
$$= 2.7 \times 10^1 \times 10^{-24} = 2.7 \times 10^{-23}$$

83. $\dfrac{(4000)(90,000)}{(0.00000012)} = \dfrac{4 \times 10^3 \cdot 9 \times 10^4}{12 \times 10^{-8}}$
$$= \dfrac{36 \times 10^7}{12 \times 10^{-8}} = 3 \times 10^{15}$$

85. $(6.3 \times 10^6)(1.45 \times 10^{-4}) = 9.135 \times 10^2$

87. $(5.36 \times 10^{-4}) + (3.55 \times 10^{-5}) = 5.715 \times 10^{-4}$

89. $\dfrac{(3.5 \times 10^5)(4.3 \times 10^{-6})}{3.4 \times 10^{-8}} = 4.426 \times 10^7$

91. $(3.56 \times 10^{85})(4.43 \times 10^{96}) = 15.77 \times 10^{181}$
$$= 1.577 \times 10^{182}$$

93. Multiply 93 million miles by 5,280 feet per mile.
9.3×10^7 miles $\cdot \dfrac{5280 \text{ feet}}{1 \text{ mile}} = 4.910 \times 10^{11}$ feet

95. Since $T = D/R$, we divide 93 million miles by 2×10^{35} miles per hour.
$\dfrac{9.3 \times 10^7 \text{ miles}}{2 \times 10^{35} \text{ miles per hour}} = 4.65 \times 10^{-28}$ hours

97. Since $C = 2\pi r$, we can divide C by 2π to get the radius.
$\dfrac{5.68 \times 10^9 \text{ feet}}{2\pi} = 9.040 \times 10^8$ feet

99. a) Copper sells for about $1 per pound and its concentration in commercial ore is about 1%.
b) If the concentration of a metal in commercial ore is 10^{-6}%, then it should sell for about 10^6 or $1,000,000 dollars per pound.
c) These four points would not be in line in our usual coordinate system.

101. $P = 50,000(1 + 0.08)^{-20} = \$10,727.41$

8.2 WARM-UPS

1. True, because of the definition of square root.

2. False, because $\sqrt[3]{2} \cdot \sqrt[3]{2} \cdot \sqrt[3]{2} = 2$.

3. True, because $(-3)^3 = -27$.

4. False, $(-5)^2 = 25$.

5. True, because $2^4 = 16$.

6. True, because $3^2 = 9$.

7. False, because $(2^3)^2 = 2^6$.

8. False, because $\sqrt{17} \cdot \sqrt{17} = 17$.

9. True, because the radical symbol always indicates the positive square root.

10. True, because $(t^3)^4 = t^{12}$.

8.2 EXERCISES

1. $\sqrt{36} = 6$ 3. $\sqrt[5]{32} = 2$

5. $\sqrt[3]{1000} = 10$

7. $\sqrt[4]{-16}$ is not a real number because the fourth power of any real number is nonnegative.

9. $\sqrt{0} = 0$ 11. $\sqrt[3]{-1} = -1$

13. $\sqrt[3]{1} = 1$

15. $\sqrt[4]{-81}$ is not a real number because the fourth power of any real number is nonnegative.

17. $\sqrt[6]{64} = 2$ 19. $\sqrt[5]{-32} = -2$

21. $-\sqrt{100} = -10$

23. $\sqrt[4]{-50}$ is not a real number because the fourth power of any real number is nonnegative.

25. $\sqrt{m^2} = m$ 27. $\sqrt[5]{y^{15}} = y^3$

29. $\sqrt[3]{y^{15}} = y^5$ 31. $\sqrt[3]{m^3} = m$

33. $\sqrt{3^6} = 3^3 = 27$ 35. $\sqrt{2^{10}} = 2^5 = 32$

37. $\sqrt[3]{5^9} = 5^3 = 125$ 39. $\sqrt{10^{20}} = 10^{10}$

41. $\sqrt{9y} = \sqrt{9}\sqrt{y} = 3\sqrt{y}$ 43. $\sqrt{4a^2} = 2a$

45. $\sqrt{x^4 y^2} = x^2 y$ 47. $\sqrt{5m^{12}} = m^6\sqrt{5}$

49. $\sqrt[3]{8y} = 2\sqrt[3]{y}$ 51. $\sqrt[3]{-27w^3} = -3w$

53. $\sqrt[4]{16s} = 2\sqrt[4]{s}$

55. $\sqrt[3]{-125a^9 y^6} = -5a^3 y^2$

57. $\sqrt{\dfrac{t}{4}} = \dfrac{\sqrt{t}}{2}$ 59. $\sqrt{\dfrac{625}{16}} = \dfrac{25}{4}$

61. $\sqrt[3]{\dfrac{t}{8}} = \dfrac{\sqrt[3]{t}}{2}$ 63. $\sqrt[3]{\dfrac{-8x^6}{y^3}} = \dfrac{-2x^2}{y}$

65. $\sqrt{\dfrac{4a^6}{9}} = \dfrac{2a^3}{3}$ 67. $\sqrt[4]{\dfrac{y}{16}} = \dfrac{\sqrt[4]{y}}{2}$

69. $\sqrt{3} + \sqrt{5} \approx 3.968$ 71. $\dfrac{\sqrt{5} + \sqrt{2}}{\sqrt{3} - 4} \approx -1.610$

73. $\sqrt{7.1^2 - 4(1.2)(3)} \approx 6.001$

75. $\dfrac{-3 + \sqrt{3^2 - 4(1)(-2.9)}}{2} \approx 0.769$

77. $E = \sqrt{\dfrac{2AS}{I}} = \sqrt{\dfrac{2(20)(5290)}{100}} = 46$

79. $V = 1.22\sqrt{A} = 1.22\sqrt{25,000} \approx 193$ miles

8.3 WARM-UPS

1. True, because $\sqrt{20} = \sqrt{4} \cdot \sqrt{5} = 2\sqrt{5}$.

2. False, because $\sqrt{18} = \sqrt{9} \cdot \sqrt{2} = 3\sqrt{2}$.

3. True, because $\dfrac{1}{\sqrt{3}} = \dfrac{1\sqrt{3}}{\sqrt{3}\sqrt{3}} = \dfrac{\sqrt{3}}{3}$.

4. False, because 9 and 4 have no common factor and 9/4 is in lowest terms.

5. True, because $\sqrt{a^3} = \sqrt{a^2} \cdot \sqrt{a} = a\sqrt{a}$.

6. False, because $(a^3)^2 = a^6$ and so the square root of a^9 is not a^3.

7. True, because $\sqrt{y^{17}} = \sqrt{y^{16}}\sqrt{y} = y^8 \cdot \sqrt{y}$.

8. False, because $\left(\dfrac{\sqrt{6}}{2}\right)^2 = \dfrac{6}{4} = \dfrac{3}{2} \neq 3$.

9. False, because $\sqrt{4} = 2$ and $2 \neq \sqrt{2}$.

10. False, because $(17)^2 = 189$.

8.3 EXERCISES

1. $\sqrt{8} = \sqrt{4}\sqrt{2} = 2\sqrt{2}$ 3. $\sqrt{24} = \sqrt{4}\sqrt{6} = 2\sqrt{6}$

5. $\sqrt{28} = \sqrt{4}\sqrt{7} = 2\sqrt{7}$

7. $\sqrt{90} = \sqrt{9}\sqrt{10} = 3\sqrt{10}$

9. $\sqrt{500} = \sqrt{100}\sqrt{5} = 10\sqrt{5}$

11. $\sqrt{150} = \sqrt{25}\sqrt{6} = 5\sqrt{6}$

13. $\dfrac{1}{\sqrt{5}} = \dfrac{1 \cdot \sqrt{5}}{\sqrt{5}\sqrt{5}} = \dfrac{\sqrt{5}}{5}$

15. $\dfrac{3}{\sqrt{2}} = \dfrac{3\sqrt{2}}{\sqrt{2}\sqrt{2}} = \dfrac{3\sqrt{2}}{2}$

17. $\dfrac{\sqrt{3}}{\sqrt{2}} = \dfrac{\sqrt{3}\sqrt{2}}{\sqrt{2}\sqrt{2}} = \dfrac{\sqrt{6}}{2}$

19. $\dfrac{-3}{\sqrt{10}} = \dfrac{-3\sqrt{10}}{\sqrt{10}\sqrt{10}} = \dfrac{-3\sqrt{10}}{10}$

21. $\dfrac{-10}{\sqrt{17}} = \dfrac{-10\sqrt{17}}{\sqrt{17}\sqrt{17}} = \dfrac{-10\sqrt{17}}{17}$

23. $\dfrac{\sqrt{11}}{\sqrt{7}} = \dfrac{\sqrt{11}\sqrt{7}}{\sqrt{7}\sqrt{7}} = \dfrac{\sqrt{77}}{7}$

25. $\sqrt{63} = \sqrt{9}\sqrt{7} = 3\sqrt{7}$

27. $\sqrt{\dfrac{3}{2}} = \dfrac{\sqrt{3}}{\sqrt{2}} = \dfrac{\sqrt{3}\sqrt{2}}{\sqrt{2}\sqrt{2}} = \dfrac{\sqrt{6}}{2}$

29. $\sqrt{\dfrac{5}{8}} = \dfrac{\sqrt{5}}{\sqrt{8}} = \dfrac{\sqrt{5}\sqrt{2}}{\sqrt{8}\sqrt{2}} = \dfrac{\sqrt{10}}{\sqrt{16}} = \dfrac{\sqrt{10}}{4}$

31. $\dfrac{\sqrt{6}}{\sqrt{10}} = \dfrac{\sqrt{6}\sqrt{10}}{\sqrt{10}\sqrt{10}} = \dfrac{\sqrt{60}}{10} = \dfrac{\sqrt{4}\sqrt{15}}{10} = \dfrac{2\sqrt{15}}{10} = \dfrac{\sqrt{15}}{5}$

33. $\dfrac{\sqrt{75}}{\sqrt{3}} = \dfrac{\sqrt{25}\sqrt{3}}{\sqrt{3}} = 5$

35. $\dfrac{\sqrt{15}}{\sqrt{10}} = \dfrac{\sqrt{3}\sqrt{5}}{\sqrt{2}\sqrt{5}} = \dfrac{\sqrt{3}}{\sqrt{2}} = \dfrac{\sqrt{3}\sqrt{2}}{\sqrt{2}\sqrt{2}} = \dfrac{\sqrt{6}}{2}$

37. $\sqrt{a^8} = a^4$

39. $\sqrt{a^9} = \sqrt{a^8}\cdot\sqrt{a} = a^4\cdot\sqrt{a}$

41. $\sqrt{8a^6} = \sqrt{4a^6}\cdot\sqrt{2} = 2a^3\cdot\sqrt{2}$

43. $\sqrt{20a^4b^9} = \sqrt{4a^4b^8}\cdot\sqrt{5b} = 2a^2b^4\cdot\sqrt{5b}$

45. $\sqrt{27x^3y^3} = \sqrt{9x^2y^2}\cdot\sqrt{3xy} = 3xy\cdot\sqrt{3xy}$

47. $\sqrt{27a^3b^8c^2} = \sqrt{9a^2b^8c^2}\cdot\sqrt{3a} = 3ab^4c\sqrt{3a}$

49. $\dfrac{1}{\sqrt{x}} = \dfrac{1\sqrt{x}}{\sqrt{x}\sqrt{x}} = \dfrac{\sqrt{x}}{x}$

51. $\dfrac{\sqrt{2}}{\sqrt{3a}} = \dfrac{\sqrt{2}\sqrt{3a}}{\sqrt{3a}\sqrt{3a}} = \dfrac{\sqrt{6a}}{3a}$

53. $\dfrac{\sqrt{3}}{\sqrt{15y}} = \dfrac{\sqrt{3}}{\sqrt{3}\sqrt{5y}} = \dfrac{1}{\sqrt{5y}} = \dfrac{1\sqrt{5y}}{\sqrt{5y}\sqrt{5y}} = \dfrac{\sqrt{5y}}{5y}$

55. $\sqrt{\dfrac{3x}{2y}} = \dfrac{\sqrt{3x}}{\sqrt{2y}} = \dfrac{\sqrt{3x}\sqrt{2y}}{\sqrt{2y}\sqrt{2y}} = \dfrac{\sqrt{6xy}}{2y}$

57. $\sqrt{\dfrac{10y}{15x}} = \sqrt{\dfrac{2\cdot5y}{3\cdot5x}} = \sqrt{\dfrac{2y}{3x}} = \dfrac{\sqrt{2y}}{\sqrt{3x}} = \dfrac{\sqrt{2y}\sqrt{3x}}{\sqrt{3x}\sqrt{3x}}$

$\qquad\qquad = \dfrac{\sqrt{6xy}}{3x}$

59. $\sqrt{\dfrac{8x^3}{y}} = \dfrac{\sqrt{4x^2}\cdot\sqrt{2x}\cdot\sqrt{y}}{\sqrt{y}\cdot\sqrt{y}} = \dfrac{2x\cdot\sqrt{2xy}}{y}$

61. $\sqrt{80x^3} = \sqrt{16x^2}\sqrt{5x} = 4x\sqrt{5x}$

63. $\sqrt{9y^9x^{15}} = \sqrt{9y^8x^{14}}\cdot\sqrt{yx} = 3y^4x^7\cdot\sqrt{yx}$

65. $\dfrac{20x^6}{\sqrt{5x^5}} = \dfrac{20x^6}{x^2\sqrt{5x}} = \dfrac{20x^6\sqrt{5x}}{x^2\sqrt{5x}\cdot\sqrt{5x}} = \dfrac{20x^6\,\sqrt{5x}}{5x^3}$

$\qquad\qquad\qquad = 4x^3\,\sqrt{5x}$

67. $\dfrac{-22p^2}{p\,\sqrt{6pq}} = \dfrac{-22p^2\cdot\sqrt{6pq}}{p\,\sqrt{6pq}\cdot\sqrt{6pq}} = \dfrac{-22p^2\cdot\sqrt{6pq}}{6p^2q}$

$\qquad\qquad = \dfrac{-11\sqrt{6pq}}{3q}$

69. $\dfrac{a^3b^7\,\sqrt{a^2b^3c^4}}{\sqrt{abc}} = a^3b^7\sqrt{ab^2c^3}$

$= a^3b^7\,\sqrt{b^2c^2}\sqrt{ac} = a^3b^7bc\sqrt{ac} = a^3b^8c\sqrt{ac}$

71. $\dfrac{\sqrt{4xy^2}}{x^9y^3\,\sqrt{6xy^3}} = \dfrac{\sqrt{4xy^2}\sqrt{6xy^3}}{x^9y^3\,\sqrt{6xy^3}\sqrt{6xy^3}}$

$= \dfrac{\sqrt{24x^2y^5}}{x^9y^3\cdot6xy^3} = \dfrac{2xy^2\,\sqrt{6y}}{6x^{10}y^6} = \dfrac{\sqrt{6y}}{3x^9y^4}$

73. $\dfrac{1}{\sqrt{2}} - \dfrac{\sqrt{2}}{2} = 0$

75. $\dfrac{\sqrt{6}}{\sqrt{2}} - \sqrt{3} = 0$

77. $E = \sqrt{\dfrac{2AS}{I}} = \dfrac{\sqrt{2AS}\cdot\sqrt{I}}{\sqrt{I}\cdot\sqrt{I}} = \dfrac{\sqrt{2AIS}}{I}$

8.4 WARM-UPS

1. False, because $\sqrt{9} + \sqrt{16} = 3 + 4 = 7$ and $\sqrt{25} = 5$.

2. True, because $\sqrt{5}\cdot\sqrt{5} = 5$.

3. False, because $2\cdot\sqrt{5} \neq \sqrt{10}$.

4. False, because $3\sqrt{2}\cdot3\sqrt{2} = 9\cdot2 = 18$.

5. True, because $3\sqrt{5}\cdot3\sqrt{2} = 3\cdot3\cdot\sqrt{5}\cdot\sqrt{2} = 9\sqrt{10}$.

6. False, because $\sqrt{5} + 3\sqrt{5} = (1+3)\sqrt{5} = 4\sqrt{5}$.

7. False, because $3 \cdot \sqrt{5} \neq \sqrt{15}$.

8. False, because $\sqrt{3} \cdot \sqrt{6} = \sqrt{18} \neq \sqrt{2}$.

9. True, because $\dfrac{\sqrt{27}}{\sqrt{3}} = \sqrt{9} = 3$.

10. True, because $(\sqrt{3} - 1)(\sqrt{3} + 1) = (\sqrt{3})^2 - 1^2$
$= 3 - 1 = 2$.

8.4 EXERCISES

1. $4\sqrt{5} + 3\sqrt{5} = (4 + 3)\sqrt{5} = 7\sqrt{5}$

3. $\sqrt[3]{2} + \sqrt[3]{2} = 2 \cdot \sqrt[3]{2}$

5. $3u\sqrt{11} + 5u\sqrt{11} = (3u + 5u)\sqrt{11} = 8u\sqrt{11}$

7. $\sqrt{2} + \sqrt{3} - 5\sqrt{2} + 3\sqrt{3} = \sqrt{3} + 3\sqrt{3} + \sqrt{2} - 5\sqrt{2}$
$$= 4\sqrt{3} - 4\sqrt{2}$$

9. $3\sqrt{y} - \sqrt{x} - 4\sqrt{y} - 3\sqrt{x}$
$= -\sqrt{x} - 3\sqrt{x} + 3\sqrt{y} - 4\sqrt{y}$
$= -4\sqrt{x} - \sqrt{y}$

11. $3x\sqrt{y} - \sqrt{a} + 2x\sqrt{y} + 3\sqrt{a}$
$= (3x + 2x)\sqrt{y} + (-1 + 3)\sqrt{a}$
$= 5x\sqrt{y} + 2\sqrt{a}$

13. $\sqrt{24} + \sqrt{54} = \sqrt{4}\sqrt{6} + \sqrt{9}\sqrt{6} = 2\sqrt{6} + 3\sqrt{6} = 5\sqrt{6}$

15. $2\sqrt{27} - 4\sqrt{75} = 2\sqrt{9}\sqrt{3} - 4\sqrt{25}\sqrt{3}$
$= 2 \cdot 3\sqrt{3} - 4 \cdot 5\sqrt{3} = 6\sqrt{3} - 20\sqrt{3} = -14\sqrt{3}$

17. $\sqrt{3a} - \sqrt{12a} = \sqrt{3a} - \sqrt{4}\sqrt{3a} = \sqrt{3a} - 2\sqrt{3a}$
$= -\sqrt{3a}$

19. $\sqrt{x^3} + x\sqrt{4x} = \sqrt{x^2}\sqrt{x} + 2x\sqrt{x} = x\sqrt{x} + 2x\sqrt{x}$
$$= 3x\sqrt{x}$$

21. $\dfrac{1}{\sqrt{3}} + \dfrac{\sqrt{2}}{\sqrt{6}} = \dfrac{1}{\sqrt{3}} + \dfrac{\sqrt{2}}{\sqrt{2}\sqrt{3}} = \dfrac{1}{\sqrt{3}} + \dfrac{1}{\sqrt{3}} = \dfrac{2}{\sqrt{3}}$
$$= \dfrac{2\sqrt{3}}{\sqrt{3}\sqrt{3}} = \dfrac{2\sqrt{3}}{3}$$

23. $\dfrac{1}{\sqrt{3}} + \sqrt{12} = \dfrac{1\sqrt{3}}{\sqrt{3}\sqrt{3}} + \sqrt{4}\sqrt{3} = \dfrac{\sqrt{3}}{3} + 2\sqrt{3}$
$$= \dfrac{\sqrt{3}}{3} + \dfrac{2\sqrt{3} \cdot 3}{3} = \dfrac{\sqrt{3}}{3} + \dfrac{6\sqrt{3}}{3} = \dfrac{7\sqrt{3}}{3}$$

25. $\sqrt{7} \cdot \sqrt{11} = \sqrt{77}$

27. $2\sqrt{6} \cdot 3\sqrt{6} = 2 \cdot 3\sqrt{6}\sqrt{6} = 2 \cdot 3 \cdot 6 = 36$

29. $-3\sqrt{5} \cdot 4\sqrt{2} = -3 \cdot 4 \cdot \sqrt{5}\sqrt{2} = -12\sqrt{10}$

31. $\sqrt{2a^3} \cdot \sqrt{6a^5} = \sqrt{12a^8} = \sqrt{4a^8} \cdot \sqrt{3} = 2a^4\sqrt{3}$

33. $\sqrt[3]{9} \cdot \sqrt[3]{3} = \sqrt[3]{27} = 3$

35. $\sqrt[3]{-4m^2} \cdot \sqrt[3]{2m} = \sqrt[3]{-8m^3} = -2m$

37. $\sqrt{2}(\sqrt{2} + \sqrt{3}) = \sqrt{2}\sqrt{2} + \sqrt{2}\sqrt{3} = 2 + \sqrt{6}$

39. $3\sqrt{2}(2\sqrt{6} + \sqrt{10}) = 6\sqrt{12} + 3\sqrt{20}$
$= 6\sqrt{4}\sqrt{3} + 3\sqrt{4}\sqrt{5} = 6 \cdot 2\sqrt{3} + 3 \cdot 2\sqrt{5} = 12\sqrt{3} + 6\sqrt{5}$

41. $2\sqrt{5}(\sqrt{5} - 3\sqrt{10}) = 2 \cdot 5 - 6\sqrt{50}$
$$= 10 - 6 \cdot \sqrt{25}\sqrt{2} = 10 - 6 \cdot 5\sqrt{2} = 10 - 30\sqrt{2}$$

43. $(\sqrt{5} - 4)(\sqrt{5} + 3) = 5 - 4\sqrt{5} + 3\sqrt{5} - 12$
$$= -7 - \sqrt{5}$$

45. $(\sqrt{3} - 1)(\sqrt{3} + 1) = (\sqrt{3})^2 - 1^2 = 3 - 1 = 2$

47. $(\sqrt{5} - \sqrt{2})(\sqrt{5} + \sqrt{2}) = 5 - \sqrt{10} + \sqrt{10} - 2 = 3$

49. $(2\sqrt{5} + 1)(3\sqrt{5} - 2) = 6 \cdot 5 + 3\sqrt{5} - 4\sqrt{5} - 2$
$$= 28 - \sqrt{5}$$

51. $(2\sqrt{3} - 3\sqrt{5})(3\sqrt{3} + 4\sqrt{5})$
$= 6 \cdot 3 - 9\sqrt{15} + 8\sqrt{15} - 12 \cdot 5 = -42 - \sqrt{15}$

53. $(2\sqrt{3} + 5)^2 = 12 + 2 \cdot 2\sqrt{3} \cdot 5 + 5^2$
$$= 12 + 20\sqrt{3} + 25 = 37 + 20\sqrt{3}$$

55. $\sqrt{10} \div \sqrt{5} = \sqrt{\dfrac{10}{5}} = \sqrt{2}$

57. $\sqrt{5} \div \sqrt{3} = \dfrac{\sqrt{5}}{\sqrt{3}} = \dfrac{\sqrt{5}\sqrt{3}}{\sqrt{3}\sqrt{3}} = \dfrac{\sqrt{15}}{3}$

59. $4\sqrt{5} \div (3\sqrt{6}) = \dfrac{4\sqrt{5}}{3\sqrt{6}} = \dfrac{4\sqrt{5}\sqrt{6}}{3\sqrt{6}\sqrt{6}} = \dfrac{4\sqrt{30}}{18} = \dfrac{2\sqrt{30}}{9}$

61. $5\sqrt{14} \div (3\sqrt{2}) = \dfrac{5\sqrt{2}\sqrt{7}}{3\sqrt{2}} = \dfrac{5\sqrt{7}}{3}$

63. $\dfrac{2 + \sqrt{8}}{2} = \dfrac{2 + \sqrt{4}\sqrt{2}}{2} = \dfrac{2 + 2\sqrt{2}}{2} = \dfrac{2(1 + \sqrt{2})}{2}$
$$= 1 + \sqrt{2}$$

65. $\dfrac{-4 + \sqrt{20}}{2} = \dfrac{-4 + \sqrt{4}\sqrt{5}}{2} = \dfrac{-4 + 2\sqrt{5}}{2}$
$$= \dfrac{2(-2 + \sqrt{5})}{2} = -2 + \sqrt{5}$$

67. $\dfrac{4-\sqrt{20}}{6} = \dfrac{4-2\sqrt{5}}{6} = \dfrac{2(2-\sqrt{5})}{2\cdot 3} = \dfrac{2-\sqrt{5}}{3}$

69. $\dfrac{-4-\sqrt{24}}{-6} = \dfrac{-4-2\sqrt{6}}{-6} = \dfrac{-2(2+\sqrt{6})}{-2\cdot 3} = \dfrac{2+\sqrt{6}}{3}$

71. $\dfrac{5}{\sqrt{3}-\sqrt{2}} = \dfrac{5(\sqrt{3}+\sqrt{2})}{(\sqrt{3}-\sqrt{2})(\sqrt{3}+\sqrt{2})} = \dfrac{5\sqrt{3}+5\sqrt{2}}{3-2}$

$$= \dfrac{5\sqrt{3}+5\sqrt{2}}{1} = 5\sqrt{3}+5\sqrt{2}$$

73. $\dfrac{\sqrt{3}}{\sqrt{5}-\sqrt{3}} = \dfrac{\sqrt{3}(\sqrt{5}+\sqrt{3})}{(\sqrt{5}-\sqrt{3})(\sqrt{5}+\sqrt{3})} = \dfrac{\sqrt{15}+3}{5-3}$

$$= \dfrac{\sqrt{15}+3}{2}$$

75. $\dfrac{2+\sqrt{3}}{5-\sqrt{3}} = \dfrac{(2+\sqrt{3})(5+\sqrt{3})}{(5-\sqrt{3})(5+\sqrt{3})} = \dfrac{10+7\sqrt{3}+3}{25-3}$

$$= \dfrac{13+7\sqrt{3}}{22}$$

77. $\dfrac{\sqrt{7}-5}{2\sqrt{7}+1} = \dfrac{(\sqrt{7}-5)(2\sqrt{7}-1)}{(2\sqrt{7}+1)(2\sqrt{7}-1)}$

$$= \dfrac{14-11\sqrt{7}+5}{28-1} = \dfrac{19-11\sqrt{7}}{27}$$

79. $\sqrt{5a}+\sqrt{20a} = \sqrt{5a}+\sqrt{4}\sqrt{5a}$

$$= \sqrt{5a}+2\sqrt{5a} = 3\sqrt{5a}$$

81. $\sqrt{75}\div\sqrt{6} = \sqrt{\dfrac{75}{6}} = \sqrt{\dfrac{25}{2}} = \dfrac{5\cdot\sqrt{2}}{\sqrt{2}\cdot\sqrt{2}} = \dfrac{5\sqrt{2}}{2}$

83. $(5+3\sqrt{5})^2 = 25+30\sqrt{5}+45 = 70+30\sqrt{5}$

85. $\sqrt{5}+\dfrac{\sqrt{20}}{3} = \dfrac{3\sqrt{5}}{3}+\dfrac{2\sqrt{5}}{3} = \dfrac{5\sqrt{5}}{3}$

87. $\dfrac{2+\sqrt{3}}{2} \approx 1.866$

89. $\dfrac{-4-\sqrt{6}}{5-\sqrt{3}} \approx -1.974$

91. $A = LW = \sqrt{18}\cdot\sqrt{8} = 3\sqrt{2}\cdot 2\sqrt{2} = 12\text{ ft}^2$

$P = 2L+2W = 2\sqrt{18}+2\sqrt{8}$

$$= 2\cdot 3\sqrt{2}+2\cdot 2\sqrt{2} = 10\sqrt{2}\text{ ft}$$

93. $V = \sqrt{2}\cdot\sqrt{3}\cdot\sqrt{6} = \sqrt{6}\cdot\sqrt{6} = 6$ cubic meters

8.5 WARM-UPS

1. False, because both 3 and -3 are solutions to $x^2 = 9$.

2. True, because the square of a real number cannot be negative.

3. False, because 0 satisfies the equation $a^2 = 0$.

4. False, because $x^2+5 = 0$ is equivalent to $x^2 = -5$ and it has no real solution.

5. True, because $-x^2 = 9$ is equivalent to $x^2 = -9$.

6. False, because the first step is to square each side.

7. False, because if we square both sides of the equation $\sqrt{x} = -2$ we get $x = 4$ and 4 is an extraneous root.

8. True, because if we square both sides of $\sqrt{x} = -1$ we get $x = 1$, but 1 is not a solution to $\sqrt{x} = -1$.

9. True, because of the square root property.

10. True, because the left side is a negative number and the right side is nonnegative for any value of the variable.

8.5 EXERCISES

1. $\quad x^2 = 16$

$\quad x = \pm\sqrt{16} = \pm 4$

The solutions to the equation are -4 and 4.

3. $\quad x^2 = 40$

$\quad x = \pm\sqrt{40} = \pm\sqrt{4}\sqrt{10} = \pm 2\sqrt{10}$

The solutions are $-2\sqrt{10}$ and $2\sqrt{10}$.

5. $\quad 3x^2 = 2$

$\quad x^2 = \dfrac{2}{3}$

$\quad x = \pm\sqrt{\dfrac{2}{3}} = \pm\dfrac{\sqrt{2}}{\sqrt{3}} = \pm\dfrac{\sqrt{2}\sqrt{3}}{\sqrt{3}\sqrt{3}} = \pm\dfrac{\sqrt{6}}{3}$

The solutions to the equation are $-\dfrac{\sqrt{6}}{3}$ and $\dfrac{\sqrt{6}}{3}$.

7. $\quad 9x^2 = -4$

$\quad x^2 = -\dfrac{4}{9}$

There is no solution to this equation.

9. $\quad (x-1)^2 = 4$

$\quad x-1 = \pm 2$

$\quad x-1 = 2 \quad\text{or}\quad x-1 = -2$

$\quad x = 3 \quad\text{or}\quad\quad x = -1$

The solutions to the equation are -1 and 3.

11.
$$2(x-5)^2 + 1 = 7$$
$$2(x-5)^2 = 6$$
$$(x-5)^2 = 3$$
$$x - 5 = \pm\sqrt{3}$$
$$x = 5 \pm \sqrt{3}$$

The solutions to the equation are $5 - \sqrt{3}$ and $5 + \sqrt{3}$.

13.
$$(x+19)^2 = 0$$
$$x + 19 = 0$$
$$x = -19$$

The solution to the equation is -19.

15.
$$\sqrt{x-9} = 9$$
$$(\sqrt{x-9})^2 = 9^2$$
$$x - 9 = 81$$
$$x = 90$$

The solution to the equation is 90.

17.
$$\sqrt{2x-3} = -4$$
$$(\sqrt{2x-3})^2 = (-4)^2$$
$$2x - 3 = 16$$
$$2x = 19$$
$$x = \frac{19}{2}$$

Check: $\sqrt{2 \cdot \frac{19}{2} - 3} = -4$

$$\sqrt{16} = -4 \quad \text{Incorrect.}$$

Since $19/2$ is an extraneous root, there is no solution to the equation.

19.
$$4 = \sqrt{x^2 - 9}$$
$$4^2 = (\sqrt{x^2 - 9})^2$$
$$16 = x^2 - 9$$
$$25 = x^2$$
$$\pm 5 = x$$

The solutions to the equation are -5 and 5.

21.
$$x = \sqrt{18 - 3x}$$
$$x^2 = (\sqrt{18 - 3x})^2$$
$$x^2 = 18 - 3x$$
$$x^2 + 3x - 18 = 0$$
$$(x+6)(x-3) = 0$$
$$x + 6 = 0 \quad \text{or} \quad x - 3 = 0$$
$$x = -6 \quad \text{or} \quad x = 3$$

Check -6: $-6 = \sqrt{18 - 3(-6)}$

$$-6 = \sqrt{36} \quad \text{Incorrect.}$$

Since -6 is an extraneous root, the only solution to the equation is 3.

23.
$$x = \sqrt{x}$$
$$x^2 = (\sqrt{x})^2$$
$$x^2 = x$$
$$x^2 - x = 0$$
$$x(x-1) = 0$$
$$x = 0 \quad \text{or} \quad x - 1 = 0$$
$$x = 1$$

The solutions to the equation are 0 and 1.

25.
$$\sqrt{x+1} = \sqrt{2x-5}$$
$$(\sqrt{x+1})^2 = (\sqrt{2x-5})^2$$
$$x + 1 = 2x - 5$$
$$6 = x$$

The solution to the equation is 6.

27.
$$3\sqrt{2x-1} + 3 = 5$$
$$\sqrt{2x-1} = \frac{2}{3}$$
$$2x - 1 = \frac{4}{9}$$
$$18x - 9 = 4$$
$$18x = 13$$
$$x = \frac{13}{18}$$

The solution to the equation is $\frac{13}{18}$.

29.
$$x - 3 = \sqrt{2x-6}$$
$$(x-3)^2 = (\sqrt{2x-6})^2$$
$$x^2 - 6x + 9 = 2x - 6$$
$$x^2 - 8x + 15 = 0$$
$$(x-3)(x-5) = 0$$
$$x - 3 = 0 \quad \text{or} \quad x - 5 = 0$$
$$x = 3 \quad \text{or} \quad x = 5$$

The solutions to the equation are 3 and 5.

31.
$$\sqrt{x+13} = x + 1$$
$$(\sqrt{x+13})^2 = (x+1)^2$$
$$x + 13 = x^2 + 2x + 1$$
$$0 = x^2 + x - 12$$
$$0 = (x+4)(x-3)$$
$$x + 4 = 0 \quad \text{or} \quad x - 3 = 0$$
$$x = -4 \quad \text{or} \quad x = 3$$

Check -4: $\sqrt{-4+13} = -4+1$

$$\sqrt{9} = -3 \quad \text{Incorrect.}$$

Since -4 is an extraneous root, the only solution to the equation is 3.

33.
$$\sqrt{10x-44} = x-2$$
$$(\sqrt{10x-44})^2 = (x-2)^2$$
$$10x-44 = x^2-4x+4$$
$$0 = x^2-14x+48$$
$$0 = (x-6)(x-8)$$
$$x-6 = 0 \quad \text{or} \quad x-8 = 0$$
$$x = 6 \quad \text{or} \quad x = 8$$
The solutions to the equation are 6 and 8.

35.
$$V = \pi r^2 h$$
$$\frac{V}{\pi h} = r^2$$
$$\pm\sqrt{\frac{V}{\pi h}} = r$$
$$r = \pm\sqrt{\frac{V}{\pi h}}$$

37.
$$a^2 + b^2 = c^2$$
$$b^2 = c^2 - a^2$$
$$b = \pm\sqrt{c^2 - a^2}$$

39.
$$b^2 - 4ac = 0$$
$$b^2 = 4ac$$
$$b = \pm\sqrt{4ac}$$
$$b = \pm 2\sqrt{ac}$$

41.
$$v = \sqrt{2pt}$$
$$v^2 = (\sqrt{2pt})^2$$
$$v^2 = 2pt$$
$$\frac{v^2}{2p} = t$$
$$t = \frac{v^2}{2p}$$

43.
$$3x^2 - 6 = 0$$
$$3x^2 = 6$$
$$x^2 = 2$$
$$x = \pm\sqrt{2}$$
The solutions to the equation are $-\sqrt{2}$ and $\sqrt{2}$.

45.
$$\sqrt{2x-3} = \sqrt{3x+1}$$
$$(\sqrt{2x-3})^2 = (\sqrt{3x+1})^2$$
$$2x-3 = 3x+1$$
$$-4 = x$$
Check:
$$\sqrt{2(-4)-3} = \sqrt{3(-4)+1}$$
$$\sqrt{-11} = \sqrt{-11}$$
Since $\sqrt{-11}$ is not a real number, there is no solution to the equation.

47.
$$(2x-1)^2 = 8$$
$$2x-1 = \pm\sqrt{8} = \pm 2\sqrt{2}$$
$$2x-1 = \sqrt{2} \quad \text{or} \quad 2x-1 = -\sqrt{2}$$
$$2x = 1+\sqrt{2} \quad \text{or} \quad 2x = 1-\sqrt{2}$$
$$x = \frac{1+\sqrt{2}}{2} \quad \text{or} \quad x = \frac{1-\sqrt{2}}{2}$$
The solutions are $\frac{1+2\sqrt{2}}{2}$ and $\frac{1-2\sqrt{2}}{2}$.

49.
$$\sqrt{2x-9} = 0$$
$$(\sqrt{2x-9})^2 = 0^2$$
$$2x-9 = 0$$
$$2x = 9$$
$$x = \frac{9}{2}$$
The solution to the equation is $\frac{9}{2}$.

51.
$$x+1 = \sqrt{2x+10}$$
$$(x+1)^2 = (\sqrt{2x+10})^2$$
$$x^2+2x+1 = 2x+10$$
$$x^2 = 9$$
$$x = \pm 3$$
Check -3:
$$-3+1 = \sqrt{2(-3)+10}$$
$$-2 = \sqrt{4} \quad \text{Incorrect.}$$
Since 3 does satisfy the original equation, the solution to the equation is 3.

53.
$$3(x+1)^2 - 27 = 0$$
$$3(x+1)^2 = 27$$
$$(x+1)^2 = 9$$
$$x+1 = \pm 3$$
$$x+1 = 3 \quad \text{or} \quad x+1 = -3$$
$$x = 2 \quad \text{or} \quad x = -4$$
Since both numbers satisfy the original equation, the solutions to the equation are -4 and 2.

55.
$$(2x-5)^2 = 0$$
$$2x-5 = 0$$
$$x = \frac{5}{2}$$
The solution to the equation is $\frac{5}{2}$.

57.
$$x^2 = 3.25$$
$$x = \pm\sqrt{3.25} = \pm 1.803$$
The solutions to the equation are -1.803 and 1.803.

59.
$$\sqrt{x+2} = 1.73$$
$$(\sqrt{x+2})^2 = (1.73)^2$$
$$x+2 = (1.73)^2$$
$$x = (1.73)^2 - 2 = 0.993$$

The solution to the equation is 0.993.

61.
$$1.3(x-2.4)^2 = 5.4$$
$$(x-2.4)^2 = \frac{5.4}{1.3}$$
$$x-2.4 = \pm\sqrt{\frac{5.4}{1.3}}$$
$$x = 2.4 \pm \sqrt{\frac{5.4}{1.3}} = 2.4 \pm 2.038$$

$$x = 2.4 + 2.038 = 4.438$$
$$\text{or } x = 2.4 - 2.038 = 0.362$$

The solutions to the equation are 0.362 and 4.438.

63. Let x = the length of a side. Since the area of a square is $A = s^2$, we can write the following equation.
$$x^2 = 18$$
$$x = \pm\sqrt{18} = \pm 3\sqrt{2}$$

The length of each side of the square is $3\sqrt{2}$ feet or 4.243 feet.

65. Let x = the length of a side of the square. Since the diagonal is 6 feet, we can use the Pythagorean theorem to write the following equation.
$$x^2 + x^2 = 6^2$$
$$2x^2 = 36$$
$$x^2 = 18$$
$$x = \pm\sqrt{18} = \pm 3\sqrt{2}$$

The length of each side of the square table is $3\sqrt{2}$ feet or 4.243 feet.

67. Let x = the length of the diagonal of the square. Since the length of each side of the square is 1 foot, we can use the Pythagorean theorem to write the following equation.
$$1^2 + 1^2 = x^2$$
$$2 = x^2$$
$$\pm\sqrt{2} = x$$

The length of the diagonal is $\sqrt{2}$ feet or 1.414 feet.

69. Let x = the length of the diagonal of the rectangle. Since the sides are length 3 feet and 4 feet, we can write the following equation.
$$3^2 + 4^2 = x^2$$
$$25 = x^2$$
$$\pm 5 = x$$

The length of the diagonal is 5 feet.

71. Use s = 100 in the formula $s = 16t^2$.
$$16t^2 = 100$$
$$t^2 = \frac{100}{16} = \frac{25}{4}$$
$$t = \pm\sqrt{\frac{25}{4}} = \pm\frac{5}{2} = \pm 2.5$$

It takes 2.5 seconds for the pine cone to fall 100 feet.

73. Let x = the length of the guy wire. Since 2/3 of 200 is 400/3, x is the length of the hypotenuse of a right triangle whose legs have lengths 400/3 feet and 200 feet.
$$x^2 = 200^2 + \left(\frac{400}{3}\right)^2$$
$$x^2 = 40000 + \frac{160000}{9}$$
$$x^2 = \frac{40000 \cdot 9}{9} + \frac{160000}{9}$$
$$x^2 = \frac{520,000}{9}$$
$$x = \pm\sqrt{\frac{520,000}{9}} = \pm\sqrt{\frac{40,000 \cdot 13}{9}}$$
$$x = \pm\frac{200\sqrt{13}}{3}$$

The length of the guy wire is $\frac{200\sqrt{13}}{3}$ feet or 240.37 feet.

8.6 WARM-UPS

1. True.   **2.** False, because $8^{5/3} = \sqrt[3]{8^5}$.

3. False, because $-16^{1/2} = -4$, while $(-16)^{1/2}$ is not a real number.

4. True.   **5.** True, because $6^{-1/2} = \frac{1}{\sqrt{6}} = \frac{\sqrt{6}}{6}$.

6. True, because $2^{1/2} \cdot 2^{1/2} = 2$

7. False, because $2^{1/2} \cdot 2^{1/2} = 2$

8. False, because $16^{-1/4} = \frac{1}{16^{1/4}} = \frac{1}{2}$.

9. True, $6^{1/6} \cdot 6^{1/6} = 6^{1/6 + 1/6} = 6^{2/6} = 6^{1/3}$.

10. True, $(2^8)^{3/4} = 2^{24/4} = 2^6$.

8.6 EXERCISES

1. $\sqrt[4]{7} = 7^{1/4}$

3. $9^{1/5} = \sqrt[5]{9}$

5. $\sqrt{5x} = (5x)^{1/2}$

7. $a^{1/2} = \sqrt{a}$

9. $25^{1/2} = \sqrt{25} = 5$

11. $(-125)^{1/3} = \sqrt[3]{-125} = -5$

13. $16^{1/4} = \sqrt[4]{16} = 2$

15. $(-4)^{1/2}$ is not a real number because it is $\sqrt{-4}$.

17. $\sqrt[3]{w^7} = w^{7/3}$

19. $\dfrac{1}{\sqrt[3]{2^{10}}} = 2^{-10/3}$

21. $w^{-3/4} = \sqrt[4]{\dfrac{1}{w^3}}$

23. $(ab)^{3/2} = \sqrt{(ab)^3}$

25. $125^{2/3} = (\sqrt[3]{125})^2 = 5^2 = 25$

27. $25^{3/2} = (\sqrt{25})^3 = 5^3 = 125$

29. $27^{-4/3} = \dfrac{1}{(\sqrt[3]{27})^4} = \dfrac{1}{3^4} = \dfrac{1}{81}$

31. $4^{-3/2} = \dfrac{1}{(\sqrt{4})^3} = \dfrac{1}{2^3} = \dfrac{1}{8}$

33. $(-27)^{-1/3} = \dfrac{1}{(-27)^{1/3}} = \dfrac{1}{-3} = -\dfrac{1}{3}$

35. $(-16)^{-1/4}$ is not a real number because it is a fourth root of a negative number.

37. $x^{1/4}x^{1/4} = x^{2/4} = x^{1/2}$

39. $n^{1/2}n^{-1/3} = n^{1/6}$

41. $\dfrac{x^2}{x^{1/2}} = x^{2-\frac{1}{2}} = x^{3/2}$

43. $\dfrac{8t^{1/2}}{4t^{1/4}} = 2t^{\frac{1}{2}-\frac{1}{4}} = 2t^{1/4}$

45. $(x^6)^{1/3} = x^{6/3} = x^2$

47. $(5^{-1/4})^{-1/2} = 5^{1/8}$

49. $(x^2y^6)^{1/2} = xy^3$

51. $(9x^{-2}y^8)^{-1/2} = 9^{-1/2}xy^{-4} = \dfrac{x}{3y^4}$

53. $16^{-1/2} + 2^{-1} = \dfrac{1}{4} + \dfrac{1}{2} = \dfrac{3}{4}$

55. $27^{-1/6} \cdot 27^{-1/2} = 27^{-4/6} = 27^{-2/3} = \dfrac{1}{9}$

57. $\dfrac{81^{5/6}}{81^{1/12}} = 81^{9/12} = 81^{3/4} = 27$

59. $(3^{-4} \cdot 6^8)^{-1/4} = 3^1 \cdot 6^{-2} = \dfrac{3}{36} = \dfrac{1}{12}$

61. $21.5 + 1.25(305.4)^{1/2} - 9.8(21.8)^{1/3} = 15.968$

So the boat does satisfy the inequality.

$$21.5 + 1.25(S)^{1/2} - 9.8(21.8)^{1/3} \le 16.296$$
$$1.25S^{1/2} \le 22.1725$$
$$S^{1/2} \le 17.738$$
$$S \le 314.637$$

Since $314.637 - 305.4 = 9.237$, about 9.2 square meters of sail area could be added.

63. $r = \left(\dfrac{S}{P}\right)^{1/n} - 1 = \left(\dfrac{52{,}796}{10{,}000}\right)^{1/10} - 1 = 0.181$

So the average annual return was 18.1%.

65. $262 \cdot 2^{1/12} = 278 \qquad 262 \cdot 2^{2/12} = 294$

$262 \cdot 2^{3/12} = 312 \qquad 262 \cdot 2^{4/12} = 330$

$262 \cdot 2^{5/12} = 350 \qquad 262 \cdot 2^{6/12} = 371$

$262 \cdot 2^{7/12} = 393 \qquad 262 \cdot 2^{8/12} = 416$

$262 \cdot 2^{9/12} = 441 \qquad 262 \cdot 2^{10/12} = 467$

$262 \cdot 2^{11/12} = 495$

The notes are tuned to 278, 294, 312, 330, 350, 371, 393, 416, 441, 467, and 495 Hz.

CHAPTER 8 REVIEW

1. $2^{-5} = \dfrac{1}{2^5} = \dfrac{1}{32}$

3. $10^{-3} = \dfrac{1}{10^3} = \dfrac{1}{1000}$

5. $x^5x^{-8} = x^{5-8} = x^{-3} = \dfrac{1}{x^3}$

7. $\dfrac{a^{-8}}{a^{-12}} = a^{-8-(-12)} = a^4$

9. $\dfrac{a^3}{a^{-7}} = a^{3-(-7)} = a^{10}$

11. $(x^{-3})^4 = x^{-3 \cdot 4} = x^{-12} = \dfrac{1}{x^{12}}$

13. $(2x^{-3})^{-3} = 2^{-3}x^9 = \dfrac{x^9}{8}$

15. $\left(\dfrac{a}{3b^{-3}}\right)^{-2} = \dfrac{a^{-2}}{(3b^{-3})^{-2}} = \dfrac{a^{-2}}{3^{-2}b^6} = \dfrac{9}{a^2 b^6}$

17. Move the decimal point 3 places to the left. Since 5000 is larger than 10 the exponent is positive.

$5000 = 5 \times 10^3$

19. Move the decimal point 5 places to the right.

$3.4 \times 10^5 = 340,000$

21. Move the decimal point 5 places to the right. Since 0.0000461 is smaller than 1, the exponent is negative.

$0.0000461 = 4.61 \times 10^{-5}$

23. Move the decimal point 6 places to the left.

$5.69 \times 10^{-6} = 0.00000569$

25. $(3.5 \times 10^8)(2.0 \times 10^{-12}) = 7 \times 10^{-4}$

27. $(2 \times 10^{-4})^4 = 2^4 \times 10^{-16} = 16 \times 10^{-16}$
$\qquad = 1.6 \times 10^1 \times 10^{-16} = 1.6 \times 10^{-15}$

29. $(0.00000004)(2,000,000,000)$
$\qquad = 4 \times 10^{-8} \cdot 2 \times 10^9 = 8 \times 10^1$

31. $(0.0000002)^5 = (2 \times 10^{-7})^5 = 32 \times 10^{-35}$
$\qquad = 3.2 \times 10^1 \times 10^{-35} = 3.2 \times 10^{-34}$

33. Since $2^5 = 32$, $\sqrt[5]{32} = 2$.

35. Since $10^3 = 1000$, $\sqrt[3]{1000} = 10$.

37. Since $(x^6)^2 = x^{12}$, $\sqrt{x^{12}} = x^6$.

39. Since $(x^2)^3 = x^6$, $\sqrt[3]{x^6} = x^2$.

41. $\sqrt{4x^2} = 2x$

43. $\sqrt[3]{125x^6} = \sqrt[3]{125} \cdot \sqrt[3]{x^6} = 5x^2$

45. $\sqrt{\dfrac{4x^{16}}{y^{14}}} = \dfrac{\sqrt{4} \cdot \sqrt{x^{16}}}{\sqrt{y^{14}}} = \dfrac{2x^8}{y^7}$

47. $\sqrt{\dfrac{w^2}{16}} = \dfrac{\sqrt{w^2}}{\sqrt{16}} = \dfrac{w}{4}$

49. $\sqrt{72} = \sqrt{36}\sqrt{2} = 6\sqrt{2}$

51. $\dfrac{1}{\sqrt{3}} = \dfrac{1\sqrt{3}}{\sqrt{3}\sqrt{3}} = \dfrac{\sqrt{3}}{3}$

53. $\sqrt{\dfrac{3}{5}} = \dfrac{\sqrt{3}}{\sqrt{5}} = \dfrac{\sqrt{3}\sqrt{5}}{\sqrt{5}\sqrt{5}} = \dfrac{\sqrt{15}}{5}$

55. $\dfrac{\sqrt{33}}{\sqrt{3}} = \sqrt{\dfrac{33}{3}} = \sqrt{11}$

57. $\dfrac{\sqrt{3}}{\sqrt{8}} = \dfrac{\sqrt{3}\sqrt{2}}{\sqrt{8}\sqrt{2}} = \dfrac{\sqrt{6}}{\sqrt{16}} = \dfrac{\sqrt{6}}{4}$

59. Since $(y^3)^2 = y^6$, $\sqrt{y^6} = y^3$.

61. $\sqrt{24t^9} = \sqrt{4t^8} \cdot \sqrt{6t} = 2t^4\sqrt{6t}$

63. $\sqrt{12m^5t^3} = \sqrt{4m^4t^2} \cdot \sqrt{3mt} = 2m^2t\sqrt{3mt}$

65. $\dfrac{\sqrt{2}}{\sqrt{x}} = \dfrac{\sqrt{2}\sqrt{x}}{\sqrt{x}\sqrt{x}} = \dfrac{\sqrt{2x}}{x}$

67. $\sqrt{\dfrac{3a^5}{2s}} = \dfrac{\sqrt{a^4}\sqrt{3a}}{\sqrt{2s}} = \dfrac{a^2\sqrt{3a}\sqrt{2s}}{\sqrt{2s}\sqrt{2s}} = \dfrac{a^2\sqrt{6as}}{2s}$

69. $2\sqrt{7} + 8\sqrt{7} = (2+8)\sqrt{7} = 10\sqrt{7}$

71. $\sqrt{12} - \sqrt{27} = \sqrt{4}\sqrt{3} - \sqrt{9}\sqrt{3}$
$\qquad = 2\sqrt{3} - 3\sqrt{3} = -\sqrt{3}$

73. $2\sqrt{3} \cdot 5\sqrt{3} = 10 \cdot 3 = 30$

75. $-3\sqrt{6} \cdot 5\sqrt{3} = -15\sqrt{18} = -15 \cdot 3\sqrt{2} = -45\sqrt{2}$

77. $-3\sqrt{3}(5 + \sqrt{3}) = (-3\sqrt{3})(5) + (-3\sqrt{3})(\sqrt{3})$
$\qquad\qquad = -15\sqrt{3} - 9$

79. $-\sqrt{3}(\sqrt{6} - \sqrt{15}) = -\sqrt{18} + \sqrt{45}$
$\qquad\quad = -\sqrt{9}\sqrt{2} + \sqrt{9}\sqrt{5} = -3\sqrt{2} + 3\sqrt{5}$

81. $(\sqrt{3} - 5)(\sqrt{3} + 5) = 3 - 25 = -22$

83. $(2\sqrt{5} - \sqrt{6})^2 = 4 \cdot 5 - 2 \cdot 2\sqrt{5}\sqrt{6} + 6$
$\qquad\qquad = 26 - 4\sqrt{30}$

85. $(4 - 3\sqrt{6})((5 - \sqrt{6}) = 20 - 19\sqrt{6} + 18$
$\quad = 38 - 19\sqrt{6}$

87. $3\sqrt{5} \div (6\sqrt{2}) = \dfrac{3\sqrt{5}}{6\sqrt{2}} = \dfrac{3\sqrt{5}\sqrt{2}}{6\sqrt{2}\sqrt{2}} = \dfrac{3\sqrt{10}}{12} = \dfrac{\sqrt{10}}{4}$

89. $\dfrac{2-\sqrt{20}}{10} = \dfrac{2-2\sqrt{5}}{2\cdot 5} = \dfrac{2(1-\sqrt{5})}{2\cdot 5} = \dfrac{1-\sqrt{5}}{5}$

91. $\dfrac{3}{1-\sqrt{5}} = \dfrac{3(1+\sqrt{5})}{(1-\sqrt{5})(1+\sqrt{5})}$

$\qquad = \dfrac{3+3\sqrt{5}}{1-5} = -\dfrac{3+3\sqrt{5}}{4}$

93. $x^2 = 400$

$\qquad x = \pm\sqrt{400} = \pm 20$

The solutions to the equation are -20 and 20.

95. $7x^2 = 3$

$\qquad x^2 = \dfrac{3}{7}$

$\qquad x = \pm\sqrt{\dfrac{3}{7}} = \pm\dfrac{\sqrt{3}}{\sqrt{7}} = \pm\dfrac{\sqrt{3}\sqrt{7}}{\sqrt{7}\sqrt{7}} = \pm\dfrac{\sqrt{21}}{7}$

The solutions to the equation are $-\dfrac{\sqrt{21}}{7}$ and $\dfrac{\sqrt{21}}{7}$.

97. $(x-4)^2 = 18$

$\qquad x-4 = \pm\sqrt{18} = \pm 3\sqrt{2}$

$\qquad x-4 = 3\sqrt{2} \qquad$ or $\quad x-4 = -3\sqrt{2}$

$\qquad\quad x = 4+3\sqrt{2} \quad$ or $\qquad x = 4-3\sqrt{2}$

The solutions to the equation are $4-3\sqrt{2}$ and $4+3\sqrt{2}$.

99. $\sqrt{x} = 9$

$\qquad (\sqrt{x})^2 = 9^2$

$\qquad\quad x = 81$

The solution to the equation is 81.

101. $x = \sqrt{36-5x}$

$\qquad x^2 = (\sqrt{36-5x})^2$

$\qquad x^2 = 36-5x$

$\quad x^2+5x-36 = 0$

$\quad (x+9)(x-4) = 0$

$\qquad x+9 = 0 \quad$ or $\quad x-4 = 0$

$\qquad\quad x = -9 \,$ or $\qquad x = 4$

Check -9: $-9 = \sqrt{36-5(-9)}$

$\qquad\qquad -9 = \sqrt{81} \quad$ Incorrect.

Since -9 does not satisfy the equation, the solution to the equation is 4.

103. $x+2 = \sqrt{52+2x}$

$\qquad (x+2)^2 = (\sqrt{52+2x})^2$

$\quad x^2+4x+4 = 52+2x$

$\quad x^2+2x-48 = 0$

$\quad (x+8)(x-6) = 0$

$\qquad x+8 = 0 \quad$ or $\quad x-6 = 0$

$\qquad\quad x = -8 \,$ or $\qquad x = 6$

Check -8: $-8+2 = \sqrt{52+2(-8)}$

$\qquad\qquad -6 = \sqrt{36} \quad$ Incorrect.

Since -8 does not satisfy the equation, the solution to the equation is 6.

105. $t^2 = 8sw$

$\qquad t = \pm\sqrt{8sw}$

$\qquad t = \pm 2\sqrt{2sw}$

107. $3a = \sqrt{bt}$

$\qquad (3a)^2 = (\sqrt{bt})^2$

$\qquad 9a^2 = bt$

$\qquad \dfrac{9a^2}{b} = t$

$\qquad t = \dfrac{9a^2}{b}$

109. $25^{-3/2} = \dfrac{1}{25^{3/2}} = \dfrac{1}{125}$

111. $25^{1/2} = \sqrt{25} = 5$

113. $64^{-1/2} = \dfrac{1}{\sqrt{64}} = \dfrac{1}{8}$

115. $x^{-3/5}x^{-2/5} = x^{-1} = \dfrac{1}{x}$

117. $(-8x^{-6})^{-1/3} = (-8)^{-1/3}x^2 = -\dfrac{1}{2}x^2$

119. $w^{-3/2} \div w^{-7/2} = w^{-3/2-(-7/2)} = w^2$

121. $\left(\dfrac{9t^{-6}}{s^{-4}}\right)^{-1/2} = \dfrac{9^{-1/2}t^3}{s^2} = \dfrac{t^3}{\sqrt{9}\cdot s^2} = \dfrac{t^3}{3s^2}$

123. $\left(\dfrac{8x^{-12}}{y^{30}}\right)^{2/3} = \dfrac{8^{2/3}x^{-8}}{y^{20}} = \dfrac{4}{x^8y^{20}}$

125. $\dfrac{1.3\times 10^{12}}{2.6\times 10^8} = \$5000$ per person

127. $r = 1-\left(\dfrac{21{,}000}{56{,}800}\right)^{1/9} \approx 0.105$ or 10.5%

129. $r = \left(\dfrac{3(0.25)}{4\pi}\right)^{1/3} \approx 0.4$ cm

131. $A = \pi r\sqrt{r^2+h^2} = \pi(2)\sqrt{2^2+6^2}$

$\qquad = 2\pi\sqrt{40} = 4\pi\sqrt{10}$ or 40 in^2

133. Let x represent the length and width of the square screen.

$$x^2 + x^2 = 40^2$$
$$2x^2 = 1600$$
$$x^2 = 800$$
$$x = \sqrt{800} = 20\sqrt{2} \approx 28.28$$

So the length and width are both 28 inches.

CHAPTER 8 TEST

1. $2^{-5} = \dfrac{1}{2^5} = \dfrac{1}{32}$

2. Since $12 \cdot 12 = 144$, $\sqrt{144} = 12$.

3. Since $(-3)^3 = -27$, $\sqrt[3]{-27} = -3$.

4. $9^{-2} = \dfrac{1}{9^2} = \dfrac{1}{81}$

5. $16^{1/4} = \sqrt[4]{16} = 2$

6. $\sqrt{24} = \sqrt{4}\sqrt{6} = 2\sqrt{6}$

7. $\sqrt{\dfrac{3}{8}} = \dfrac{\sqrt{3}}{\sqrt{8}} = \dfrac{\sqrt{3}\sqrt{2}}{\sqrt{8}\sqrt{2}} = \dfrac{\sqrt{6}}{\sqrt{16}} = \dfrac{\sqrt{6}}{4}$

8. $(-4)^{3/2}$ is not a real number because it is a square root of a negative number.

9. $\sqrt{8} + \sqrt{2} = \sqrt{4}\sqrt{2} + \sqrt{2} = 2\sqrt{2} + \sqrt{2} = 3\sqrt{2}$

10. $(2 + \sqrt{3})^2 = 2^2 + 2 \cdot 2\sqrt{3} + 3 = 7 + 4\sqrt{3}$

11. $(3\sqrt{2} - \sqrt{7})(3\sqrt{2} + \sqrt{7}) = 9 \cdot 2 - 7 = 11$

12. $\sqrt{21} \div \sqrt{3} = \dfrac{\sqrt{21}}{\sqrt{3}} = \sqrt{\dfrac{21}{3}} = \sqrt{7}$

13. $\sqrt{20} \div \sqrt{3} = \dfrac{\sqrt{20}}{\sqrt{3}} = \dfrac{\sqrt{4}\sqrt{5}}{\sqrt{3}} = \dfrac{2\sqrt{5}\sqrt{3}}{\sqrt{3}\sqrt{3}} = \dfrac{2\sqrt{15}}{3}$

14. $\dfrac{2 + \sqrt{8}}{2} = \dfrac{2 + \sqrt{4}\sqrt{2}}{2} = \dfrac{2 + 2\sqrt{2}}{2} = 1 + \sqrt{2}$

15. $27^{4/3} = (\sqrt[3]{27})^4 = 3^4 = 81$

16. $\sqrt{3}(\sqrt{6} - \sqrt{3}) = \sqrt{18} - 3 = \sqrt{9}\sqrt{2} - 3 = 3\sqrt{2} - 3$

17. $3x^{-2} \cdot 5x^7 = 15x^{-2+7} = 15x^5$

18. $(2x^{-6})^3 = 2^3(x^{-6})^3 = 8x^{-18} = \dfrac{8}{x^{18}}$

19. $(-3x^{-5}y^2)^{-3} = (-3)^{-3}(x^{-5})^{-3}(y^2)^{-3}$

$= -\dfrac{1}{27}x^{15}y^{-6} = \dfrac{-x^{15}}{27y^6}$

20. $\dfrac{2y^{-5}}{8y^9} = \dfrac{1}{4y^{9-(-5)}} = \dfrac{1}{4y^{14}}$

21. $\dfrac{t^{-7}}{t^{-3}} = t^{-7-(-3)} = t^{-4} = \dfrac{1}{t^4}$

22. $(x^3y^9)^{1/3} = xy^3$

23. $(-2s^{-3}t^2)^{-2} = (-2)^{-2}s^6t^{-4} = \dfrac{s^6}{(-2)^2t^4}$

$= \dfrac{s^6}{4t^4}$

24. $\left(\dfrac{125w^3}{u^{-12}}\right)^{-1/3} = \dfrac{125^{-1/3}w^{-1}}{u^4} = \dfrac{1}{5u^4w}$

25. $\sqrt{\dfrac{3}{t}} = \dfrac{\sqrt{3}}{\sqrt{t}} = \dfrac{\sqrt{3}\sqrt{t}}{\sqrt{t}\sqrt{t}} = \dfrac{\sqrt{3t}}{t}$

26. $\sqrt{4y^6} = \sqrt{4}\sqrt{y^6} = 2y^3$

27. $\sqrt[3]{8y^{12}} = \sqrt[3]{8} \cdot \sqrt[3]{y^{12}} = 2y^4$

28. $\sqrt{18t^7} = \sqrt{9t^6} \cdot \sqrt{2t} = 3t^3 \cdot \sqrt{2t}$

29. $\quad (x+3)^2 = 36$

$\quad\quad x + 3 = \pm 6$

$\quad x + 3 = 6 \quad$ or $\quad x + 3 = -6$

$\quad\quad x = 3 \quad$ or $\quad\quad x = -9$

The solutions to the equation are -9 and 3.

30. $\quad\quad \sqrt{x+7} = 5$

$\quad (\sqrt{x+7})^2 = 5^2$

$\quad\quad\quad x + 7 = 25$

$\quad\quad\quad\quad x = 18$

The solution to the equation is 18.

31.
$$5x^2 = 2$$
$$x^2 = \frac{2}{5}$$
$$x = \pm\sqrt{\frac{2}{5}} = \pm\frac{\sqrt{2}}{\sqrt{5}} = \pm\frac{\sqrt{2}\sqrt{5}}{\sqrt{5}\sqrt{5}} = \pm\frac{\sqrt{10}}{5}$$

The solutions to the equation are $-\frac{\sqrt{10}}{5}$ and $\frac{\sqrt{10}}{5}$.

32.
$$(3x - 4)^2 = 0$$
$$3x - 4 = 0$$
$$3x = 4$$
$$x = \frac{4}{3}$$

The solution to the equation is $\frac{4}{3}$.

33.
$$x - 3 = \sqrt{5x + 9}$$
$$(x - 3)^2 = 5x + 9$$
$$x^2 - 6x + 9 = 5x + 9$$
$$x^2 - 11x = 0$$
$$x(x - 11) = 0$$
$$x = 0 \quad \text{or} \quad x - 11 = 0$$
$$x = 11$$

Since 0 does not check in the original equation, the only solution to the equation is 11.

34.
$$S = \pi r^2 h$$
$$r^2 = \frac{S}{\pi h}$$
$$r = \pm\sqrt{\frac{S}{\pi h}}$$

35. Move the decimal point 6 places to the left. Since 5,433,000 is larger than 10, the exponent is positive.

$$5,433,000 = 5.433 \times 10^6$$

36. Move the decimal point 6 places to the right. Since 0.0000065 is smaller than 1, the exponent is negative.

$$0.0000065 = 6.5 \times 10^{-6}$$

37. $(80,000)(0.000006) = (8 \times 10^4)(6 \times 10^{-6})$
$$= 48 \times 10^{-2} = 4.8 \times 10^1 \times 10^{-2} = 4.8 \times 10^{-1}$$

38. $(0.0000003)^4 = (3 \times 10^{-7})^4 = 81 \times 10^{-28}$
$$= 8.1 \times 10^1 \times 10^{-28} = 8.1 \times 10^{-27}$$

39. Let $x =$ the length of a side of the square. Since the diagonal is the hypotenuse of a right triangle, we can write the following equation.
$$x^2 + x^2 = 5^2$$
$$2x^2 = 25$$
$$x^2 = \frac{25}{2}$$
$$x = \pm\sqrt{\frac{25}{2}} = \pm\frac{5}{\sqrt{2}} = \pm\frac{5\sqrt{2}}{\sqrt{2}\sqrt{2}} = \pm\frac{5\sqrt{2}}{2}$$

The length of each side of the square is $\frac{5\sqrt{2}}{2}$ meters.

40. Since $A = \pi r^2$, we have $100,000 = \pi r^2$ or
$$r^2 = \frac{100,000}{\pi}$$
$$r = \sqrt{\frac{100,000}{\pi}} \approx 178.4 \text{ meters}$$

Tying It All Together Chapters 1-8

1.
$$2x + 3 = 0$$
$$2x = -3$$
$$x = -\frac{3}{2}$$

The solution to the equation is $-\frac{3}{2}$.

2.
$$2x = 3$$
$$x = \frac{3}{2}$$

The solution to the equation is $\frac{3}{2}$.

3.
$$2x + 3 > 0$$
$$2x > -3$$
$$x > -\frac{3}{2}$$

4.
$$-2x + 3 > 0$$
$$-2x > -3$$
$$x < \frac{3}{2}$$

5. $2(x+3) = 0$

$\qquad x + 3 = 0$

$\qquad\quad x = -3$

The solution to the equation is -3.

6. $2x^2 = 3$

$\qquad x^2 = \frac{3}{2}$

$\qquad\; x = \pm\sqrt{\frac{3}{2}} = \pm\frac{\sqrt{6}}{2}$

The solutions to the equation are $-\frac{\sqrt{6}}{2}$ and $\frac{\sqrt{6}}{2}$.

7. $\frac{x}{3} = \frac{2}{x}$

$\qquad x^2 = 6$

$\qquad\; x = \pm\sqrt{6}$

The solutions to the equation are $-\sqrt{6}$ and $\sqrt{6}$.

8. $\frac{x-1}{x} = \frac{x}{x-2}$

$\qquad x^2 = (x-1)(x-2)$

$\qquad x^2 = x^2 - 3x + 2$

$\qquad\; 0 = -3x + 2$

$\qquad 3x = 2$

$\qquad\; x = \frac{2}{3}$

The solution to the equation is $\frac{2}{3}$.

9. $(2x+3)^2 = 0$

$\qquad 2x + 3 = 0$

$\qquad\quad 2x = -3$

$\qquad\quad\; x = -\frac{3}{2}$

The solution to the equation is $-\frac{3}{2}$.

10. $(2x+3)(x-3) = 0$

$\quad 2x + 3 = 0 \quad$ or $\quad x - 3 = 0$

$\qquad 2x = -3 \quad$ or $\qquad x = 3$

$\qquad\; x = -\frac{3}{2}$

The solutions to the equation are $-\frac{3}{2}$ and 3.

11. $2x^2 + 3 = 0$

$\qquad 2x^2 = -3$

$\qquad\; x^2 = -\frac{3}{2}$

Since the square of any real number is nonnegative, there is no solution to the equation.

12. $(2x+3)^2 = 1$

$\qquad 2x + 3 = \pm 1$

$2x + 3 = 1 \quad$ or $\quad 2x + 3 = -1$

$\quad 2x = -2 \quad$ or $\qquad 2x = -4$

$\qquad x = -1 \quad$ or $\qquad\; x = -2$

The solutions to the equation are -2 and -1.

13. $(2x+3)^2 = -1$

Since the square of any real number is nonnegative, there is no solution to this equation.

14. $\sqrt{2x^2 - 14} = x - 1$

$\quad (\sqrt{2x^2 - 14})^2 = (x-1)^2$

$\qquad 2x^2 - 14 = x^2 - 2x + 1$

$\qquad x^2 + 2x - 15 = 0$

$\qquad (x+5)(x-3) = 0$

$\qquad x + 5 = 0 \quad$ or $\quad x - 3 = 0$

$\qquad\quad x = -5 \;$ or $\qquad x = 3$

Check -5: $\sqrt{2(-5)^2 - 14} = -5 - 1$

$\qquad\qquad\qquad\quad \sqrt{36} = -6 \quad$ Incorrect.

Since 3 does satisfy the original equation, the only solution to the equation is 3.

15. Use $b = -3$ in the expression b^2.

$b^2 = (-3)^2 = 9$

16. Use $a = 2$ and $c = -9$ in the expression $-4ac$.

$-4ac = -4(2)(-9) = 72$

17. Use $a = 2$, $b = -3$, and $c = -9$ in the expression $b^2 - 4ac$.

$b^2 - 4ac = (-3)^2 - 4(2)(-9) = 9 - (-72) = 81$

18. Use the value of $b^2 - 4ac$ from exercise 17 in the expression $\sqrt{b^2 - 4ac}$.

$$\sqrt{b^2 - 4ac} = \sqrt{81} = 9$$

19. Use the value of $\sqrt{b^2 - 4ac}$ from exercise 18 in the expression $-b + \sqrt{b^2 - 4ac}$.

$$-b + \sqrt{b^2 - 4ac} = -(-3) + 9 = 12$$

20. Use the value of $b^2 - 4ac$ from exercise 17 in the expression $-b - \sqrt{b^2 - 4ac}$.

$$-b - \sqrt{b^2 - 4ac} = -(-3) - \sqrt{81} = 3 - 9 = -6$$

21. From exercise 19 the value of the numerator is 12.

$$\frac{-b + \sqrt{b^2 - 4ac}}{2a} = \frac{12}{2(2)} = 3$$

22. From exercise 20 the value of the numerator is -6.

$$\frac{-b - \sqrt{b^2 - 4ac}}{2a} = \frac{-6}{2(2)} = -\frac{6}{4} = -\frac{3}{2}$$

23. $x^2 - 6x + 9 = x^2 - 2 \cdot x \cdot 3 + 3^2 = (x - 3)^2$

24. $x^2 + 10x + 25 = x^2 + 2 \cdot x \cdot 5 + 5^2 = (x + 5)^2$

25. $x^2 + 12x + 36 = x^2 + 2 \cdot x \cdot 6 + 6^2 = (x + 6)^2$

26. $x^2 - 20x + 100 = x^2 - 2 \cdot x \cdot 10 + 10^2$

$$= (x - 10)^2$$

27. $2x^2 - 8x + 8 = 2(x^2 - 4x + 4)$

$$= 2(x - 2)^2$$

28. $3x^2 + 6x + 3 = 3(x^2 + 2x + 1)$

$$= 3(x + 1)^2$$

29. $(3 + 2x) - (6 - 5x) = 3 + 2x - 6 + 5x$

$$= 7x - 3$$

30. $(5 + 3t)(4 - 5t) = 20 + 12t - 25t - 15t^2$

$$= -15t^2 - 13t + 20$$

31. $(8 - 6j)(3 + 4j) = 24 - 18j + 32j - 24j^2$

$$= -24j^2 + 14j + 24$$

32. $(1 - u) + (5 + 7u) = 6u + 6$

33. $(3 - 4v) - (2 - 5v) = 3 - 4v - 2 + 5v$

$$= v + 1$$

34. $t^2 + 4t + 4$

35. $t^2 - 49$

36. $-4n^2 + 9$

37. $m^2 - 2m + 1$

38. $2t - 1$

39. $-4r^2 - r + 3$

40. $-18y^2 + 2$

41. $3j - 5$

42. $-6j + 2$

43. $2 - 3x$

44. $-1 - 3p$

45. $-2 + 3q$

46. $-4 + z$

47. a) $P = M(av - b)$

$$P = 60(1.02 \times 10^{-3}(300) - 2.62 \times 10^{-2})$$

$P \approx 16.788$ kilocalories per minute

b) $55(1.02 \times 10^{-3}v - 2.62 \times 10^{-2}) = 14$

$$1.02 \times 10^{-3}v - 2.62 \times 10^{-2} = \frac{14}{55}$$

$$v = \frac{\frac{14}{55} + 2.62 \times 10^{-2}}{1.02 \times 10^{-3}} \approx 275.24 \text{ m/min}$$

9.1 WARM-UPS

1. True, because $4^2 - 16 = 0$ and $(-4)^2 - 16 = 0$.

2. False, because $(x-3)^2 = 8$ is equivalent to $x - 3 = \pm 2\sqrt{2}$.

3. False, because some second degree polynomials cannot be factored.

4. True, because $(-5-4)(-5+5) = 0$ and $(4-4)(4+5) = 0$ are both correct.

5. True, because the square of every real number is nonnegative.

6. False, because 0 is a solution to $x^2 = 0$.

7. False, because $(2x+3)(4x-5) = 0$ is equivalent to $x = -\frac{3}{2}$ or $x = \frac{5}{4}$.

8. True, because $(x+2)^2 = 0$ is equivalent to $x + 2 = 0$.

9. False, because $(x-3)$ and $(x-5)$ could be any numbers with a product of 4, not just 2 and 2.

10. False, because the equation $(x-1)^2 = 0$ is a quadratic equation with only one solution.

9.1 EXERCISES

1.
$$x^2 - 36 = 0$$
$$x^2 = 36$$
$$x = \pm 6$$
The solutions to the equation are -6 and 6.

3.
$$x^2 + 10 = 0$$
$$x^2 = -10$$

Since no real number has a square that is negative, there is no solution to the equation.

5.
$$5x^2 = 50$$
$$x^2 = 10$$
$$x = \pm\sqrt{10}$$
The solutions to the equation are $-\sqrt{10}$ and $\sqrt{10}$.

7.
$$3t^2 - 5 = 0$$
$$3t^2 = 5$$
$$t^2 = \frac{5}{3}$$
$$t = \pm\sqrt{\frac{5}{3}} = \pm\frac{\sqrt{15}}{3}$$
The solutions to the equation are $-\frac{\sqrt{15}}{3}$ and $\frac{\sqrt{15}}{3}$.

9.
$$-3y^2 + 8 = 0$$
$$-3y^2 = -8$$
$$y^2 = \frac{8}{3}$$
$$y = \pm\sqrt{\frac{8}{3}} = \pm\frac{2\sqrt{6}}{3}$$

The solutions to the equation are $-\frac{2\sqrt{6}}{3}$ and $\frac{2\sqrt{6}}{3}$.

11.
$$(x-3)^2 = 4$$
$$x - 3 = \pm 2$$
$$x - 3 = 2 \quad \text{or} \quad x - 3 = -2$$
$$x = 5 \quad \text{or} \quad x = 1$$
The solutions to the equation are 1 and 5.

13.
$$(y-2)^2 = 18$$
$$y - 2 = \pm\sqrt{18} = \pm 3\sqrt{2}$$
$$y - 2 = 3\sqrt{2} \quad \text{or} \quad y - 2 = -3\sqrt{2}$$
$$y = 2 + 3\sqrt{2} \quad \text{or} \quad y = 2 - 3\sqrt{2}$$
The solutions to the equation are $2 - 3\sqrt{2}$ and $2 + 3\sqrt{2}$.

15.
$$2(x+1)^2 = \frac{1}{2}$$
$$(x+1)^2 = \frac{1}{4}$$
$$x + 1 = \pm\sqrt{\frac{1}{4}} = \pm\frac{1}{2}$$
$$x + 1 = \frac{1}{2} \quad \text{or} \quad x + 1 = -\frac{1}{2}$$
$$x = -\frac{1}{2} \quad \text{or} \quad x = -\frac{3}{2}$$

The solutions to the equation are $-\frac{3}{2}$ and $-\frac{1}{2}$.

17.
$$(x-1)^2 = \frac{1}{2}$$
$$x - 1 = \pm\sqrt{\frac{1}{2}} = \pm\frac{\sqrt{2}}{2}$$
$$x - 1 = \frac{\sqrt{2}}{2} \quad \text{or} \quad x - 1 = -\frac{\sqrt{2}}{2}$$
$$x = 1 + \frac{\sqrt{2}}{2} \quad \text{or} \quad x = 1 - \frac{\sqrt{2}}{2}$$
$$x = \frac{2 + \sqrt{2}}{2} \quad \text{or} \quad x = \frac{2 - \sqrt{2}}{2}$$

The solutions to the equation are $\frac{2 - \sqrt{2}}{2}$ and $\frac{2 + \sqrt{2}}{2}$.

19.
$$\left(x+\tfrac{1}{2}\right)^2=\tfrac{1}{2}$$
$$x+\tfrac{1}{2}=\pm\sqrt{\tfrac{1}{2}}=\pm\tfrac{\sqrt{2}}{2}$$
$$x+\tfrac{1}{2}=\tfrac{\sqrt{2}}{2} \quad \text{or} \quad x+\tfrac{1}{2}=-\tfrac{\sqrt{2}}{2}$$
$$x=-\tfrac{1}{2}+\tfrac{\sqrt{2}}{2} \quad \text{or} \quad x=-\tfrac{1}{2}-\tfrac{\sqrt{2}}{2}$$

The solutions to the equation are $\dfrac{-1-\sqrt{2}}{2}$ and $\dfrac{-1+\sqrt{2}}{2}$.

21.
$$(x-11)^2=0$$
$$x-11=0$$
$$x=11$$

The solution to the equation is 11.

23.
$$x^2-2x-15=0$$
$$(x-5)(x+3)=0$$
$$x-5=0 \quad \text{or} \quad x+3=0$$
$$x=5 \quad \text{or} \quad x=-3$$

The solutions to the equation are -3 and 5.

25.
$$x^2+6x+9=0$$
$$(x+3)^2=0$$
$$x+3=0$$
$$x=-3$$

The solution to the equation is -3.

27.
$$4x^2-4x=8$$
$$4x^2-4x-8=0$$
$$4(x^2-x-2)=0$$
$$4(x-2)(x+1)=0$$
$$x-2=0 \quad \text{or} \quad x+1=0$$
$$x=2 \quad \text{or} \quad x=-1$$

The solutions to the equation are -1 and 2.

29.
$$3x^2-6x=0$$
$$3x(x-2)=0$$
$$3x=0 \quad \text{or} \quad x-2=0$$
$$x=0 \quad \text{or} \quad x=2$$

The solutions to the equation are 0 and 2.

31.
$$-4t^2+6t=0$$
$$-2t(2t-3)=0$$
$$-2t=0 \quad \text{or} \quad 2t-3=0$$
$$t=0 \quad \text{or} \quad t=\tfrac{3}{2}$$

The solutions to the equation are 0 and $\tfrac{3}{2}$.

33.
$$2x^2+11x-21=0$$
$$(2x-3)(x+7)=0$$
$$2x-3=0 \quad \text{or} \quad x+7=0$$
$$x=\tfrac{3}{2} \quad \text{or} \quad x=-7$$

The solutions to the equation are -7 and $\tfrac{3}{2}$.

35.
$$x^2-10x+25=0$$
$$(x-5)^2=0$$
$$x-5=0$$
$$x=5$$

The solution to the equation is 5.

37.
$$2x^2-7x=30$$
$$2x^2-7x-30=0$$
$$(2x+5)(x-6)=0$$
$$2x+5=0 \quad \text{or} \quad x-6=0$$
$$x=-\tfrac{5}{2} \quad \text{or} \quad x=6$$

The solutions to the equation are $-\tfrac{5}{2}$ and 6.

39.
$$\tfrac{1}{10}a^2-a+\tfrac{12}{5}=0$$
$$a^2-10a+24=0$$
$$(a-4)(a-6)=0$$
$$a-4=0 \quad \text{or} \quad a-6=0$$
$$a=4 \quad \text{or} \quad a=6$$

The solutions to the equation are 4 and 6.

41.
$$x^2-2x=2(3-x)$$
$$x^2-2x=6-2x$$
$$x^2=6$$
$$x=\pm\sqrt{6}$$

The solutions to the equation are $-\sqrt{6}$ and $\sqrt{6}$.

43.
$$x=\tfrac{27}{12-x}$$
$$x(12-x)=27$$
$$12x-x^2=27$$
$$0=x^2-12x+27$$
$$0=(x-9)(x-3)$$
$$x-9=0 \quad \text{or} \quad x-3=0$$
$$x=9 \quad \text{or} \quad x=3$$

The solutions to the equation are 3 and 9.

45.
$$\sqrt{3x-8} = x-2$$
$$(\sqrt{3x-8})^2 = (x-2)^2$$
$$3x-8 = x^2-4x+4$$
$$0 = x^2-7x+12$$
$$0 = (x-3)(x-4)$$
$$x-3=0 \quad \text{or} \quad x-4=0$$
$$x=3 \quad \text{or} \quad x=4$$

The solutions to the equation are 3 and 4.

47. Let x = the length of a side of the square.

Since the diagonal is the hypotenuse of a right triangle, we can write the following equation.

$$x^2+x^2 = 5^2$$
$$2x^2 = 25$$
$$x^2 = \frac{25}{2}$$
$$x = \pm\sqrt{\frac{25}{2}} = \pm\frac{5\sqrt{2}}{2}$$

The length of each side is $\frac{5\sqrt{2}}{2}$ meters.

49. Let x = the distance from where Howard started to where he finished walking. The line joining his starting point to his finish point is the hypotenuse of a right triangle whose legs are 8 blocks and 4 blocks. We can use the Pythagorean theorem to write the following equation.

$$x^2 = 8^2+4^2$$
$$x^2 = 80$$
$$x = \pm\sqrt{80} = \pm 4\sqrt{5}$$

Howard is $4\sqrt{5}$ blocks from where he started.

51.
$$d^2 = 10^2+12^2$$
$$d^2 = 244$$
$$d = \sqrt{244} = 2\sqrt{61}$$

So the length of the brace is $2\sqrt{61}$ feet.

53.
$$500(1+r)^2 = 565$$
$$(1+r)^2 = 1.13$$
$$1+r = \pm\sqrt{1.13}$$
$$r = -1 \pm \sqrt{1.13}$$
$$r \approx -2.063 \quad \text{or} \quad r = 0.063$$

Since the rate is not negative, the rate is 6.3%.

55.
$$-16t^2+80t+6 = 102$$
$$-16t^2+80t-96 = 0$$
$$t^2-5t+6 = 0$$
$$(t-2)(t-3) = 0$$
$$t-2=0 \quad \text{or} \quad t-3=0$$
$$t=2 \quad \text{or} \quad t=3$$

The stone is 102 feet above the ground for t = 2 sec and t = 3 sec.

57. Use 45 in place of S in the formula and solve for n.

$$\frac{n^2+n}{2} = 45$$
$$n^2+n = 90$$
$$n^2+n-90 = 0$$
$$(n+10)(n-9) = 0$$
$$n+10=0 \quad \text{or} \quad n-9=0$$
$$n=-10 \quad \text{or} \quad n=9$$

Since the formula is for the sum of the first n positive integers, the sum is 45 when n is 9.

9.2 WARM-UPS

1. False, because completing the square is used to solve a quadratic equation.

2. False, because one-half of 2/3 is 1/3 and 1/3 squared is 1/9.

3. False, because some quadratic polynomials cannot be factored.

4. False, because one-half of −1 is −1/2 and −1/2 squared is 1/4.

5. True, because the process works for any coefficients of the terms.

6. True, because if we add 2 to each side of $x-2 = \pm\sqrt{3}$ we get $x = 2\pm\sqrt{3}$.

7. True, because the square of any real number is nonnegative.

8. True, because one-half of −5 is −5/2 and −5/2 squared is 25/4.

9. True, because $\frac{1}{2} \cdot \frac{4}{5} = \frac{4}{10} = \frac{2}{5}$.

10. True, because $\frac{1}{2} \cdot \frac{3}{4} = \frac{3}{8}$.

9.2 EXERCISES

1. One-half of 6 is 3 and $3^2 = 9$. So the perfect square trinomial is $x^2 + 6x + 9$ and $x^2 + 6x + 9 = (x + 3)^2$.

3. One-half of 14 is 7 and $7^2 = 49$. So $x^2 + 14x + 49 = (x + 7)^2$

5. One-half of -16 is -8 and $(-8)^2 = 64$. So $x^2 - 16x + 64 = (x - 8)^2$.

7. One-half of -18 is -9 and $(-9)^2 = 81$. So $t^2 - 18t + 81 = (t - 9)^2$.

9. One-half of 3 is $3/2$ and $(3/2)^2 = 9/4$. So $m^2 + 3m + \frac{9}{4} = \left(m + \frac{3}{2}\right)^2$.

11. One-half of 1 is $1/2$ and $(1/2)^2 = 1/4$. So $z^2 + z + \frac{1}{4} = \left(z + \frac{1}{2}\right)^2$.

13. One-half of $-1/2$ is $-1/4$ and $(-1/4)^2 = 1/16$. So the perfect square trinomial is $x^2 - \frac{1}{2}x + \frac{1}{16} = \left(x - \frac{1}{4}\right)^2$.

15. One-half of $1/4$ is $1/8$ and $(1/8)^2 = 1/64$. So $y^2 + \frac{1}{4}y + \frac{1}{64} = \left(y + \frac{1}{8}\right)^2$.

17. $x^2 + 10x + 25 = x^2 + 2 \cdot 5x + 5^2 = (x + 5)^2$

19. $m^2 - 2m + 1 = m^2 - 2 \cdot m \cdot 1 + 1^2 = (m - 1)^2$

21. $x^2 + x + \frac{1}{4} = x^2 + 2 \cdot \frac{1}{2}x + \left(\frac{1}{2}\right)^2 = (x + \frac{1}{2})^2$

23. $t^2 + \frac{1}{3}t + \frac{1}{36} = t^2 + 2 \cdot \frac{1}{6}t + \left(\frac{1}{6}\right)^2 = (t + \frac{1}{6})^2$

25. $x^2 + \frac{2}{5}x + \frac{1}{25} = x^2 + 2 \cdot \frac{1}{5}x + \left(\frac{1}{5}\right)^2 = (x + \frac{1}{5})^2$

27.
$$x^2 + 2x - 15 = 0$$
$$x^2 + 2x \qquad = 15$$
$$x^2 + 2x + 1 = 15 + 1$$
$$(x + 1)^2 = 16$$
$$x + 1 = \pm 4$$
$$x = -1 \pm 4$$
$$x = -1 + 4 \quad \text{or} \quad x = -1 - 4$$
$$x = 3 \qquad \text{or} \quad x = -5$$

The solutions to the equation are -5 and 3.

29.
$$x^2 - 4x - 21 = 0$$
$$x^2 - 4x \qquad = 21$$
$$x^2 - 4x + 4 = 21 + 4$$
$$(x - 2)^2 = 25$$
$$x - 2 = \pm 5$$
$$x = 2 \pm 5$$
$$x = 2 + 5 \quad \text{or} \quad x = 2 - 5$$
$$x = 7 \qquad \text{or} \quad x = -3$$

The solutions to the equation are -3 and 7.

31.
$$x^2 + 6x + 9 = 0$$
$$(x + 3)^2 = 0$$
$$x + 3 = 0$$
$$x = -3$$

The solution to the equation is -3.

33.
$$2t^2 - 3t + 1 = 0$$
$$t^2 - \frac{3}{2}t + \frac{1}{2} = 0$$
$$t^2 - \frac{3}{2}t \qquad = -\frac{1}{2}$$
$$t^2 - \frac{3}{2}t + \frac{9}{16} = \frac{9}{16} - \frac{1}{2}$$
$$(t - \frac{3}{4})^2 = \frac{1}{16}$$
$$t - \frac{3}{4} = \pm \frac{1}{4}$$
$$t = \frac{3}{4} \pm \frac{1}{4}$$
$$t = \frac{3}{4} + \frac{1}{4} = 1 \quad \text{or} \quad t = \frac{3}{4} - \frac{1}{4} = \frac{1}{2}$$

The solutions to the equation are $\frac{1}{2}$ and 1.

35.
$$2w^2 - 7w + 6 = 0$$
$$w^2 - \frac{7}{2}w + 3 = 0$$
$$w^2 - \frac{7}{2}w \qquad = -3$$
$$w^2 - \frac{7}{2}w + \frac{49}{16} = \frac{49}{16} - 3$$
$$(w - \frac{7}{4})^2 = \frac{1}{16}$$
$$w - \frac{7}{4} = \pm \frac{1}{4}$$
$$w = \frac{7}{4} \pm \frac{1}{4}$$
$$w = \frac{7}{4} + \frac{1}{4} = 2 \quad \text{or} \quad w = \frac{7}{4} - \frac{1}{4} = \frac{3}{2}$$

The solutions to the equation are $\frac{3}{2}$ and 2.

37.
$$3x^2 + 2x - 1 = 0$$
$$x^2 + \frac{2}{3}x - \frac{1}{3} = 0$$
$$x^2 + \frac{2}{3}x \quad = \frac{1}{3}$$
$$x^2 + \frac{2}{3}x + \frac{1}{9} = \frac{1}{3} + \frac{1}{9}$$
$$\left(x + \frac{1}{3}\right)^2 = \frac{4}{9}$$
$$x + \frac{1}{3} = \pm\frac{2}{3}$$
$$x = -\frac{1}{3} \pm \frac{2}{3}$$
$$x = -\frac{1}{3} + \frac{2}{3} = \frac{1}{3} \quad \text{or} \quad x = -\frac{1}{3} - \frac{2}{3} = -1$$

The solutions to the equation are -1 and $\frac{1}{3}$.

39.
$$x^2 + 2x - 6 = 0$$
$$x^2 + 2x \quad = 6$$
$$x^2 + 2x + 1 = 6 + 1$$
$$(x+1)^2 = 7$$
$$x + 1 = \pm\sqrt{7}$$
$$x = -1 \pm \sqrt{7}$$

The solutions to the equation are $-1 - \sqrt{7}$ and $-1 + \sqrt{7}$.

41.
$$x^2 + 6x + 1 = 0$$
$$x^2 + 6x \quad = -1$$
$$x^2 + 6x + 9 = -1 + 9$$
$$(x+3)^2 = 8$$
$$x + 3 = \pm\sqrt{8}$$
$$x = -3 \pm 2\sqrt{2}$$

The solutions to the equation are $-3 - 2\sqrt{2}$ and $-3 + 2\sqrt{2}$.

43.
$$y^2 - y - 3 = 0$$
$$y^2 - y \quad = 3$$
$$y^2 - y + \frac{1}{4} = 3 + \frac{1}{4}$$
$$\left(y - \frac{1}{2}\right)^2 = \frac{13}{4}$$
$$y - \frac{1}{2} = \pm\sqrt{\frac{13}{4}}$$
$$y = \frac{1}{2} \pm \frac{\sqrt{13}}{2}$$

The solutions to the equation are $\frac{1 - \sqrt{13}}{2}$ and $\frac{1 + \sqrt{13}}{2}$.

45.
$$v^2 + 3v - 3 = 0$$
$$v^2 + 3v \quad = 3$$
$$v^2 + 3v + \frac{9}{4} = 3 + \frac{9}{4}$$
$$\left(v + \frac{3}{2}\right)^2 = \frac{21}{4}$$
$$v + \frac{3}{2} = \pm\sqrt{\frac{21}{4}}$$
$$v = -\frac{3}{2} \pm \frac{\sqrt{21}}{2}$$

The solutions to the equation are $\frac{-3 - \sqrt{21}}{2}$ and $\frac{-3 + \sqrt{21}}{2}$.

47.
$$2m^2 - m - 4 = 0$$
$$m^2 - \frac{1}{2}m - 2 = 0$$
$$m^2 - \frac{1}{2}m \quad = 2$$
$$m^2 - \frac{1}{2}m + \frac{1}{16} = 2 + \frac{1}{16}$$
$$\left(m - \frac{1}{4}\right)^2 = \frac{33}{16}$$
$$m - \frac{1}{4} = \pm\sqrt{\frac{33}{16}}$$
$$m = \frac{1}{4} \pm \frac{\sqrt{33}}{4}$$

The solutions to the equation are $\frac{1 - \sqrt{33}}{4}$ and $\frac{1 + \sqrt{33}}{4}$.

49.
$$(x-5)^2 = 7$$
$$x - 5 = \pm\sqrt{7}$$
$$x = 5 \pm \sqrt{7}$$

The solutions to the equation are $5 - \sqrt{7}$ and $5 + \sqrt{7}$.

51.
$$3n^2 - 5 = 0$$
$$3n^2 = 5$$
$$n^2 = \frac{5}{3}$$
$$n = \pm\sqrt{\frac{5}{3}} = \pm\frac{\sqrt{15}}{3}$$

The solutions to the equation are $-\frac{\sqrt{15}}{3}$ and $\frac{\sqrt{15}}{3}$.

53.
$$3x^2 + 1 = 0$$
$$3x^2 = -1$$
$$x^2 = -\frac{1}{3}$$

Since the square of any real number is nonnegative, there is no solution to the equation.

55.
$$x^2 + 5 = 8x - 3$$
$$x^2 + 8 = 8x$$
$$x^2 - 8x + 8 = 0$$
$$x^2 - 8x = -8$$
$$x^2 - 8x + 16 = -8 + 16$$
$$(x - 4)^2 = 8$$
$$x - 4 = \pm\sqrt{8}$$
$$x = 4 \pm 2\sqrt{2}$$

The solutions to the equation are $4 - 2\sqrt{2}$ and $4 + 2\sqrt{2}$.

57.
$$(2x - 7)^2 = 0$$
$$2x - 7 = 0$$
$$2x = 7$$
$$x = \frac{7}{2}$$

The solution to the equation is $\frac{7}{2}$.

59.
$$y^2 + 6y = 11$$
$$y^2 + 6y + 9 = 11 + 9$$
$$(y + 3)^2 = 20$$
$$y + 3 = \pm\sqrt{20}$$
$$y = -3 \pm 2\sqrt{5}$$

The solutions to the equation are $-3 - 2\sqrt{5}$ and $-3 + 2\sqrt{5}$.

61.
$$\frac{1}{4}w^2 + \frac{1}{2} = w$$
$$\frac{1}{4}w^2 - w + \frac{1}{2} = 0$$
$$w^2 - 4w + 2 = 0$$
$$w^2 - 4w + 4 = 2$$
$$(w - 2)^2 = 2$$
$$w - 2 = \pm\sqrt{2}$$
$$w = 2 \pm\sqrt{2}$$

The solutions to the equation are $2 \pm\sqrt{2}$.

63.
$$t^2 + 0.2t = 0.24$$
$$t^2 + 0.2t + 0.01 = 0.24 + 0.01$$
$$(t + 0.1)^2 = 0.25$$
$$t + 0.1 = \pm\sqrt{0.25}$$
$$t = -0.1 \pm 0.5$$
$$t = -0.1 + 0.5 = 0.4$$
or $\quad t = -0.1 - 0.5 = -0.6$

The solutions to the equation are -0.6 and 0.4.

65. The sum of the base and height is 10 inches. Let x = the base and $10 - x$ = the height. Since the area is 11 square inches, we can write the following equation.

$$\frac{1}{2}x(10 - x) = 11$$
$$x(10 - x) = 22$$
$$10x - x^2 = 22$$
$$-x^2 + 10x - 22 = 0$$
$$x^2 - 10x + 22 = 0$$
$$x^2 - 10x = -22$$
$$x^2 - 10x + 25 = -22 + 25$$
$$(x - 5)^2 = 3$$
$$x - 5 = \pm\sqrt{3}$$
$$x = 5 \pm\sqrt{3}$$

If $x = 5 - \sqrt{3}$ then
$10 - x = 10 - (5 - \sqrt{3}) = 5 + \sqrt{3}$.
If $x = 5 + \sqrt{3}$ then
$10 - x = 10 - (5 + \sqrt{3}) = 5 - \sqrt{3}$.
The base and height are $5 - \sqrt{3}$ inches and $5 + \sqrt{3}$ inches.

67. Let x represent one number and $12 - x$ represent the other number. Since their product is 34, we have

$$x(12 - x) = 34$$
$$-x^2 + 12x = 34$$
$$x^2 - 12x + 36 = -34 + 36$$
$$(x - 6)^2 = 2$$
$$x - 6 = \pm\sqrt{2}$$
$$x = 6 \pm\sqrt{2}$$

If $x = 6 + \sqrt{2}$, then
$12 - x = 12 - (6 + \sqrt{2}) = 6 - \sqrt{2}$.
If $x = 6 - \sqrt{2}$, then
$12 - x = 12 - (6 - \sqrt{2}) = 6 + \sqrt{2}$.

So the numbers are $6 - \sqrt{2}$ and $6 + \sqrt{2}$.

69. The total number of candles she has is the sum of the positive integers from 1 to n for some value n and n is her age. Use 78 in place of S in the formula from exercise 47 of Section 7.1.

$$\frac{n^2 + n}{2} = 78$$
$$n^2 + n = 156$$
$$n^2 + n - 156 = 0$$
$$(n - 12)(n + 13) = 0$$
$$n - 12 = 0 \quad \text{or} \quad n + 13 = 0$$
$$n = 12 \quad \text{or} \quad x = -13$$

Joan is 12 years old.

9.3 WARM-UPS

1. True, because we found the quadratic formula by completing the square on the equation $ax^2 + bx + c = 0$.

2. False, because $b = -1$.

3. False, because when the equation is in the proper form, $x^2 - 5x - 3 = 0$, we have $a = 1$, $b = -5$, and $c = -3$.

4. False, because $-b$ is to be divided by $2a$ as well as the square root.

5. True, because if $a = 2$, $b = -6$, and $c = 0$, $b^2 - 4ac = (-6)^2 - 4(2)(0) = 36$.

6. False, because some have one solution and some have no real solutions.

7. False, because every quadratic equation can be solved by the quadratic formula.

8. True, because $2x^2 - 6x = 0$ is equivalent to $2x(x - 3) = 0$.

9. True, because 0 and 1 are the solutions to both equations.

10. True, because -3 is a solution to $x^2 + 6x + 9 = 0$ and $b^2 - 4ac = 6^2 - 4(1)(9) = 0$.

9.3 EXERCISES

1. For $x^2 + 2x - 15 = 0$, $a = 1$, $b = 2$, and $c = -15$.

$$x = \frac{-2 \pm \sqrt{2^2 - 4(1)(-15)}}{2(1)} = \frac{-2 \pm \sqrt{64}}{2} = \frac{-2 \pm 8}{2}$$

$$x = \frac{-2 + 8}{2} = 3 \quad \text{or} \quad x = \frac{-2 - 8}{2} = -5$$

The solutions to the equation are -5 and 3.

3. For $x^2 + 10x + 25 = 0$, $a = 1$, $b = 10$, and $c = 25$.

$$x = \frac{-10 \pm \sqrt{10^2 - 4(1)(25)}}{2(1)} = \frac{-10 \pm \sqrt{0}}{2} = -5$$

The solution to the equation is -5.

5. For $2x^2 + x - 6 = 0$, $a = 2$, $b = 1$, and $c = -6$.

$$x = \frac{-1 \pm \sqrt{1^2 - 4(2)(-6)}}{2(2)} = \frac{-1 \pm \sqrt{49}}{4} = \frac{-1 \pm 7}{4}$$

$$x = \frac{-1 + 7}{4} = \frac{3}{2} \quad \text{or} \quad x = \frac{-1 - 7}{4} = \frac{-8}{4} = -2$$

The solutions to the equation are -2 and $\frac{3}{2}$.

7. For $4x^2 + 4x - 3 = 0$, $a = 4$, $b = 4$, and $c = -3$.

$$x = \frac{-4 \pm \sqrt{4^2 - 4(4)(-3)}}{2(4)} = \frac{-4 \pm \sqrt{64}}{8} = \frac{-4 \pm 8}{8}$$

$$x = \frac{-4 + 8}{8} = \frac{1}{2} \quad \text{or} \quad x = \frac{-4 - 8}{8} = \frac{-12}{8} = -\frac{3}{2}$$

The solutions to the equation are $-\frac{3}{2}$ and $\frac{1}{2}$.

9. For $2y^2 - 6y + 3 = 0$, $a = 2$, $b = -6$, and $c = 3$.

$$y = \frac{-(-6) \pm \sqrt{(-6)^2 - 4(2)(3)}}{2(2)} = \frac{6 \pm \sqrt{12}}{4} = \frac{6 \pm 2\sqrt{3}}{4}$$

$$= \frac{2(3 \pm \sqrt{3})}{2 \cdot 2} = \frac{3 \pm \sqrt{3}}{2}$$

The solutions to the equation are $\frac{3 - \sqrt{3}}{2}$ and $\frac{3 + \sqrt{3}}{2}$.

11. For $2t^2 + 4t + 1 = 0$, $a = 2$, $b = 4$, and $c = 1$.

$$t = \frac{-4 \pm \sqrt{4^2 - 4(2)(1)}}{2(2)} = \frac{-4 \pm \sqrt{8}}{4} = \frac{-4 \pm 2\sqrt{2}}{4}$$

$$= \frac{2(-2 \pm \sqrt{2})}{2 \cdot 2} = \frac{-2 \pm \sqrt{2}}{2}$$

The solutions to the equation are $\frac{-2 - \sqrt{2}}{2}$ and $\frac{-2 + \sqrt{2}}{2}$.

13. For $2x^2 - 2x + 3 = 0$, $a = 2$, $b = -2$, and $c = 3$.

$$x = \frac{2 \pm \sqrt{(-2)^2 - 4(2)(3)}}{2(2)} = \frac{2 \pm \sqrt{-20}}{2}$$

Because $\sqrt{-20}$ is not a real number, the equation has no real solution.

15. For $8x^2 - 4x = 0$, $a = 8$, $b = -4$, and $c = 0$.

$$x = \frac{4 \pm \sqrt{(-4)^2 - 4(8)(0)}}{2(8)} = \frac{4 \pm \sqrt{16}}{16} = \frac{4 \pm 4}{16}$$

$$x = \frac{1}{2} \quad \text{or} \quad w = 0$$

The solutions to the equation are 0 and $\frac{1}{2}$.

17. For $5w^2 - 3 = 0$, $a = 5$, $b = 0$, and $c = -3$.

$$w = \frac{0 \pm \sqrt{(0)^2 - 4(5)(-3)}}{2(5)} = \frac{\pm\sqrt{60}}{10} = \frac{\pm 2\sqrt{15}}{10}$$

$$w = \pm\frac{\sqrt{15}}{5}$$

The solutions to the equation are $-\frac{\sqrt{15}}{5}$ and $\frac{\sqrt{15}}{5}$.

19. For $\frac{1}{2}h^2 + 7h + \frac{1}{2} = 0$, $a = \frac{1}{2}$, $b = 7$, and $c = \frac{1}{2}$.

$$h = \frac{-7 \pm \sqrt{(7)^2 - 4(\frac{1}{2})(\frac{1}{2})}}{2(\frac{1}{2})} = \frac{-7 \pm \sqrt{48}}{1}$$

$$h = -7 \pm 4\sqrt{3}$$

The solutions to the equation are $-7 + 4\sqrt{3}$ and $-7 - 4\sqrt{3}$.

21. For $4x^2 - 4x + 1 = 0$, $a = 4$, $b = -4$, and $c = 1$.

$b^2 - 4ac = (-4)^2 - 4(4)(1) = 0$

Since the discriminant has a value of 0, there is one real solution to the equation.

23. For $6x^2 - 7x + 4 = 0$, $a = 6$, $b = -7$, and $c = 4$.
$b^2 - 4ac = (-7)^2 - 4(6)(4) = -47$
Since the discriminant has a negative value, there are no real solutions to the equation.

25. For $-5t^2 - t + 9 = 0$, $a = -5$, $b = -1$, and $c = 9$.
$b^2 - 4ac = (-1)^2 - 4(-5)(9) = 181$
Since the discriminant has a positive value, there are two real solutions to the equation.

27. For $4x^2 - 12x + 9 = 0$, $a = 4$, $b = -12$, and $c = 9$.
$b^2 - 4ac = (-12)^2 - 4(4)(9) = 0$

Since the discriminant has a value of 0, there is one real solution to the equation.

29. For $x^2 + x + 4 = 0$, $a = 1$, $b = 1$, and $c = 4$.
$b^2 - 4ac = (1)^2 - 4(1)(4) = -15$
Since the discriminant has a negative value, there are no real solutions to the equation.

31. For $3x^2 - x + 5 = 0$, $a = 3$, $b = -1$, and $c = 5$.
$b^2 - 4ac = (-1)^2 - 4(3)(5) = -59$
Since the discriminant has a negative value, there are no real solutions to the equation.

33. $$x^2 + \frac{3}{2}x = 1$$
$$2x^2 + 3x = 2$$
$$2x^2 + 3x - 2 = 0$$
$$(2x - 1)(x + 2) = 0$$
$$2x - 1 = 0 \quad \text{or} \quad x + 2 = 0$$
$$x = \frac{1}{2} \quad \text{or} \quad x = -2$$

The solutions to the equation are -2 and $\frac{1}{2}$.

35. $$(x - 1)^2 + (x - 2)^2 = 5$$
$$x^2 - 2x + 1 + x^2 - 4x + 4 = 5$$
$$2x^2 - 6x = 0$$
$$2x(x - 3) = 0$$
$$2x = 0 \quad \text{or} \quad x - 3 = 0$$
$$x = 0 \quad \text{or} \quad x = 3$$

The solutions to the equation are 0 and 3.

37. $$\frac{1}{x} + \frac{1}{x + 2} = \frac{5}{12}$$
$$12x(x + 2)\left(\frac{1}{x} + \frac{1}{x + 2}\right) = 12x(x + 2)\frac{5}{12}$$
$$12(x + 2) + 12x = x(x + 2)5$$
$$12x + 24 + 12x = 5x^2 + 10x$$
$$0 = 5x^2 - 14x - 24$$
$$0 = (5x + 6)(x - 4)$$
$$5x + 6 = 0 \quad \text{or} \quad x - 4 = 0$$
$$x = -\frac{6}{5} \quad \text{or} \quad x = 4$$

The solutions to the equation are $-\frac{6}{5}$ and 4.

39. $$x^2 + 6x + 8 = 0$$
$$(x + 4)(x + 2) = 0$$
$$x + 4 = 0 \quad \text{or} \quad x + 2 = 0$$
$$x = -4 \quad \text{or} \quad x = -2$$

The solutions to the equation are -4 and -2.

41. $$x^2 - 9x = 0$$
$$x(x - 9) = 0$$
$$x = 0 \quad \text{or} \quad x - 9 = 0$$
$$x = 9$$

The solutions to the equation are 0 and 9.

43. $$(x + 5)^2 = 9$$
$$x + 5 = \pm 3$$
$$x = -5 \pm 3$$
$$x = -5 + 3 = -2 \quad \text{or} \quad x = -5 - 3 = -8$$

The solutions to the equation are -8 and -2.

45.
$$x(x-3) = 2 - 3(x+4)$$
$$x^2 - 3x = 2 - 3x - 12$$
$$x^2 = -10$$

Since the square of any real number is nonnegative, there is no solution to the equation.

47.
$$\frac{x}{3} = \frac{x+2}{x}$$

$$x^2 = 3x + 6$$
$$x^2 - 3x - 6 = 0$$

$$x = \frac{-(-3) \pm \sqrt{(-3)^2 - 4(1)(-6)}}{2(1)} = \frac{3 \pm \sqrt{33}}{2}$$

The solutions to the equation are $\frac{3 - \sqrt{33}}{2}$ and $\frac{3 + \sqrt{33}}{2}$.

49.
$$2x^2 - 3x = 0$$
$$x(2x - 3) = 0$$
$$x = 0 \quad \text{or} \quad 2x - 3 = 0$$
$$x = 0 \quad \text{or} \quad x = \frac{3}{2}$$

The solutions to the equation are 0 and $\frac{3}{2}$.

51.
$$x^2 - 3x - 3 = 0$$

$$x = \frac{-(-3) \pm \sqrt{(-3)^2 - 4(1)(-3)}}{2(1)} = \frac{3 \pm \sqrt{21}}{2}$$

$$x = \frac{3 + \sqrt{21}}{2} = 3.79 \quad \text{or} \quad x = \frac{3 - \sqrt{21}}{2} = -0.79$$

The solutions to the equation are -0.79 and 3.79.

53.
$$x^2 - x - 3.2 = 0$$

$$x = \frac{-(-1) \pm \sqrt{(-1)^2 - 4(1)(-3.2)}}{2(1)} = \frac{1 \pm \sqrt{13.8}}{2}$$

$$x = \frac{-1 - \sqrt{13.8}}{2} = -1.36$$

or $x = \frac{-1 + \sqrt{13.8}}{2} = 2.36$

The solutions to the equation are -1.36 and 2.36.

55.
$$5.29x^2 - 3.22x + 0.49 = 0$$

$$x = \frac{-(-3.22) \pm \sqrt{(-3.22)^2 - 4(5.29)(.49)}}{2(5.29)}$$

$$= \frac{3.22 \pm \sqrt{0}}{2(5.29)} = 0.30$$

The solution to the equation is 0.30.

57. $-0.000625x^2 + 0.025x + 0.50 = 0.70$

$$-0.000625x^2 + 0.025x - 0.20 = 0$$

$$x = \frac{-0.025 \pm \sqrt{0.025^2 - 4(-0.000625)(-0.20)}}{2(-0.000625)}$$

$$x = \frac{-0.025 \pm \sqrt{0.000125}}{-0.00125} \approx 29 \text{ or } 11$$

The percentage of young women in the workforce was 70% in 1981 and will again be 70% in 1999.

9.4 WARM-UPS

1. False, because if $x = 4$ then $x + 10 = 14$ and $4 + 14 \neq 10$.

2. True, because in a right triangle if one leg is the base, then the other leg is the height.

3. False, because the speed of the boat with the current is $x + 5$ mph.

4. False, because he eats $1/x$ of the bag per hour.

5. True, because $R = D/T$.

6. True, because $7 - \sqrt{50}$ is negative ($\sqrt{50}$ is slightly larger than 7).

7. True, because $\sqrt{27}$ is slightly larger than 5.

8. False, because for a square $A = s^2 = (x + 9)^2$
$$= x^2 + 18x + 81.$$

9. True, because if she did the whole job in 3 hours, then she does $1/3$ of the job per hour.

10. True, because the current will reduce the speed of the boat by 5 miles per hour.

9.4 EXERCISES

1. Let $x =$ the width and $x + 2 =$ the length. Since the area is 10 square meters, we can write the following equation.

$$x(x+2) = 10$$
$$x^2 + 2x - 10 = 0$$

$$x = \frac{-2 \pm \sqrt{2^2 - 4(1)(-10)}}{2(1)} = \frac{-2 \pm \sqrt{44}}{2}$$

$$= \frac{-2 \pm 2\sqrt{11}}{2} = -1 \pm \sqrt{11}$$

Since $-1 - \sqrt{11}$ is negative, the width is $-1 + \sqrt{11}$ meters. The length is $-1 + \sqrt{11} + 2 = 1 + \sqrt{11}$ meters.

3. Let $x =$ the length of a side of the square. Since the diagonal is the hypotenuse of a right triangle, we can write the following equation.

$$x^2 + x^2 = 8^2$$
$$2x^2 = 64$$
$$x^2 = 32$$
$$x = \pm\sqrt{32} = \pm 4\sqrt{2}$$

The length of each side is $4\sqrt{2}$ feet.

5. Let $x =$ the height and $x + 6 =$ the length of the base. Since the area of a parallelogram is the base times the height, we can write the following equation.

$$x(x+6) = 10$$
$$x^2 + 6x - 10 = 0$$

$$x = \frac{-6 \pm \sqrt{6^2 - 4(1)(-10)}}{2(1)} = \frac{-6 \pm \sqrt{76}}{2}$$

$$= \frac{-6 \pm 2\sqrt{19}}{2} = -3 \pm \sqrt{19}$$

Since $-3 - \sqrt{19}$ is a negative number, the height is $-3 + \sqrt{19}$ inches. The base is $-3 + \sqrt{19} + 6 = 3 + \sqrt{19}$ inches long.

7. Let $x =$ the number of hours for Alberta to pick the berries by herself and $x + 2 =$ the number of hours for Ernie to pick the berries by himself. Since they pick 1/2 of the berries per hour when they work together, we can write the following equation.

$$\frac{1}{x} + \frac{1}{x+2} = \frac{1}{2}$$

$$2x(x+2)\frac{1}{x} + 2x(x+2)\frac{1}{x+2} = 2x(x+2)\frac{1}{2}$$

$$2x + 4 + 2x = x^2 + 2x$$
$$-x^2 + 2x + 4 = 0$$
$$x^2 - 2x - 4 = 0$$

$$x = \frac{-(-2) \pm \sqrt{(-2)^2 - 4(1)(-4)}}{2(1)} = \frac{2 \pm \sqrt{20}}{2}$$

$$= \frac{2 \pm 2\sqrt{5}}{2} = 1 \pm \sqrt{5}$$

Alberta's time is $1 + \sqrt{5}$, so Ernie's time is $3 + \sqrt{5}$ or 5.24 hours.

9. Let $x =$ the time required for Tasha to complete the job and $x + 8 =$ the time required for Tena to complete the job. In 20 hours, Tasha does 20/x of the job and Tena does 20/(x+8) of the job. Since the job is completed in 20 hours these fractions have a sum of 1.

$$\frac{20}{x} + \frac{20}{x+8} = 1$$

$$x(x+8)\frac{20}{x} + x(x+8)\frac{20}{x+8} = 1 \cdot x(x+8)$$

$$20x + 160 + 20x = x^2 + 8x$$
$$-x^2 + 32x + 160 = 0$$
$$x^2 - 32x - 160 = 0$$

$$x = \frac{32 \pm \sqrt{(-32)^2 - 4(1)(-160)}}{2(1)}$$

$$= \frac{32 \pm \sqrt{1664}}{2} = \frac{32 \pm 8\sqrt{26}}{2} = 16 \pm 4\sqrt{26}$$

Since $16 - 4\sqrt{26}$ is negative, Tasha's time alone is $16 + 4\sqrt{26}$ hours or 36.4 hours.

11. The height of the ball above the earth is given by the formula $S = -16t^2 + v_0 t + s_0$ or $S = -16t^2 + 60t + 4$. To find the time it takes for the ball to reach the earth, let $S = 0$ and solve the equation for t.

$$-16t^2 + 60t + 4 = 0$$
$$4t^2 - 15t - 1 = 0$$

$$t = \frac{15 \pm \sqrt{(-15)^2 - 4(4)(-1)}}{2(4)} = \frac{15 \pm \sqrt{241}}{8}$$

It will take $\frac{15 + \sqrt{241}}{8}$ or 3.82 seconds for the ball to reach the earth.

13. $S = -16t^2 + 20t + 5$

$$-16t^2 + 20t + 5 = 0$$

$$t = \frac{-20 \pm \sqrt{20^2 - 4(-16)(5)}}{2(-16)} = \frac{-20 \pm \sqrt{720}}{-32}$$

$$= \frac{-20 \pm 12\sqrt{5}}{-32} = \frac{5 \pm 3\sqrt{5}}{8} \approx -0.2 \text{ or } 1.5$$

The ball reaches the ground 1.5 seconds after it is tossed.

15. Let $x =$ the speed of the boat in still water. With the current the boat goes $x + 2$ miles per hour and against the current the boat goes $x - 2$ miles per hour. Her time upstream is $6/(x-2)$ hours and her time downstream is $6/(x+2)$ miles per hour. Since her time upstream was 1/3 hour longer, we can write the following equation.

$$\frac{6}{x-2} = \frac{6}{x+2} + \frac{1}{3}$$

$$3(x-2)(x+2)\frac{6}{x-2} = 3(x-2)(x+2)\left(\frac{6}{x+2} + \frac{1}{3}\right)$$

$$18x + 36 = 18(x-2) + (x-2)(x+2)$$
$$18x + 36 = 18x - 36 + x^2 - 4$$
$$76 = x^2$$
$$x = \pm\sqrt{76} = \pm 2\sqrt{19}$$

The speed of her boat is $2\sqrt{19}$ or 8.72 mph.

17. Let $x =$ the amount of increase. The new dimensions for the garden will be $5 + x$ feet by $8 + x$ feet. Since the area of the new garden is to be 100 square feet, we can write the following equation.

$$(5+x)(8+x) = 100$$
$$40 + 13x + x^2 = 100$$
$$x^2 + 13x - 60 = 0$$

$$x = \frac{-13 \pm \sqrt{13^2 - 4(1)(-60)}}{2(1)} = \frac{-13 \pm \sqrt{409}}{2}$$

He should increase each dimension by $\frac{-13 + \sqrt{409}}{2}$ or 3.61 feet.

9.5 WARM-UPS

1. True, because $(3 + i) + (2 - 4i)$
$= 3 + 2 + i - 4i = 5 - 3i$.
2. True, $(4 - 2i)(3 - 5i) = 12 - 26i + 10i^2$
$= 12 - 26i - 10 = 2 - 26i$
3. True, because $(4 - i)(4 + i) = 16 - i^2$
$= 16 - (-1) = 17$.
4. True, because $i^4 = i^2 \cdot i^2 = (-1)(-1) = 1$.
5. False, because $\sqrt{-5} = i\sqrt{5}$.
6. False, because $\sqrt{-36} = 6i$ but $\sqrt{-36} \neq -6i$.
7. False, the complex conjugate of $-2 + 3i$ is $-2 - 3i$.
8. False, because all real numbers are complex numbers.
9. False, because the only solutions to $x^2 = 4$ are 2 and -2.
10. True, because the complex solutions can always be found with the quadratic formula.

9.5 EXERCISES

1. $(3 + 5i) + (2 + 4i) = 3 + 2 + 5i + 4i = 5 + 9i$

3. $(-1 + i) + (2 - i) = -1 + 2 + i - i = 1$

5. $(4 - 5i) - (2 + 3i) = 4 - 2 - 5i - 3i = 2 - 8i$

7. $(-3 - 5i) - (-2 - i) = -3 + 2 - 5i + i$
$$= -1 - 4i$$

9. $(8 - 3i) - (9 - 3i) = 8 - 9 - 3i + 3i = -1$

11. $\left(\frac{1}{2} + i\right) + \left(\frac{1}{4} - \frac{1}{2}i\right) = \frac{3}{4} + \frac{1}{2}i$

13. $3(2 - 3i) = 6 - 9i$

15. $(6i)^2 = 6^2 i^2 = 36(-1) = -36$

17. $(-6i)^2 = (-6)^2 i^2 = 36(-1) = -36$

19. $(2 + 3i)(3 - 5i) = 6 + 9i - 10i - 15i^2$
$= 6 - i - 15(-1) = 6 - i + 15 = 21 - i$

21. $(5 - 2i)^2 = 25 - 2(5)(2i) + (2i)^2$
$= 25 - 20i + 4i^2 = 25 - 20i - 4 = 21 - 20i$

23. $(4 - 3i)(4 + 3i) = 16 - 9i^2 = 16 - 9(-1) = 25$

25. $(1 - i)(1 + i) = 1 - i^2 = 1 - (-1) = 2$

27. $(2 + 5i)(2 - 5i) = 4 - 25i^2 = 29$

29. $(4 - 6i)(4 + 6i) = 16 - 36i^2 = 52$

31. $(-3 + 2i)(-3 - 2i) = 9 - 4i^2 = 13$

33. $i(-i) = -i^2 = 1$

35. $(2 - 6i) \div 2 = \frac{2}{2} - \frac{6i}{2} = 1 - 3i$

37. $\frac{-2 + 8i}{2} = \frac{-2}{2} + \frac{8i}{2} = -1 + 4i$

39. $\frac{4 + 6i}{-2i} = \frac{(4 + 6i)(2i)}{(-2i)(2i)} = \frac{8i + 12i^2}{-4i^2}$
$= \frac{8i - 12}{4} = 2i - 3 = -3 + 2i$

41. $\frac{4i}{3 + 2i} = \frac{4i(3 - 2i)}{(3 + 2i)(3 - 2i)} = \frac{12i - 8i^2}{9 - 4i^2}$
$= \frac{12i + 8}{9 + 4} = \frac{8 + 12i}{13} = \frac{8}{13} + \frac{12}{13}i$

43. $\frac{2 + i}{2 - i} = \frac{(2 + i)(2 + i)}{(2 - i)(2 + i)} = \frac{4 + 4i + i^2}{4 - i^2}$
$= \frac{4 + 4i - 1}{4 - (-1)} = \frac{3 + 4i}{5} = \frac{3}{5} + \frac{4}{5}i$

45. $\frac{4 - 12i}{3 + i} = \frac{(4 - 12i)(3 - i)}{(3 + i)(3 - i)} = \frac{12 - 40i + 12i^2}{9 - i^2}$
$= \frac{-40i}{10} = -4i$

47. $5 + \sqrt{-9} = 5 + i\sqrt{9} = 5 + 3i$

49. $-3 - \sqrt{-7} = -3 - i\sqrt{7}$

51. $\frac{-2 + \sqrt{-12}}{2} = \frac{-2 + i\sqrt{4}\sqrt{3}}{2} = \frac{-2 + 2i\sqrt{3}}{2}$
$= -1 + i\sqrt{3}$

53. $\frac{-8 - \sqrt{-20}}{-4} = \frac{-8 - i\sqrt{4}\sqrt{5}}{-4} = \frac{-8 - 2i\sqrt{5}}{-4}$
$= \frac{-2(4 + i\sqrt{5})}{-2 \cdot 2} = \frac{4 + i\sqrt{5}}{2} = 2 + \frac{1}{2}i\sqrt{5}$

55. $\frac{-4 + \sqrt{-28}}{6} = \frac{-4 + 2i\sqrt{7}}{6} = -\frac{4}{6} + \frac{2i\sqrt{7}}{6}$
$= -\frac{2}{3} + \frac{1}{3}i\sqrt{7}$

57. $\frac{-2 + \sqrt{-100}}{-10} = \frac{-2 + 10i}{-10} = \frac{-2}{-10} + \frac{10i}{-10} = \frac{1}{5} - i$

59. $x^2 + 81 = 0$
$x^2 = -81$
$x = \pm\sqrt{-81}$
$x = \pm 9i$
The solutions to the equation are $-9i$ and $9i$.

61. $x^2 + 5 = 0$
$x^2 = -5$
$x = \pm\sqrt{-5} = \pm i\sqrt{5}$
The solutions to the equation are $-i\sqrt{5}$ and $i\sqrt{5}$.

63. $3y^2 + 2 = 0$
$3y^2 = -2$
$y^2 = -\frac{2}{3}$
$y = \pm\sqrt{-\frac{2}{3}} = \pm i\sqrt{\frac{2}{3}} = \pm i\frac{\sqrt{6}}{3}$
The solutions to the equation are $-i\frac{\sqrt{6}}{3}$ and $i\frac{\sqrt{6}}{3}$.

65. $x^2 - 4x + 5 = 0$
$x = \frac{-(-4) \pm \sqrt{(-4)^2 - 4(1)(5)}}{2(1)} = \frac{4 \pm \sqrt{-4}}{2}$
$= \frac{4 \pm 2i}{2} = 2 \pm i$
The solutions to the equation are $2 - i$ and $2 + i$.

67. $y^2 + 13 = 6y$
$y^2 - 6y + 13 = 0$
$y = \frac{-(-6) \pm \sqrt{(-6)^2 - 4(1)(13)}}{2(1)} = \frac{6 \pm \sqrt{-16}}{2}$
$= \frac{6 \pm 4i}{2} = 3 \pm 2i$
The solutions to the equation are $3 - 2i$ and $3 + 2i$.

168

69. $x^2 - 4x + 7 = 0$

$$x = \frac{-(-4) \pm \sqrt{(-4)^2 - 4(1)(7)}}{2(1)} = \frac{4 \pm \sqrt{-12}}{2}$$

$$= \frac{4 \pm 2i\sqrt{3}}{2} = 2 \pm i\sqrt{3}$$

The solutions to the equation are $2 - i\sqrt{3}$ and $2 + i\sqrt{3}$.

71. $9y^2 - 12y + 5 = 0$

$$y = \frac{-(-12) \pm \sqrt{(-12)^2 - 4(9)(5)}}{2(9)} = \frac{12 \pm \sqrt{-36}}{18}$$

$$= \frac{12 \pm 6i}{18} = \frac{6(2 \pm i)}{6 \cdot 3} = \frac{2 \pm i}{3}$$

The solutions to the equation are $\frac{2-i}{3}$ and $\frac{2+i}{3}$.

73. $x^2 - x + 1 = 0$

$$x = \frac{-(-1) \pm \sqrt{(-1)^2 - 4(1)(1)}}{2(1)} = \frac{1 \pm \sqrt{-3}}{2}$$

$$= \frac{1 \pm i\sqrt{3}}{2}$$

The solutions to the equation are $\frac{1 - i\sqrt{3}}{2}$ and $\frac{1 + i\sqrt{3}}{2}$.

75. $-4x^2 + 8x - 9 = 0$

$$x = \frac{-8 \pm \sqrt{(8)^2 - 4(-4)(-9)}}{2(-4)} = \frac{-8 \pm \sqrt{-80}}{-8}$$

$$= \frac{-8 \pm 4i\sqrt{5}}{-8} = \frac{-4(2 \pm i\sqrt{5})}{-4 \cdot 2} = \frac{2 \pm i\sqrt{5}}{2}$$

The solutions to the equation are $\frac{2 - i\sqrt{5}}{2}$ and $\frac{2 + i\sqrt{5}}{2}$.

77. $(2 - 3i)^2 + 4(2 - 3i) - 9$

$$= 4 - 12i + 9i^2 + 8 - 12i - 9$$

$$= 4 - 12i - 9 + 8 - 12i - 9 = -6 - 24i$$

79. Use $4 - i$ in place of x in $x^2 - 8x + 17$.

$x^2 - 8x + 17 = (4 - i)^2 - 8(4 - i) + 17$

$$= 16 - 8i + i^2 - 32 + 8i + 17$$

$$= 16 - 8i - 1 - 32 + 8i + 17$$

$$= 0$$

81. $[x - (6 - i)][x - (6 + i)]$

$$= [x - 6 + i][x - 6 - i]$$

$$= [x - 6 + i]x - [x - 6 + i]6 - [x - 6 + i]i$$

$$= x^2 - 6x + ix - 6x + 36 - 6i - ix + 6i - i^2$$

$$= x^2 - 12x + ix - ix - 6i + 6i + 36 - (-1)$$

$$= x^2 - 12x + 37$$

83. $(x - 3i)(x + 3i) = 0$

$$x^2 - 9i^2 = 0$$

$$x^2 + 9 = 0$$

9.6 WARM-UPS

1. True, because $-1 = (-2)^2 - 5$ is correct.

2. False, the y-intercept is $(0, 9)$.

3. True, because the solution to $x^2 - 5 = 0$ is $x = \pm\sqrt{5}$.   **4.** True, because $a > 0$.

5. False, because in $y = x^2 + 4$, $a > 0$.

6.   True,   because $x = \frac{-2}{2(1)} = -1$ and $(-1)^2 + 2(-1) = -1$.

7. True, because $x^2 + 1 = 0$ has no real solution.

8. True, because if $x = 0$, then $y = c$.

9. True, because the parabola opens downward from the vertex $(0, 9)$

10. False, the minimum value of y occurs when $x = 7/6$.

9.6 EXERCISES

1. If $x = 3$, then $y = 3^2 - 3 - 12 = -6$.

If $y = 0$, then $x^2 - x - 12 = 0$.

$$(x - 4)(x + 3) = 0$$

$$x - 4 = 0 \quad \text{or} \quad x + 3 = 0$$

$$x = 4 \quad \text{or} \quad x = -3$$

So the ordered pairs are $(3, -6)$, $(4, 0)$, and $(-3, 0)$.

3. If $t = 4$, then $s = -16 \cdot 4^2 + 32(4) = -128$.
It $s = 0$, then $-16t^2 + 32t = 0$.
$$-16t(t-2) = 0$$
$$-16t = 0 \quad \text{or} \quad t - 2 = 0$$
$$t = 0 \quad \text{or} \quad t = 2$$
The ordered pairs are $(4, -128)$, $(0, 0)$ and $(2, 0)$.

5. The ordered pairs $(-2, 6)$, $(-1, 3)$, $(0, 2)$, $(1, 3)$, and $(2, 6)$ satisfy $y = x^2 + 2$. The domain is R and the range is $\{y \mid y \geq 2\}$.

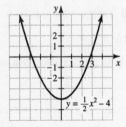

7. The ordered pairs $(-4, 4)$, $(-2, -2)$, $(0, -4)$, $(2, -2)$, and $(4, 4)$ satisfy $y = \frac{1}{2}x^2 - 4$. The domain is R and the range is $\{y \mid y \geq -4\}$.

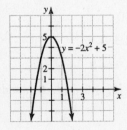

9. The ordered pairs $(-2, -3)$, $(-1, 3)$, $(0, 5)$, $(1, 3)$, and $(2, -3)$ satisfy $y = -2x^2 + 5$. The domain is R and the range is $\{y \mid y \leq 5\}$.

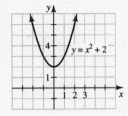

11. The ordered pairs $(-6, -7)$, $(-3, 2)$, $(0, 5)$, $(3, 2)$, and $(6, -7)$ satisfy $y = -\frac{1}{3}x^2 + 5$. The domain is R and the range is $\{y \mid y \leq 5\}$.

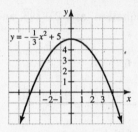

13. The ordered pairs $(0, 4)$, $(1, 1)$, $(2, 0)$, $(3, 1)$, and $(4, 4)$ satisfy $y = (x - 2)^2$. The domain is R and the range is $\{y \mid y \geq 0\}$.

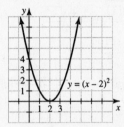

15. $x = \dfrac{-b}{2a} = \dfrac{1}{2(1)} = \dfrac{1}{2}$.

$$y = \left(\frac{1}{2}\right)^2 - \frac{1}{2} - 2 = -\frac{9}{4}$$

The vertex is $\left(\frac{1}{2}, -\frac{9}{4}\right)$. The y-intercept is $(0, -2)$.
$$x^2 - x - 2 = 0$$
$$(x - 2)(x + 1) = 0$$
$$x - 2 = 0 \quad \text{or} \quad x + 1 = 0$$
$$x = 2 \quad \text{or} \quad x = -1$$
The x-intercepts are $(-1, 0)$ and $(2, 0)$.
The domain is R and the range is $\{y \mid y \geq -\frac{9}{4}\}$.

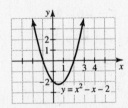

17. $x = \dfrac{-b}{2a} = \dfrac{-2}{2(1)} = -1$

$y = (-1)^2 + 2(-1) - 8 = -9$

The vertex is $(-1, -9)$. The y-intercept is $(0, -8)$.

$$x^2 + 2x - 8 = 0$$
$$(x - 2)(x + 4) = 0$$
$$x - 2 = 0 \quad \text{or} \quad x + 4 = 0$$
$$x = 2 \quad \text{or} \quad x = -4$$

The x-intercepts are $(-4, 0)$ and $(2, 0)$.
The domain is R and the range is $\{y \mid y \geq -9\}$.

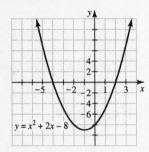

19. $x = \dfrac{-b}{2a} = \dfrac{4}{2(-1)} = -2$

$y = -(-2)^2 - 4(-2) - 3 = 1$

Vertex $(-2, 1)$, y-intercept $(0, -3)$.

$$-x^2 - 4x - 3 = 0$$
$$x^2 + 4x + 3 = 0$$
$$(x + 1)(x + 3) = 0$$
$$x + 1 = 0 \quad \text{or} \quad x + 3 = 0$$
$$x = -1 \quad \text{or} \quad x = -3$$

The x-intercepts are $(-1, 0)$ and $(-3, 0)$.
The domain is R and the range is $\{y \mid y \leq 1\}$.

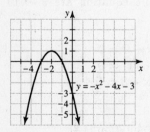

21. $y = -x^2 + 3x + 4$

$x = \dfrac{-b}{2a} = \dfrac{-3}{2(-1)} = \dfrac{3}{2}$

$y = -\left(\dfrac{3}{2}\right)^2 + 3\left(\dfrac{3}{2}\right) + 4 = \dfrac{25}{4}$

Vertex $\left(\dfrac{3}{2}, \dfrac{25}{4}\right)$, y-intercept $(0, 4)$.

$$-x^2 + 3x + 4 = 0$$
$$x^2 - 3x - 4 = 0$$
$$(x - 4)(x + 1) = 0$$
$$x - 4 = 0 \quad \text{or} \quad x + 1 = 0$$
$$x = 4 \quad \text{or} \quad x = -1$$

The x-intercepts are $(4, 0)$ and $(-1, 0)$.
The domain is R and the range is $\{y \mid y \leq \dfrac{25}{4}\}$.

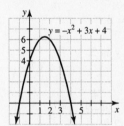

23. $a = b^2 - 6b - 16$

$b = \dfrac{6}{2(1)} = 3$

$a = 3^2 - 6(3) - 16 = -25$

Vertex $(3, -25)$, y-intercept $(0, -16)$.

$$b^2 - 6b - 16 = 0$$
$$(b - 8)(b + 2) = 0$$
$$b - 8 = 0 \quad \text{or} \quad b + 2 = 0$$
$$b = 8 \quad \text{or} \quad b = -2$$

The b-intercepts are $(8, 0)$ and $(-2, 0)$.
The domain is R and the range is $\{a \mid a \geq -25\}$.

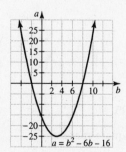

171

25. $y = x^2 - 8$
$$x = \frac{-b}{2a} = \frac{0}{2(1)} = 0$$
$y = 0^2 - 8 = -8$

The minimum value of y is −8.

27. $y = -3x^2 + 14$
$$x = \frac{-b}{2a} = \frac{0}{2(-3)} = 0$$
$y = -3 \cdot 0^2 + 14 = 14$

The maximum value of y is 14.

29. $y = x^2 + 2x + 3$
$$x = \frac{-b}{2a} = \frac{-2}{2(1)} = -1$$
$y = (-1)^2 + 2(-1) + 3 = 2$

The minimum value of y is 2.

31. $y = -2x^2 - 4x$
$$x = \frac{-b}{2a} = \frac{4}{2(-2)} = -1$$
$y = -2(-1)^2 - 4(-1) = 2$

The maximum value of y is 2.

33. $s = -16t^2 + 64t$
$$t = \frac{-64}{2(-16)} = 2$$
$s = -16 \cdot 2^2 + 64(2) = 64$

Maximum height is 64 feet.

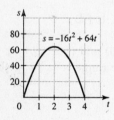

35. $A = -w^2 + 50w$
$$w = \frac{-50}{2(-1)} = 25$$
$A = -25^2 + 50(25) = 625$

Maximum area is 625 square meters.

37. $A = -2t^2 + 32t + 12$
$$t = \frac{-32}{2(-2)} = 8$$
The nitrogen dioxide is at its maximum 8 hours after 6 A.M. or at 2 P.M.

39. Since the x-coordinate of the tower on the right is 20, we get $y = 0.0375(20)^2 = 15$. So the height of the towers is 15 meters. Use the Pythagorean theorem to find z:
$$15^2 + 20^2 = z^2$$
$$z^2 = 625$$
$$z = 25$$

So the length of the cable marked z is 25 meters.

CHAPTER 9 REVIEW

1. $\quad x^2 - 9 = 0$
$$x^2 = 9$$
$$x = \pm 3$$

The solutions to the equation are −3 and 3.

3. $\quad x^2 - 9x = 0$
$$x(x - 9) = 0$$
$$x = 0 \quad \text{or} \quad x - 9 = 0$$
$$x = 9$$

The solutions to the equation are 0 and 9.

5. $\quad x^2 - x = 2$
$$x^2 - x - 2 = 0$$
$$(x - 2)(x + 1) = 0$$
$$x - 2 = 0 \quad \text{or} \quad x + 1 = 0$$
$$x = 2 \quad \text{or} \quad x = -1$$

The solutions to the equation are −1 and 2.

7. $\quad (x - 9)^2 = 10$
$$x - 9 = \pm\sqrt{10}$$
$$x = 9 \pm \sqrt{10}$$

The solutions to the equation are $9 - \sqrt{10}$ and $9 + \sqrt{10}$.

9. $\quad 4x^2 - 12x + 9 = 0$
$$(2x - 3)^2 = 0$$
$$2x - 3 = 0$$
$$x = \frac{3}{2}$$

The solution to the equation is $\frac{3}{2}$.

11. $t^2 - 9t + 20 = 0$
$(t-4)(t-5) = 0$
$t - 4 = 0$ or $t - 5 = 0$
$t = 4$ or $t = 5$
The solutions to the equation are 4 and 5.

13. $\dfrac{x}{2} = \dfrac{7}{x+5}$

$x(x+5) = 14$
$x^2 + 5x - 14 = 0$
$(x+7)(x-2) = 0$
$x + 7 = 0$ or $x - 2 = 0$
$x = -7$ or $x = 2$
The solutions to the equation are -7 and 2.

15. $\dfrac{1}{2}x^2 + \dfrac{7}{4}x = 1$
$2x^2 + 7x = 4$
$2x^2 + 7x - 4 = 0$
$(2x - 1)(x + 4) = 0$
$2x - 1 = 0$ or $x + 4 = 0$
$x = \dfrac{1}{2}$ or $x = -4$
The solutions to the equation are -4 and $\frac{1}{2}$.

17. $x^2 + 4x - 7 = 0$
$x^2 + 4x \qquad = 7$
$x^2 + 4x + 4 = 7 + 4$
$(x+2)^2 = 11$
$x + 2 = \pm\sqrt{11}$
$x = -2 \pm \sqrt{11}$
The solutions to the equation are $-2 - \sqrt{11}$ and $-2 + \sqrt{11}$.

19. $x^2 + 3x - 28 = 0$
$x^2 + 3x \qquad = 28$
$x^2 + 3x + \dfrac{9}{4} = 28 + \dfrac{9}{4}$
$\left(x + \dfrac{3}{2}\right)^2 = \dfrac{121}{4}$
$x + \dfrac{3}{2} = \pm\dfrac{11}{2}$
$x = -\dfrac{3}{2} \pm \dfrac{11}{2}$
$x = -\dfrac{3}{2} - \dfrac{11}{2} = -7$ or
$x = -\dfrac{3}{2} + \dfrac{11}{2} = \dfrac{8}{2} = 4$

The solutions to the equation are -7 and 4.

21. $x^2 + 3x - 5 = 0$
$x^2 + 3x \qquad = 5$
$x^2 + 3x + \dfrac{9}{4} = 5 + \dfrac{9}{4}$
$\left(x + \dfrac{3}{2}\right)^2 = \dfrac{29}{4}$
$x + \dfrac{3}{2} = \pm\sqrt{\dfrac{29}{4}} = \pm\dfrac{\sqrt{29}}{2}$
$x = -\dfrac{3}{2} \pm \dfrac{\sqrt{29}}{2}$
The solutions to the equation are $\dfrac{-3 - \sqrt{29}}{2}$ and $\dfrac{-3 + \sqrt{29}}{2}$.

23. $2x^2 + 9x - 5 = 0$
$x^2 + \dfrac{9}{2}x - \dfrac{5}{2} = 0$
$x^2 + \dfrac{9}{2}x \qquad = \dfrac{5}{2}$
$x^2 + \dfrac{9}{2}x + \dfrac{81}{16} = \dfrac{5}{2} + \dfrac{81}{16}$
$\left(x + \dfrac{9}{4}\right)^2 = \dfrac{121}{16}$
$x + \dfrac{9}{4} = \pm\dfrac{11}{4}$
$x = -\dfrac{9}{4} \pm \dfrac{11}{4}$
$x = -\dfrac{9}{4} - \dfrac{11}{4} = -5$ or
$x = -\dfrac{9}{4} + \dfrac{11}{4} = \dfrac{2}{4} = \dfrac{1}{2}$
The solutions to the equation are -5 and $\frac{1}{2}$.

25.   For $25t^2 - 10t + 1 = 0$, $a = 25$, $b = -10$, and $c = 1$.
$b^2 - 4ac = (-10)^2 - 4(25)(1) = 0$
Since the discriminant is zero, there is one real solution to the equation.

27.   For $-3w^2 + 4w - 5 = 0$, $a = -3$, $b = 4$, and $c = -5$.
$b^2 - 4ac = (4)^2 - 4(-3)(-5) = -44$
Since the discriminant is negative, there are no real solutions to the equation.

29.   For $-3v^2 + 4v + 5 = 0$, $a = -3$, $b = 4$, and $c = 5$.
$b^2 - 4ac = (4)^2 - 4(-3)(5) = 76$
Since the discriminant is positive, there are two real solutions to the equation.

31. For $6x^2 + x - 2 = 0$, $a = 6$, $b = 1$, and $c = -2$.

$$x = \frac{-1 \pm \sqrt{1^2 - 4(6)(-2)}}{2(6)} = \frac{-1 \pm \sqrt{49}}{12} = \frac{-1 \pm 7}{12}$$

$$x = \frac{-1 - 7}{12} = \frac{-8}{12} = -\frac{2}{3} \quad \text{or} \quad x = \frac{-1 + 7}{12} = \frac{6}{12} = \frac{1}{2}$$

The solutions to the equation are $-\frac{2}{3}$ and $\frac{1}{2}$.

33. For $x^2 - x - 4 = 0$, $a = 1$, $b = -1$, and $c = -4$.

$$x = \frac{-(-1) \pm \sqrt{(-1)^2 - 4(1)(-4)}}{2(1)} = \frac{1 \pm \sqrt{17}}{2}$$

The solutions to the equation are $\frac{1 - \sqrt{17}}{2}$ and $\frac{1 + \sqrt{17}}{2}$.

35. For $5x^2 - 6x - 1 = 0$, $a = 5$, $b = -6$, and $c = -1$.

$$x = \frac{-(-6) \pm \sqrt{(-6)^2 - 4(5)(-1)}}{2(5)} = \frac{6 \pm \sqrt{56}}{10}$$

$$= \frac{6 \pm 2\sqrt{14}}{10} = \frac{3 \pm \sqrt{14}}{5}$$

The solutions to the equation are $\frac{3 - \sqrt{14}}{5}$ and $\frac{3 + \sqrt{14}}{5}$.

37. For $3x^2 - 5x = 0$, $a = 3$, $b = -5$, and $c = 0$.

$$x = \frac{-(-5) \pm \sqrt{(-5)^2 - 4(3)(0)}}{2(3)} = \frac{5 \pm \sqrt{25}}{6} = \frac{5 \pm 5}{6}$$

$$x = \frac{5 - 5}{6} = 0 \quad \text{or} \quad x = \frac{5 + 5}{6} = \frac{10}{6} = \frac{5}{3}$$

The solutions to the equation are 0 and $\frac{5}{3}$.

39. Let $x =$ the distance from Chuck's eyes to the nest. The distance from Chuck's eyes to the nest is the hypotenuse of a right triangle with sides of 5 feet and 12 feet.

$$x^2 = 5^2 + 12^2$$
$$x^2 = 169$$
$$x = \pm 13$$

The distance from Chuck's eyes to the nest is 13 meters.

41. Let $x =$ the length of one leg and $x + 2 =$ the length of the other leg. Since the hypotenuse is 5 meters, we can write the following equation.

$$x^2 + (x + 2)^2 = 5^2$$
$$x^2 + x^2 + 4x + 4 = 25$$
$$2x^2 + 4x - 21 = 0$$

$$x = \frac{-(4) \pm \sqrt{(4)^2 - 4(2)(-21)}}{2(2)} = \frac{-4 \pm \sqrt{184}}{4}$$

$$= \frac{-4 \pm 2\sqrt{46}}{4} = \frac{-2 \pm \sqrt{46}}{2}$$

If $x = \frac{-2 + \sqrt{46}}{2}$, then

$$x + 2 = \frac{-2 + \sqrt{46}}{2} + 2 = \frac{-2 + \sqrt{46}}{2} + \frac{4}{2} = \frac{2 + \sqrt{46}}{2}.$$

The length of the shorter leg is $\frac{-2 + \sqrt{46}}{2} \approx$ 2.391 meters and the length of the longer leg is $\frac{2 + \sqrt{46}}{2} \approx$ 4.391 meters.

43. Let $x =$ the length of the height and $x + 4 =$ the length of the base. Since the area is 20 square inches, we can write the following equation.

$$\frac{1}{2}(x + 4)x = 20$$
$$(x + 4)x = 40$$
$$x^2 + 4x - 40 = 0$$

$$x = \frac{-(4) \pm \sqrt{(4)^2 - 4(1)(-40)}}{2(1)} = \frac{-4 \pm \sqrt{176}}{2}$$

$$= \frac{-4 \pm 4\sqrt{11}}{2} = -2 \pm \sqrt{11}$$

If $x = -2 + \sqrt{11}$, then
$x + 4 = -2 + \sqrt{11} + 4 = 2 + \sqrt{11}$.
The height is $-2 + 2\sqrt{11} \approx$ 4.633 inches and the base is $2 + 2\sqrt{11} \approx$ 8.633 inches.

45. Since the sum of the numbers is 6, we can let $x =$ one number and $6 - x =$ the other. Since their product is 7, we can write the following equation.

$$x(6 - x) = 7$$
$$6x - x^2 = 7$$
$$-x^2 + 6x - 7 = 0$$
$$x^2 - 6x + 7 = 0$$

$$x = \frac{-(-6) \pm \sqrt{(-6)^2 - 4(1)(7)}}{2(1)} = \frac{6 \pm \sqrt{8}}{2}$$

$$= \frac{6 \pm 2\sqrt{2}}{2} = 3 \pm \sqrt{2}$$

If $x = 3 + \sqrt{2}$, then $6 - x = 6 - (3 + \sqrt{2}) = 3 - \sqrt{2}$.
If $x = 3 - \sqrt{2}$, then $6 - x = 6 - (3 - \sqrt{2}) = 3 + \sqrt{2}$.

The numbers are $3 + \sqrt{2} \approx 4.414$ and $3 - \sqrt{2} \approx 1.586$.

47. Let $x =$ the number of hours for the new printer and $x + 2 =$ the number of hours for the old printer. The new printer does $1/x$ of the job per hour, the old printer does $1/(x+2)$ of the job per hour, and together they do $1/8$ of the job per hour.

$$\frac{1}{x} + \frac{1}{x+2} = \frac{1}{8}$$

$$8x(x+2)\left(\frac{1}{x} + \frac{1}{x+2}\right) = 8x(x+2)\frac{1}{8}$$

$$8x + 16 + 8x = x^2 + 2x$$
$$-x^2 + 14x + 16 = 0$$
$$x^2 - 14x - 16 = 0$$

$$x = \frac{-(-14) \pm \sqrt{(-14)^2 - 4(1)(-16)}}{2(1)} = \frac{14 \pm \sqrt{260}}{2}$$

$$= \frac{14 \pm 2\sqrt{65}}{2} = 7 \pm \sqrt{65}$$

If $x = 7 + \sqrt{65}$, then $x + 2 = 9 + \sqrt{65}$.

The new printer takes $7 + \sqrt{65} \approx 15.062$ hours to do the job alone and the old printer takes $9 + \sqrt{65} \approx 17.062$ hours to do the job alone.

49. $(2 + 3i) + (5 - 6i) = 2 + 5 + 3i - 6i$
$$= 7 - 3i$$

51. $(-5 + 4i) - (-2 - 3i) = -5 + 4i + 2 + 3i$
$$= -3 + 7i$$

53. $(2 - 9i)(3 + i) = 6 - 27i + 2i - 9i^2$
$$= 6 - 25i + 9 = 15 - 25i$$

55. $(3 + 8i)^2 = 9 + 2 \cdot 3 \cdot 8i + 64i^2$
$$= 9 + 48i - 64 = -55 + 48i$$

57. $\dfrac{-2 - \sqrt{-8}}{2} = \dfrac{-2 - 2i\sqrt{2}}{2} = -1 - i\sqrt{2}$

59. $\dfrac{1 + 3i}{6 - i} = \dfrac{(1 + 3i)(6 + i)}{(6 - i)(6 + i)} = \dfrac{6 + 19i + 3i^2}{36 - i^2}$

$$= \frac{6 + 19i - 3}{36 - (-1)} = \frac{3 + 19i}{37} = \frac{3}{37} + \frac{19}{37}i$$

61. $\dfrac{5 + i}{4 - i} = \dfrac{(5 + i)(4 + i)}{(4 - i)(4 + i)} = \dfrac{20 + 9i + i^2}{16 - i^2}$

$$= \frac{20 + 9i - 1}{16 - (-1)} = \frac{19 + 9i}{17} = \frac{19}{17} + \frac{9}{17}i$$

63. $x^2 + 121 = 0$
$$x^2 = -121$$
$$x = \pm\sqrt{-121} = \pm 11i$$

The solutions to the equation are $-11i$ and $11i$.

65. $x^2 - 16x + 65 = 0$

$$x = \frac{-(-16) \pm \sqrt{(-16)^2 - 4(1)(65)}}{2(1)} = \frac{16 \pm \sqrt{-4}}{2}$$

$$= \frac{16 \pm 2i}{2} = 8 \pm i$$

The solutions to the equation are $8 - i$ and $8 + i$.

67. $2x^2 - 3x + 9 = 0$

$$x = \frac{-(-3) \pm \sqrt{(-3)^2 - 4(2)(9)}}{2(2)} = \frac{3 \pm \sqrt{-63}}{4}$$

$$= \frac{3 \pm i\sqrt{9}\sqrt{7}}{2} = \frac{3 \pm 3i\sqrt{7}}{2}$$

The solutions to the equation are $\dfrac{3 - 3i\sqrt{7}}{4}$ and $\dfrac{3 + 3i\sqrt{7}}{4}$.

69. $y = x^2 - 6x$

$x = \dfrac{-b}{2a} = \dfrac{6}{2(1)} = 3$, $y = 3^2 - 6(3) = -9$

The vertex is $(3, -9)$. The y-intercept is $(0, 0)$.
$$x^2 - 6x = 0$$
$$x(x - 6) = 0$$
$$x = 0 \quad \text{or} \quad x - 6 = 0$$
$$x = 0 \quad \text{or} \qquad x = 6$$
The x-intercepts are $(0, 0)$ and $(6, 0)$.

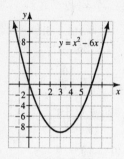

71. $y = x^2 - 4x - 12$

$x = \dfrac{-b}{2a} = \dfrac{4}{2(1)} = 2$, $y = 2^2 - 4(2) - 12 = -16$

The vertex is $(2, -16)$. The y-intercept is $(0, -12)$. The solutions to $x^2 - 4x - 12 = 0$ are 6 and -2. So the x-intercepts are $(-2, 0)$ and $(6, 0)$.

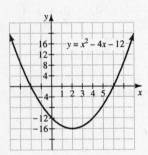

73. $y = -2x^2 + 8x$

$x = \dfrac{-b}{2a} = \dfrac{-8}{2(-2)} = 2$, $\quad y = -2 \cdot 2^2 + 8(2) = 8$

The vertex is $(2, 8)$. The y-intercept is $(0, 0)$.

The solutions to $-2x^2 + 8x = 0$ are 0 and 4. So

the x-intercepts are $(0, 0)$ and $(4, 0)$.

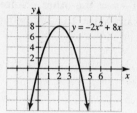

75. $y = -x^2 + 2x + 3$

$x = \dfrac{-b}{2a} = \dfrac{-2}{2(-1)} = 1$, $y = -1^2 + 2(1) + 3 = 4$

The vertex is $(1, 4)$. The y-intercept is $(0, 3)$.
The solutions to $-x^2 + 2x + 3 = 0$ are -1 and 3.
So the x-intercepts are $(-1, 0)$ and $(3, 0)$.

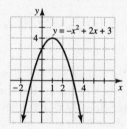

77. $y = x^2 + 4x + 1$

$x = \dfrac{-b}{2a} = \dfrac{-4}{2(1)} = -2$,

$y = (-2)^2 + 4(-2) + 1 = -3$

The domain is R and the range is $\{y \mid y \geq -3\}$.

79. $y = -2x^2 - x + 4$

$x = \dfrac{-b}{2a} = \dfrac{1}{2(-2)} = -\dfrac{1}{4}$,

$y = -2(-\tfrac{1}{4})^2 - (-\tfrac{1}{4}) + 4 = 4.125$

The domain is R and the range is

$\{y \mid y \leq 4.125\}$.

81. $C = 0.004n^2 - 3.2n + 660$
If $n = 390$, then
$C = 0.004(390)^2 - 3.2(390) + 660 = \$20.40$

$n = \dfrac{3.2}{2(0.004)} = 400$

The unit cost is at a minimum for 400 starters.

176

CHAPTER 9 TEST

1. For $9x^2 - 12x + 4 = 0$, $a = 9$, $b = -12$, and $c = 4$.

$b^2 - 4ac = (-12)^2 - 4(9)(4) = 0$
Since the discriminant has a value of 0, there is only one real solution to the equation.

2. For $-2x^2 + 3x - 5 = 0$, $a = -2$, $b = 3$, and $c = -5$.
$b^2 - 4ac = (3)^2 - 4(-2)(-5) = -31$
Since the value of the discriminant is negative, there are no real solutions to the equation.

3. For $-2x^2 + 5x - 1 = 0$, $a = -2$, $b = 5$, and $c = -1$.
$b^2 - 4ac = (5)^2 - 4(-2)(-1) = 17$
Since the value of the discriminant is positive, there are two real solutions to the equation.

4. For $5x^2 + 2x - 3 = 0$, $a = 5$, $b = 2$, and $c = -3$.

$$x = \frac{-(2) \pm \sqrt{(2)^2 - 4(5)(-3)}}{2(5)} = \frac{-2 \pm \sqrt{64}}{10} = \frac{-2 \pm 8}{10}$$

$$x = \frac{-2 + 8}{10} = \frac{6}{10} = \frac{3}{5} \text{ or } x = \frac{-2 - 8}{10} = -1$$

The solutions to the equation are -1 and $\frac{3}{5}$.

5. For $2x^2 - 4x - 3 = 0$, $a = 2$, $b = -4$, and $c = -3$.

$$x = \frac{-(-4) \pm \sqrt{(-4)^2 - 4(2)(-3)}}{2(2)} = \frac{4 \pm \sqrt{40}}{4}$$

$$= \frac{4 \pm 2\sqrt{10}}{4} = \frac{2 \pm \sqrt{10}}{2}$$

The solutions to the equation are $\frac{2 - \sqrt{10}}{2}$ and $\frac{2 + \sqrt{10}}{2}$.

6.
$$x^2 + 4x - 21 = 0$$
$$x^2 + 4x = 21$$
$$x^2 + 4x + 4 = 21 + 4$$
$$(x + 2)^2 = 25$$
$$x + 2 = \pm 5$$
$$x = -2 \pm 5$$
$$x = -2 + 5 = 3 \text{ or } x = -2 - 5 = -7$$

The solutions to the equation are -7 and 3.

7.
$$x^2 + 3x - 5 = 0$$
$$x^2 + 3x = 5$$
$$x^2 + 3x + \frac{9}{4} = 5 + \frac{9}{4}$$
$$\left(x + \frac{3}{2}\right)^2 = \frac{29}{4}$$
$$x + \frac{3}{2} = \pm\sqrt{\frac{29}{4}} = \pm\frac{\sqrt{29}}{2}$$
$$x = -\frac{3}{2} \pm \frac{\sqrt{29}}{2}$$

The solutions to the equation are $\frac{-3 - \sqrt{29}}{2}$ and $\frac{-3 + \sqrt{29}}{2}$.

8.
$$x(x + 1) = 20$$
$$x^2 + x = 20$$
$$x^2 + x - 20 = 0$$
$$(x + 5)(x - 4) = 0$$
$$x + 5 = 0 \text{ or } x - 4 = 0$$
$$x = -5 \text{ or } x = 4$$

The solutions to the equation are -5 and 4.

9.
$$x^2 - 28x + 75 = 0$$
$$(x - 3)(x - 25) = 0$$
$$x - 3 = 0 \text{ or } x - 25 = 0$$
$$x = 3 \text{ or } x = 25$$

The solutions to the equation are 3 and 25.

10.
$$\frac{x - 1}{3} = \frac{x + 1}{2x}$$

$$2x(x - 1) = 3(x + 1)$$
$$2x^2 - 2x = 3x + 3$$
$$2x^2 - 5x - 3 = 0$$
$$(2x + 1)(x - 3) = 0$$
$$2x + 1 = 0 \text{ or } x - 3 = 0$$
$$x = -\frac{1}{2} \text{ or } x = 3$$

The solutions to the equation are $-\frac{1}{2}$ and 3.

11. $(2 - 3i) + (8 + 6i) = 10 + 3i$

12. $(-2 - 5i) - (4 - 12i) = -2 - 5i - 4 + 12i$
$$= -6 + 7i$$

13. $(-6i)^2 = 36i^2 = 36(-1) = -36$

14. $(3 - 5i)(4 + 6i) = 12 - 2i - 30i^2$
$$= 12 - 2i + 30 = 42 - 2i$$

15. $(8 - 2i)(8 + 2i) = 64 - 4i^2 = 64 + 4 = 68$

16. $(4 - 6i) \div 2 = \frac{4}{2} - \frac{6i}{2} = 2 - 3i$

17. $\dfrac{-2+\sqrt{-12}}{2} = \dfrac{-2+2i\sqrt{3}}{2} = -1+i\sqrt{3}$

18. $\dfrac{6-\sqrt{-18}}{-3} = \dfrac{6-3i\sqrt{2}}{-3} = -2+i\sqrt{2}$

19. $\dfrac{5i}{4+3i} = \dfrac{5i(4-3i)}{(4+3i)(4-3i)} = \dfrac{20i-15i^2}{16-9i^2}$

$= \dfrac{20i+15}{16+9} = \dfrac{15+20i}{25} = \dfrac{3}{5}+\dfrac{4}{5}i$

20. For $x^2+6x+12=0$, $a=1$, $b=6$, and $c=12$.

$x = \dfrac{-(6) \pm \sqrt{(6)^2-4(1)(12)}}{2(1)} = \dfrac{-6 \pm \sqrt{-12}}{2}$

$\qquad = \dfrac{-6 \pm 2i\sqrt{3}}{2} = -3 \pm i\sqrt{3}$

The solutions to the equation are $-3-i\sqrt{3}$ and $-3+i\sqrt{3}$.

21. For $-5x^2+6x-5=0$, $a=-5$, $b=6$, and $c=-5$.

$x = \dfrac{-(6) \pm \sqrt{(6)^2-4(-5)(-5)}}{2(-5)} = \dfrac{-6 \pm \sqrt{-64}}{-10}$

$= \dfrac{-6 \pm 8i}{-10} = \dfrac{-6}{-10} \pm \dfrac{8i}{-10} = \dfrac{3}{5} \pm \dfrac{4}{5}i$

The solutions are $\dfrac{3}{5}-\dfrac{4}{5}i$ and $\dfrac{3}{5}+\dfrac{4}{5}i$.

22. The graph of $y=16-x^2$ goes through $(-2, 12)$, $(-1, 15)$, $(0, 16)$, $(1, 15)$, and $(2, 12)$. The vertex is $(0, 16)$. The domain is R and the range is $\{y \mid y \leq 16\}$.

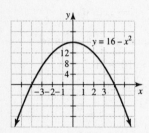

23. The graph of $y=x^2-3x$ goes through $(0, 0)$, $(1, -2)$, $(2, -2)$, and $(3, 0)$. The vertex is $\left(\dfrac{3}{2}, -\dfrac{9}{4}\right)$. The domain is R and the range is $\left\{y \mid y \geq -\dfrac{9}{4}\right\}$

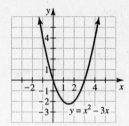

24.
$$x^2-6x+5=0$$
$$(x-5)(x-1)=0$$
$$x-5=0 \quad \text{or} \quad x-1=0$$
$$x=5 \quad \text{or} \quad x=1$$

The x-intercepts are $(1, 0)$ and $(5, 0)$.

25. $s=-16t^2+48t$

$t = \dfrac{-48}{2(-16)} = \dfrac{3}{2}$, $s = -16\left(\dfrac{3}{2}\right)^2 + 48\left(\dfrac{3}{2}\right) = 36$

The maximum height reached by the ball is 36 feet.

26. Let $x=$ one of the numbers and $10-x=$ the other number. Since their product is 23, we can write the following equation.

$$x(10-x)=23$$
$$10x-x^2=23$$
$$-x^2+10x-23=0$$
$$x^2-10x+23=0$$

$$x = \dfrac{-(-10) \pm \sqrt{(-10)^2-4(1)(23)}}{2(1)} = \dfrac{10 \pm \sqrt{8}}{2}$$

$$= \dfrac{10 \pm 2\sqrt{2}}{2} = 5 \pm \sqrt{2}$$

If $x=5+\sqrt{2}$, then $10-x = 10-(5+\sqrt{2})$

$= 5-\sqrt{2}$.

If $x=5-\sqrt{2}$, then $10-x = 10-(5-\sqrt{2})$

$= 5+\sqrt{2}$.

The numbers are $5-\sqrt{2}$ and $5+\sqrt{2}$.

Tying It All Together Chapters 1-9

1. $2x - 1 = 0$

$2x = 1$

$x = \frac{1}{2}$

The solution to the equation is $\frac{1}{2}$.

2. $2(x - 1) = 0$

$2x - 2 = 0$

$2x = 2$

$x = 1$

The solution to the equation is 1.

3. $2x^2 - 1 = 0$

$2x^2 = 1$

$x^2 = \frac{1}{2}$

$x = \pm\sqrt{\frac{1}{2}} = \pm\frac{\sqrt{2}}{2}$

The solutions to the equation are $-\frac{\sqrt{2}}{2}$ and $\frac{\sqrt{2}}{2}$.

4. $(2x - 1)^2 = 8$

$2x - 1 = \pm\sqrt{8}$

$2x = 1 \pm 2\sqrt{2}$

$x = \frac{1 \pm 2\sqrt{2}}{2}$

The solutions to the equation are $\frac{1 - 2\sqrt{2}}{2}$ and $\frac{1 + 2\sqrt{2}}{2}$.

5. $2x^2 - 4x - 1 = 0$

$x = \frac{-(-4) \pm \sqrt{(-4)^2 - 4(2)(-1)}}{2(2)} = \frac{4 \pm \sqrt{24}}{4}$

$= \frac{4 \pm 2\sqrt{6}}{4} = \frac{2(2 \pm \sqrt{6})}{2 \cdot 2} = \frac{2 \pm \sqrt{6}}{2}$

The solutions to the equation are $\frac{2 - \sqrt{6}}{2}$ and $\frac{2 + \sqrt{6}}{2}$.

6. $2x^2 - 4x = 0$

$2x(x - 2) = 0$

$2x = 0$ or $x - 2 = 0$

$x = 0$ or $x = 2$

The solutions to the equation are 0 and 2.

7. $2x^2 + x = 1$

$2x^2 + x - 1 = 0$

$(2x - 1)(x + 1) = 0$

$2x - 1 = 0$ or $x + 1 = 0$

$x = \frac{1}{2}$ or $x = -1$

The solutions to the equation are -1 and $\frac{1}{2}$.

8. $x - 2 = \sqrt{2x - 1}$

$(x - 2)^2 = (\sqrt{2x - 1})^2$

$x^2 - 4x + 4 = 2x - 1$

$x^2 - 6x + 5 = 0$

$(x - 5)(x - 1) = 0$

$x - 5 = 0$ or $x - 1 = 0$

$x = 5$ or $x = 1$

Check $x = 1$: $1 - 2 = \sqrt{2(1) - 1}$

$-1 = \sqrt{1}$ Incorrect.

Since 5 does check in the original equation, the solution to the equation is 5.

9. $\frac{1}{x} = \frac{x}{2x - 15}$

$x^2 = 2x - 15$

$x^2 - 2x + 15 = 0$

$x = \frac{-(-2) \pm \sqrt{(-2)^2 - 4(1)(15)}}{2(1)} = \frac{2 \pm \sqrt{-56}}{2}$

$= \frac{2 \pm 2i\sqrt{14}}{2} = 1 \pm i\sqrt{14}$

The solutions to the equation are $1 - i\sqrt{14}$ and $1 + i\sqrt{14}$.

10. $\frac{1}{x} - \frac{1}{x - 1} = -\frac{1}{2}$

$2x(x - 1)\left(\frac{1}{x} - \frac{1}{x - 1}\right) = 2x(x - 1)\left(-\frac{1}{2}\right)$

$2x - 2 - 2x = x^2 - x$

$-x^2 + x - 2 = 0$

$x^2 - x + 2 = 0$

$(x - 2)(x + 1) = 0$

$x - 2 = 0$ or $x + 1 = 0$

$x = 2$ or $x = -1$

The solutions to the equation are -1 and 2.

11.
$$5x - 4y = 8$$
$$-4y = -5x + 8$$
$$y = \frac{-5x}{-4} + \frac{8}{-4}$$
$$y = \frac{5}{4}x - 2$$

12.
$$3x - y = 9$$
$$3x = 9 + y$$
$$3x - 9 = y$$
$$y = 3x - 9$$

13.
$$\frac{y-4}{x+2} = \frac{2}{3}$$
$$y - 4 = \frac{2}{3}(x+2)$$
$$y - 4 = \frac{2}{3}x + \frac{4}{3}$$
$$y = \frac{2}{3}x + \frac{4}{3} + 4$$
$$y = \frac{2}{3}x + \frac{16}{3}$$

14.
$$ay + b = 0$$
$$ay = -b$$
$$y = -\frac{b}{a}$$

15.
$$ay^2 + by + c = 0$$
$$y = \frac{-b \pm \sqrt{b^2 - 4ac}}{2a}$$

16.
$$y - 1 = -\frac{2}{3}(x - 9)$$
$$y - 1 = -\frac{2}{3}x + 6$$
$$y = -\frac{2}{3}x + 7$$

17.
$$\frac{2}{3}x + \frac{1}{2}y = \frac{1}{9}$$
$$\frac{1}{2}y = -\frac{2}{3}x + \frac{1}{9}$$
$$2 \cdot \frac{1}{2}y = 2\left(-\frac{2}{3}x + \frac{1}{9}\right)$$
$$y = -\frac{4}{3}x + \frac{2}{9}$$

18.
$$x^2 + y^2 = a^2$$
$$y^2 = a^2 - x^2$$
$$y = \pm\sqrt{a^2 - x^2}$$

19. $P = 4s$ **20.** $A = s^2$

21. Since $s^2 + s^2 = d^2$, $2s^2 = d^2$, $s\sqrt{2} = d$, or $s = \frac{d}{\sqrt{2}}$.

$$P = 4s = 4 \cdot \frac{d}{\sqrt{2}} = \frac{4d\sqrt{2}}{\sqrt{2}\sqrt{2}} = 2d\sqrt{2}$$
So $P = 2d\sqrt{2}$.

22. Since $A = s^2$ and $s^2 = \frac{d^2}{2}$, we have $A = \frac{d^2}{2}$.
So $d^2 = 2A$ or $d = \sqrt{2A}$.

23.
$$5(3x - 2y) = 5 \cdot 12$$
$$2(2x + 5y) = 2(-11)$$

$$15x - 10y = 60$$
$$\underline{4x + 10y = -22}$$
$$19x \qquad = 38$$
$$x = 2$$
$$3(2) - 2y = 12$$
$$-2y = 6$$
$$y = -3$$
The solution to the system is $(2, -3)$.

24. Use substitution to eliminate y.
$$3x - 0.6(3x + 1) = 3$$
$$3x - 1.8x - 0.6 = 3$$
$$1.2x = 3.6$$
$$x = 3$$
$$y = 3(3) + 1 = 10$$
The solution to the system is $(3, 10)$.

25. The graph of $y = x - 3$ is a straight line with y-intercept $(0, -3)$ and slope 1.

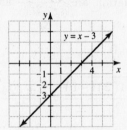

26. The graph of $y = 2 - x$ is a straight line with y-intercept $(0, 2)$ and slope -1.

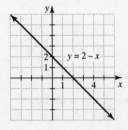

180

27. The graph of $y = x^2 - 3$ is a parabola with vertex $(0, -3)$. It passes through the points $(2, 1)$ and $(-2, 1)$.

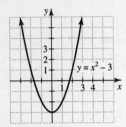

28. The graph of $y = 2 - x^2$ is a parabola with vertex $(0, 2)$. It passes through $(2, -2)$ and $(-2, -2)$.

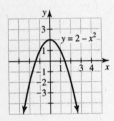

29. The graph of $y = \frac{2}{3}x - 4$ is a line with y-intercept $(0, -4)$ and slope $2/3$.

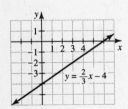

30. The graph of $y = -\frac{4}{3}x + 5$ is a straight line with slope $-4/3$ and y-intercept $(0, 5)$.

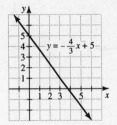

31. a) $m = \dfrac{8000 - 7500}{10 - 12} = -250$

$s - 8000 = -250(p - 10)$

$s - 8000 = -250p + 2500$

$s = -250p + 10,500$

b) At \$10 the revenue is \$80,000. At \$12 the revenue is \$90,000. At \$16 the revenue is \$104,000. So the revenue has increased as the price was raised.

c) $R = sp = (-250p + 10,500)p$

$R = -250p^2 + 10,500p$

d) $p = \dfrac{-10500}{2(-250)} = 21$

A price of \$21 would produce the maximum revenue.

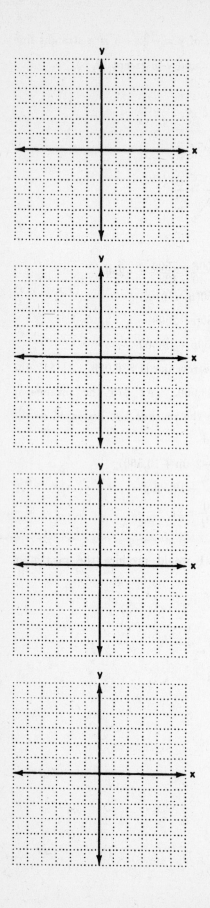

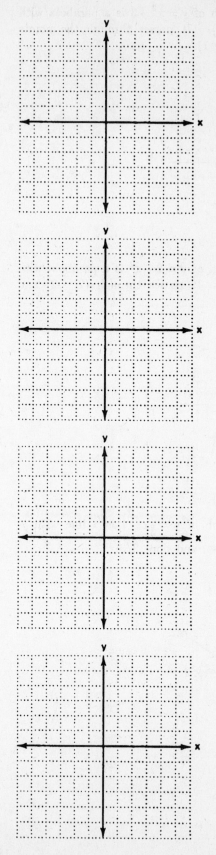

182

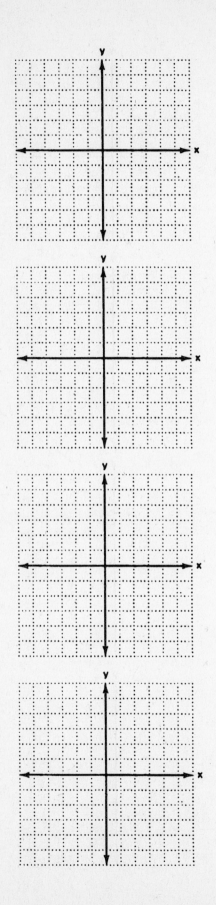

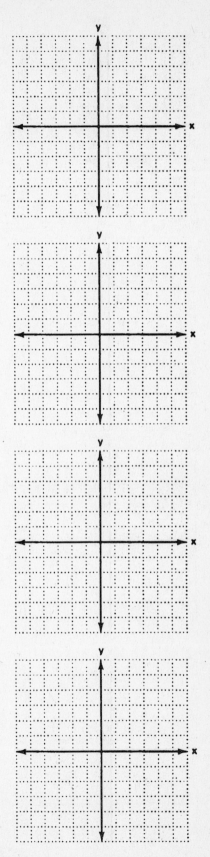

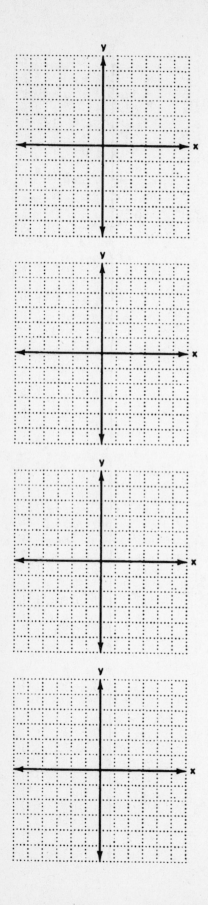

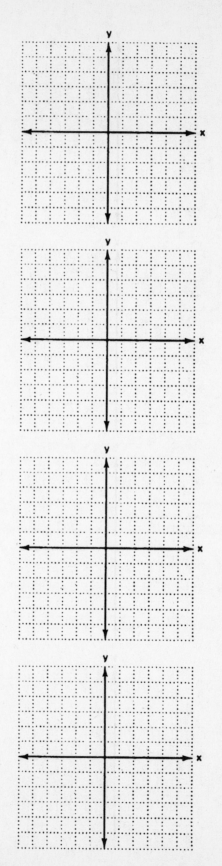

184

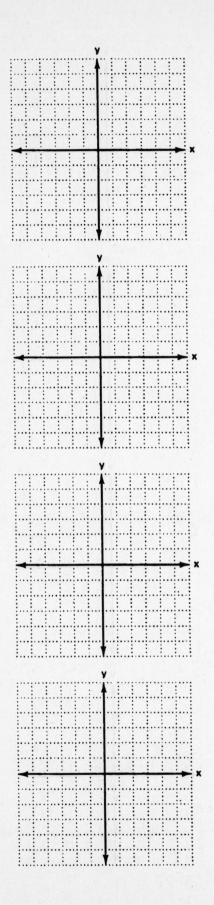
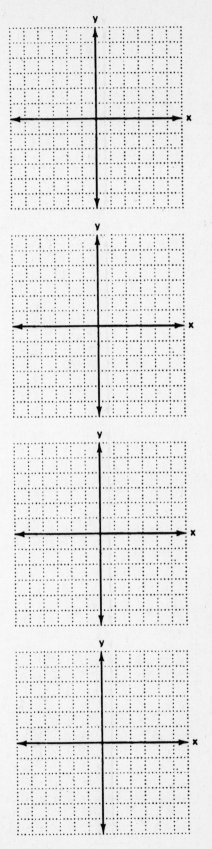

185

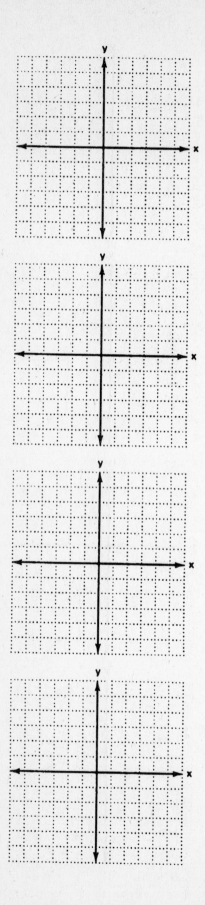
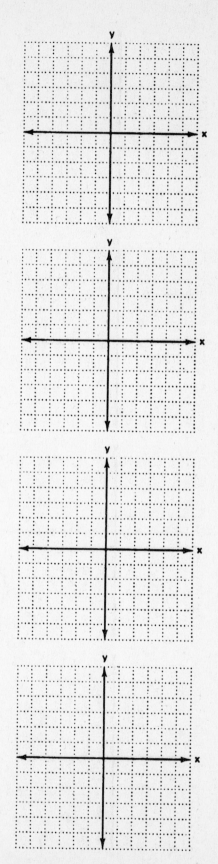

186

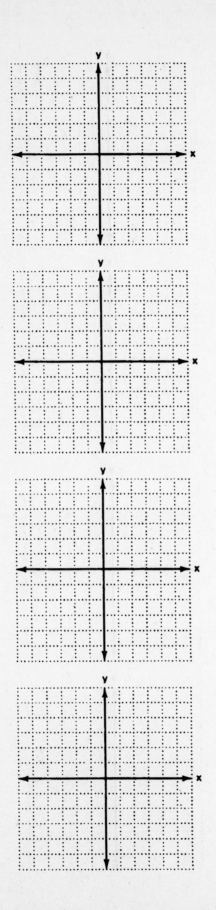
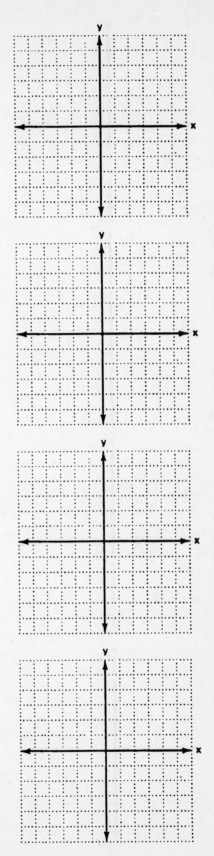

187

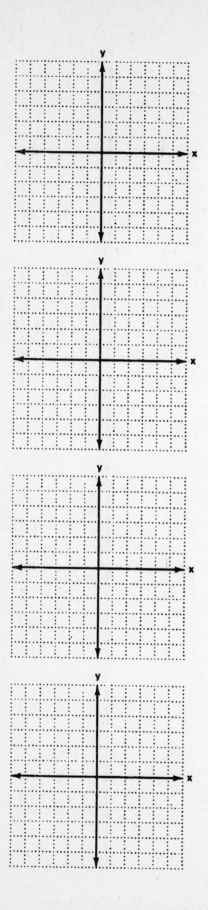
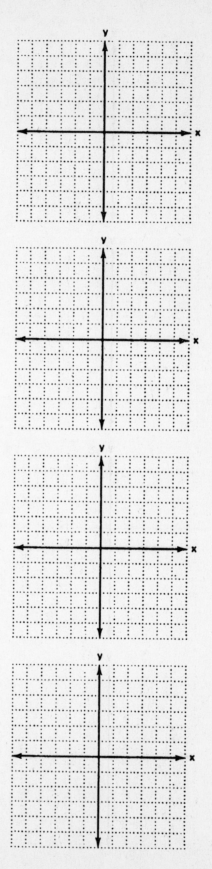

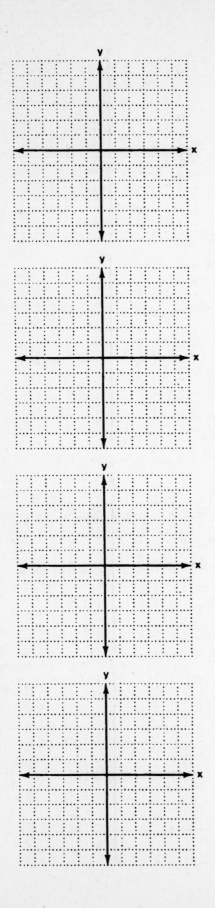
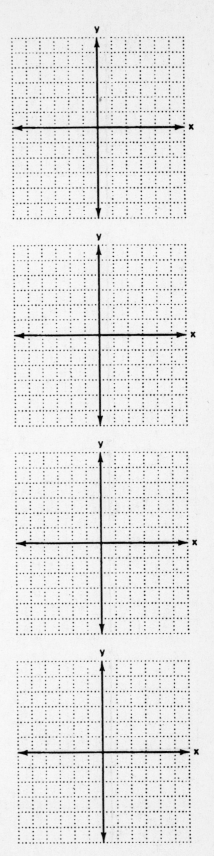

189

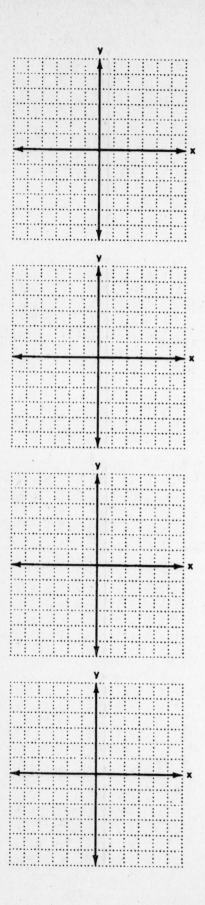

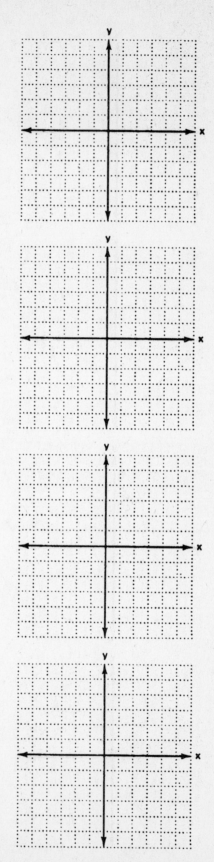

190